CPSS Power Electronics Series

This series comprises advanced textbooks, research monographs, professional books, and reference works covering different aspects of power electronics, such as Variable Frequency Power Supply, DC Power Supply, Magnetic Technology, New Energy Power Conversion, Electromagnetic Compatibility as well as Wireless Power Transfer Technology and Equipment. The series features leading Chinese scholars and researchers and publishes authored books as well as edited compilations. It aims to provide critical reviews of important subjects in the field, publish new discoveries and significant progress that has been made in development of applications and the advancement of principles, theories and designs, and report cutting-edge research and relevant technologies. The CPSS Power Electronics series has an editorial board with members from the China Power Supply Society and a consulting editor from Springer.

Readership: Research scientists in universities, research institutions and the industry, graduate students, and senior undergraduates.

More information about this series at http://www.springer.com/series/15422

Yanfeng Chen · Bo Zhang

Equivalent-Small-Parameter Analysis of DC/DC Switched-Mode Converter

Springer

Yanfeng Chen
School of Electric Power
South China University of Technology
Guangzhou, Guangdong, China

Bo Zhang
School of Electric Power
South China University of Technology
Guangzhou, Guangdong, China

ISSN 2520-8853 ISSN 2520-8861 (electronic)
CPSS Power Electronics Series
ISBN 978-981-13-2573-1 ISBN 978-981-13-2574-8 (eBook)
https://doi.org/10.1007/978-981-13-2574-8

Library of Congress Control Number: 2018954618

This Springer imprint is published by the registered company Springer Nature Singapore Pte Ltd.
The registered company address is: 152 Beach Road, #21-01/04 Gateway East, Singapore 189721, Singapore

Preface

This book focuses on the nonlinear analysis of DC/DC power switching converter systems.

As it is known that DC/DC switched-mode converter systems are typical strongly nonlinear and time-varying ill-conditioned systems, and thus it is one of the focuses and difficulties in the field of power electronics to understand and reveal the operating mechanisms from the perspective of nonlinear theory, and to analyze the stability of closed-loop systems under large signal disturbances.

After a comprehensive review of the typical large-signal methods for analyzing switching converter systems (Chap. 1), we believe that the Equivalent-Small-Parameter (ESP) method, which combines the characteristics of perturbation method and harmonic balance principle, has the advantage of relatively simple calculation process and is easy to be understood. It is a symbol algorithm suitable for strong nonlinear systems. Therefore, it is of practical significance to apply it to the modeling and analysis of DC/DC switching converter systems. Thus, we first establish a unified nonlinear mathematical description of the DC/DC switched-mode converters (Chap. 2). Then in Chap. 3, the basic principles of the symbolic analysis method for strong nonlinear systems, the equivalent small-parameter method, are introduced. Furthermore, in the following chapters, the ESP method is applied to the steady-state or transient analysis of different kinds of DC/DC switched-mode converter systems, wherein the converters with PWM (pulse-width-modulation, belongs to hard-switching technology) or PFM (pulse-frequency-modulation, belongs to soft-switching technology), current-mode control or voltage-mode control, continuous-conduction-mode (CCM) and discontinuous-conduction-mode (DCM), and so on, are all considered. Moreover, the stability problem of closed-loop systems is studied based on the ESP method, and finally a further extension to the modeling and analysis of fractional-order converters is introduced in detail. The details about the chapters are as follows.

General Nonlinear Modeling

By introducing a switching function that characterizes the on/off state of the switch, a unified nonlinear large-signal equivalent circuit model and mathematical description of the basic PWM converters operating in CCM and DCM are established in Chap. 2, which can be easily extended to the modeling and analysis of PFM type converters, such as quasi-resonant converters (Chap. 8).

Steady-State and Transient Analysis of Open-Loop Converter Systems

Chapter 4 introduces the general method of analyzing DC/DC switching converters using ESP and its applications in the steady-state and transient-state of open-loop switching converters. The analytical expressions of steady-state and transient solutions can be obtained. It can be seen that the ESP method has the advantages of higher precision and lower computational workload, which are especially evident in the transient analysis of the switching converter. Using the results of the steady-state analysis based on ESP method, an ESP algorithm suitable for transient analysis can be obtained. This method only needs to solve the linear differential equations with lower orders to overcome the shortcomings of the general average method, and can also be applied to the analysis of closed-loop switching converters and resonant switching converters.

Analysis of Steady-State of Closed-Loop Converter Systems

In Chaps. 5 and 6, mathematical models and the state-variable-based duty cycle equations are built for voltage-mode controlled (duty-cycle programmed) PWM converters in both CCM and DCM operation, in which state variables of both the power stage main circuit and feedback control compensation networks are considered. Meanwhile, a single-iterative symbol algorithm and a double-iterative symbol algorithm are developed for the analysis of closed-loop systems based on the ESP method, by which the analytical solutions for the whole closed-loop systems are obtained, wherein state variables and duty cycle ripples can be represented by symbols.

Chapter 7 presents an extended application case for the analysis of a constant frequency current-mode controlled converter, wherein the solving methods are put forward for the exact solution and simplified solution of duty ratio. Among them, the former has broader applicability since it does not rely on the hypothesis of linear inductor current waveform and small ripple, whereas the latter is of more significance in engineering applications since the calculating procedure is simplified.

Both the order-reduced and full-order model of the frequency-modulated quasi-resonant converters (QRCs) are presented in Chap. 8, wherein the analysis process of the full-order model is absolutely the same with typical PWM converters. As a result, the analyzing procedures for PWM converters and resonant converters are unified based on the full-order model. Moreover, as can be seen from the solution obtained by the full-wave model, the higher harmonic component of the resonant state variable has no significant effect on the output filter state variable, and thus can be neglected. Therefore, when using the ESP algorithm, more accurate steady-state periodic solutions of the low-frequency state variables can be obtained by very few steps of iterations. Hence, it is reasonable for the order-reduced model to ignore the dynamic characteristics of the resonant variables and only analyze the low-frequency state variables of the circuit.

Stability Analysis

Chapter 9 proposes an analytical method, i.e., Characteristic-Equation-Root (CER) method, for the stability of the equilibrium point of a closed-loop system based on the steady-state periodic solution by ESP method. The ESP-based CER method considers the correction effect of higher harmonics on DC components, and hence performs better than the stability analysis method based on State-Space-Averaging (SSA). Meanwhile, taking the Buck regulator as an example, some useful explorations are carried out on the analytical analysis and numerical analysis of the typical nonlinear dynamic phenomenon. The author believes that the low-frequency large-signal oscillation phenomenon that occurs in the Buck regulator is mainly caused by the periodic switching between different operating modes, which also have a significant impact on the large signal stability of other kinds of closed-loop converter systems.

Extension of ESP Method in Fractional-Order DC/DC Converter

Chapter 10 extends the ESP method to modeling and analysis of fractional-order DC/DC converters operating in continuous conduction mode (CCM). As an example, a fractional-order Boost converter is studied in detail. Being different from those modeling works based on fractional calculus, the converter is described by a general state vector differential equation first. Then by combining the principle of harmonic balance and equivalent small parameter method, the approximate analytical steady-state solutions of state variables could be obtained. Subsequently, the CCM-operating criterion of the fractional-order converter is discussed based on the proposed method, and the appropriate parameters are provided to make sure that the fractional-order converter operates in CCM. In addition, a statistical analysis of harmonics is made, from which influences of fractional orders on harmonics could be seen. Moreover, numerical simulations are performed, which are based on the Adams-Bashforth-Moulton typed predictor-corrector method and Oustaloup's filter-based approximation method. Both DC components and ripples of state variables from these two methods are put together with those obtained by the ESP method, and they are in good agreements with each other. Finally, the fractional-order capacitor and inductor are designed and verified by simulations, and on the basis of the designed fractional-order capacitor and inductor, experiments of the fractional-order Boost converter are carried out for further verification of the proposed method.

In summary, this book uses the ESP method for the analysis of voltage-mode and current-mode controlled PWM-type and PFM-type quasi-resonant DC/DC converters, as well as fractional-order DC/DC converter systems. All the work in the book has been verified by Pspice simulation or numerical simulation. The analysis of some circuits, such as integer-order and fractional-order duty-cycle controlled PWM Boost converter, has been experimentally verified. The obtained results are very consistent with the theoretical analysis, which shows that the ESP method has wide applicability, and still has the advantages of both simplicity and accuracy in analyzing such strong nonlinear systems as DC/DC converter systems. Moreover, the obtained results are in symbolic form, by which the working mechanism of the circuit can be reflected, and the output ripples can be calculated easily. Therefore, it has obvious applications in engineering design and computer symbol analysis.

For the analysis of closed-loop integer-order DC/DC converter systems, the principle of the method and the application cases show that: (a) The ripples have little effect on the duty cycle of PWM switched-mode converter with a higher switching frequency. (b) The DC offset of the system can be suppressed by adding appropriate integral compensations to the feedback control circuit. (c) The single iterative algorithm has a higher accuracy for the analysis of closed-loop systems with feedback compensation (the actual application circuit generally belongs to this situation). While for the closed-loop systems with large ripple components, the analysis results of the double iterative algorithm have an obvious improvement than those of the single iterative algorithm.

For the fractional-order converter, the steady-state periodic solutions of the fractional-order converter system can be obtained by using the ESP method without considering the complex definition of fractional-order calculus. And the analysis results show that the order of the fractional components not only affects the ripple component of the converter but also affects the DC component of the converter, which is the significant difference between the ESP and SSA methods since the DC solution based on state-space-averaging (SSA) is independent of the order of the fractional components.

At last, the authors would hope that this book can be referred by the researchers and engineers engaged in power electronics and nonlinear circuits, as well as the senior and graduate students with related majors in colleges and universities.

Guangzhou, China

Yanfeng Chen
Bo Zhang

Acknowledgements

This book brings together the authors' research results on nonlinear modeling and analysis of switching converters for many years. For the successful completion of this book, we are indebted to a number of people, institutions, and organizations.

First of all, the authors would like to miss and thank our respectable Professor Shuisheng Qiu for his careful guidance during the past years. Although Prof. Qiu has passed away, his voice and smile, his noble personality and rigorous research style would always live in our hearts!

Second, we must thank National Natural Science Foundation of China under Grant 51437005 and 51077057, and the Natural Science Foundation of Guangdong Province for its funding of the Team Program under Grant 2017B030312001.

We would also like to thank Dr. Botao Lin and Dr. Yanzhen Wu, as part of the examples in this book refers to their papers; and thank our students, Mr. Xi Chen, Mr. Mingjie Liu, Mr. Zhu Chen, Ms. XiaoXiao Li, Mr. Lei Wang, and Mr. Zongqi Jiang for their wonderful work for verifying our proposed algorithm.

Last, but by no means least, we wish to express our sincere appreciation to the editors of this book, and to the staff of Springer for their professional and enthusiastic support of this project.

Contents

About the Authors

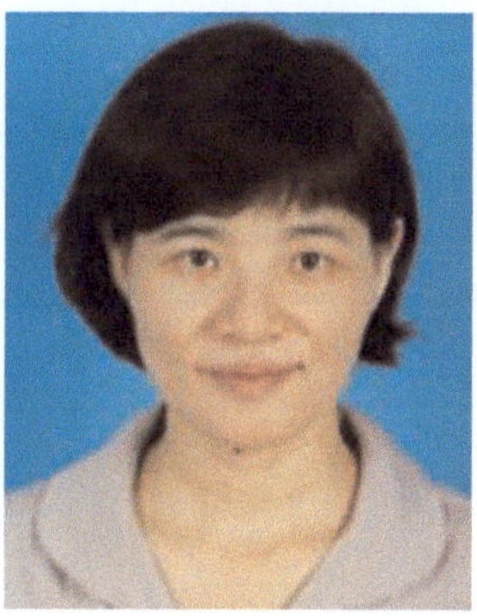

Yanfeng Chen received MS in power electronics technology from Wuhan University, Wuhan, China, in 1995, and Ph.D. in circuits and systems from South China University of Technology, Guangzhou, China, in 2000. From November 2005 to December 2006, she was a Research Associate with the Department of Electronic and Information Engineering, Hong Kong Polytechnic University, Hong Kong.

She is currently a Professor of the School of Electric Power, South China University of Technology. She has authored or coauthored two books, and more than 50 papers and 50 patents. Her main research interests are modeling and analysis of nonlinear systems and power electronics.

Bo Zhang was born in Shanghai, China, in 1962. He received a BS in electrical engineering from Zhejiang University, Hangzhou, China, in 1982, an MS in power electronics from Southwest Jiaotong University, Chengdu, China, in 1988, and a Ph.D. in power electronics from the Nanjing University of Aeronautics and Astronautics, Nanjing, China, in 1994.

He is currently a Professor of the School of Electric Power, South China University of Technology, Guangzhou, China. He has authored or coauthored four books, more than 450 technical papers and 100 patents. His current research interests include nonlinear analysis and control of power electronic converters and wireless power transfer applications.

Chapter 1
Introduction

1.1 Introduction

Power Electronics is a subject that studies electric power conversion using power semiconductor devices under high speed information control [1–6]. As an interdisciplinary of Electronics, Electrodynamics, and Control Theory, it mainly includes three types of technology: power electronic devices, power conversion topologies and automatic control. Power conversion technology with devices as core can realize the transformation of voltage, current, waveform, frequency and phase in power circuit or system by applying modern control theory and computer control technology. As the main component of power conversion systems, power converters can be divided into DC-DC converters, DC-AC inverters, AC-AC converters, and AC-DC rectifiers according to the conversion functionality, and they have been widely used in the fields such as electronics, electricity, machinery, metallurgy, mining, and transportation, etc. [7, 8].

The research field of Power Electronics contains the following parts. (1) Devices, including the design, test, model analysis, manufacture and simulation of power semiconductor devices. (2) Circuit, including topology analysis and synthesis, modeling, simulation and control of various converters. (3) Application. This book takes DC-DC converters as an example to study the modeling and analysis of converter.

Power regulator is a closed-loop system composed of DC-DC switching converter and the feedback control network for voltage or current stabilization. Figure 1.1 presents the typical diagram of closed-loop DC-DC converter with voltage-programming PWM feedback control. Differently from current-programming control, converter with such traditional PWM feedback control can be called as duty-ratio programmed converter or voltage-programming converter. As shown in the figure, the power stage is a combination of large inertia energy storage elements, such as filter inductors and capacitors and it can play a part in energy transferring, power processing and dc voltage conversion. The feedback control circuit can control the on-off states of the main circuit

Y. Chen and B. Zhang, *Equivalent-Small-Parameter Analysis of DC/DC Switched-Mode Converter*, CPSS Power Electronics Series,
https://doi.org/10.1007/978-981-13-2574-8_1

switch by modulating the continuous error signal into a series of discrete pulse signals. The control network is composed of integrated circuits with small-capacity capacitors and thus small time constant. The duty-ratio is defined as the ratio of conduction time t_{on} to switching period T, i.e., $d(t) = t_{on}/T$, and it can directly control the amount of energy transmitted from the input side to the output side. Therefore, the power regulator is actually a system that can regulate the output by applying the principle of negative feedback to control the pulse signal $d(t)$. There are three ways to regulate $d(t)$, including Pulse Width Modulation (PWM) with the constant T and variable t_{on}, Pulse Frequency Modulation (PFM) with the constant t_{on} and variable T, and the hybrid modulation with both variable t_{on} and T. The latter two ways are mostly used in resonant converters.

The time constant of the main circuit is different from that of the control circuit in several orders of magnitude. And the topology of the converter changes periodically according to the on-off state of the nonlinear switching devices, which makes DC-DC converter with closed-loop control a strongly nonlinear, time-varying, discrete and ill conditioned system. Therefore, establishing the model of the system and obtaining its analytical solution of transient response in theory have become one of the difficulties and hotspots in the field of power electronics.

The methods of analyzing the power electronics converter can be divided into numerical method and analytic method. Numerical method (also called discrete time domain simulation method) can obtain the large-signal characteristic of the converter based on a certain algorithm with the help of computer processing. In spite of accuracy, the method has no obvious physical meaning and cannot provide a guidance to the improvement of the converter performance. On the other hand, analytic method focuses on the operation principle of the circuit and makes some reasonable approximation to obtain its analytic solutions, which is of clearer significance to the design of the circuit. A further breakdown of analytic method leads to discrete analytic method and continuous analytic method. Difference equation and z-domain transform technique are applied in discrete analytic method while the Laplace transform technique is often adopted in analog circuit design, which means that although the discrete analytic method is accurate, it is not convenient for circuit design.

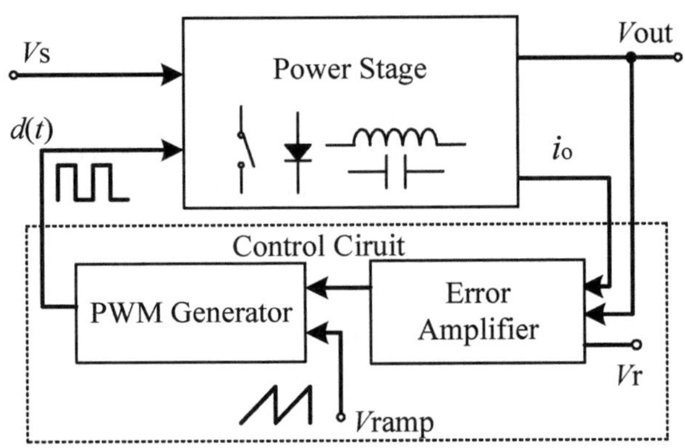

Fig. 1.1 Basic diagram of PWM control

Since DC-DC converter is essentially a nonlinear and time-varying system, the method of obtaining the analytical solution of its dynamic characteristics is relatively complex. Generally speaking, the analytical methods for power electronics converter can be divided into two categories: small-signal analysis method and large-signal analysis method. Small-signal analysis method can approximately linearize the system after assuming that there is a small disturbance near the equilibrium point. So it has the advantage of using the mature linear system theory to analyze and design the system, and the disadvantage of failing to analyze the transient characteristics of the large signal disturbances such as ripple and input/load mutation, since it neglects the nonlinear nature of the system. The large signal analysis method is a nonlinear analysis method without assuming small signal during the analysis, so as to preserve the nonlinear information of the original system to the maximum extent.

1.2 Small-Signal Linearized Averaging Methods

The state space averaging method proposed by R.D.Middlebrook and S.Cuk of California Institute of Technology in 1976, provides an effective way to the analysis of steady-state and dynamic low-frequency small-signal characteristics of the PWM DC-DC converters, which has inaugurated a new epoch in the modeling and analysis of power electronics [9, 10]. Since then, some methods with the same precision have been published, such as the current injection equivalent circuit method [11], the equivalent controlled source method [12, 13], the three terminal switch device method [14–16] (these three all belong to the circuit averaging methods). Also, other proposed methods with higher precision are suitable for occasions with higher signal frequency, including sampling data method [17, 18], discrete averaging method [19, 20], and other methods on the foundation of nonlinear theory like KBM ((Krylov-Bogoliubov-Mitropolsk) method, improved averaging method [21–23] and general averaging method [24–26], all of which have overcome the limitations in the state space averaging method of inability to analyze ripples, incapacity of the analysis of resonant converters, and the inaccuracy of stability analysis.

Nowadays, the state space averaging method [10] and circuit averaging method [11–16] are still playing major roles in the small-signal analysis methods with the basic assumption as follow.

(1) The amplitude of the disturbance signal is much smaller than the average value of state variable in a switching period.
(2) The frequency of the disturbance signal is much smaller than that of the switching frequency.

Under the two assumptions, the linear state differential equation can be used to describe the circuit and can be solved in the frequency domain.

The circuit averaging method and state space averaging method are the same in essence, but different in the processing. The former method focus on the topology, and the voltage/current across the nonlinear elements, such as switch and diode, for example, is averaged over a switching period by replacing the instantaneous value with the average value. Since the average value over the switch is pertinent to that of the voltage/current in other branches in the circuit, the switching device can be approximately represented by a controlled source in relation of duty ratio, and thus the converter can be transformed into a time-continuous linear circuit with a controlled source. Further, the solution of the linear circuit can be applied to steady state analysis and dynamic analysis of AC small signal, the result of which is the same as that of the state space averaging method.

On the other hand, the state space averaging method is from the perspective of mathematics. Suppose that over the switching period the total count of the topologies is j, each of which can be described by a linear state differential equation,

$$\dot{x} = A_i x + B_i v_s(t), \quad y = C_i^T x \quad t_{i-1} < t < t_i \quad (i = 1, 2..., j) \tag{1.1}$$

where x is the state variable, y is the output variable and A_i, B_i, C_i are the coefficient matrix related to the topology i respectively.

The duty ratio of topology i can be represented by

$$d_i = (t_i - t_{i-1})/T \tag{1.2}$$

where T is the switching period and $\sum d_i = 1$.

The weighted average of each state equation with the coefficient d_i leads to the state space average equation,

$$\dot{x} = \left(\sum_{i=1}^{j} d_i A_i\right) x + \left(\sum_{i=1}^{j} d_i B_i\right) v_s(t), \quad y = \left(\sum_{i=1}^{j} d_i C_i^T\right) x \tag{1.3}$$

Note that the state variable and output variable in (1.3) are actually the averaged value during the switching period. The low-frequency small-signal model in frequency domain can be obtained after perturbation and linearization of (1.3). Take the CCM operation as an example. The perturbation and linearization can lead to

$$\begin{aligned} (sI - A)\hat{x}(s) &= B\hat{v}_s(s) + F\hat{d}(s) \\ \hat{y}(s) &= C^T \hat{x}(s) + K\hat{d}(s) \end{aligned} \tag{1.4}$$

where

$$A = DA_1 + (1 - D)A_2 \quad B = DB_1 + (1 - D)B_2 \quad C = DC_1 + (1 - D)C_2$$
$$F = (A_1 - A_2)X + (B_1 - B_2)V_s \quad K = (C_1 - C_2)^T X$$

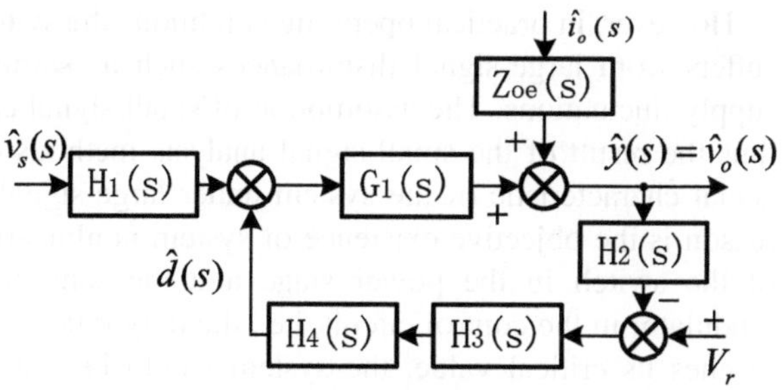

Fig. 1.2 Small-signal model of switching power regulators

The small-signal model [10] of switching power regulator derived from (1.4) is illustrated in Fig. 1.2, where $G_1(s)$ is the transfer function of the main circuit, $H_1(s)$ is the transfer function of the input, $Z_{oe}(s)$ is the equivalent output impedance when the system is open-loop, $H_2(s)$ is the transfer function of the sampling circuit, $H_3(s)$ is the transfer function of the feedback compensation network, and $H_4(s)$ is the transfer function of the pulse width modulator. The fluctuation of the input or load can lead to the change of the output, and thus the change of the duty ratio to adjust the system stable. Figure 2.2 can be simplified as a typical negative feedback closed-loop control system, as shown in Fig. 1.3 where $G(s) = H_3(s) \cdot H_4(s) \cdot G_1(s), H(s) = H_2(s)$. Define $G(s) \cdot H(s)$ as open loop transfer function (or loop gain). According to the automatic control theory, the stability of the system can be assessed by the root of the characteristic equation $1 + G(s) \cdot H(s) = 0$, or the Bode plot obtained by the open loop transfer function.

1.3 Large-Signal Analysis Methods

The essentially nonlinear characteristic of switching power regulator and the difficulty of precisely obtaining its solution make it hard to design a stable feedback loop in an exact way [27, 28], which is also the reason that small-signal analysis methods are commonly used in practical engineering design. The small-signal linearization method is an approximate equivalent mathematical model that postulates small disturbances near the quiescent point and neglects the nonlinearity of the system. Therefore, linear circuit theory can be simply used to analyze and design such system.

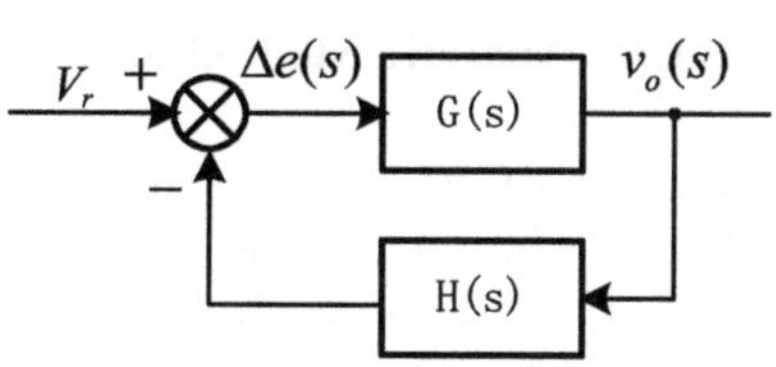

Fig. 1.3 Diagram of closed-loop control

However, in practical operating conditions the switching power regulator always suffers from large-signal disturbances such as startup, load mutation and power supply fluctuations. The assumption of small signal can be no more applicable and thus the result of the small-signal analysis methods cannot be used to predict the actual characteristic of the system under large-signal disturbances [28]. The main reason is the objective existence of system nonlinearity, including the nonlinearity of the switch in the power stage and the saturation nonlinearity of the pulse modulator in the control circuit, i.e., the duty ratio is bounded. When the duty ratio reaches its critical value, the system would be out of control and transient over voltage or over current phenomena would occur across semiconductor devices.

The analysis of large signal transient characteristics is difficult and complex. The present analytical methods can be approximately divided into two categories. One is the large-signal average modeling method based on the state space averaging method or the circuit averaging method such as the analytic method from the direct solution of differential equations [29, 30], the phase plane method [28], the large-signal equivalent circuit model method [31, 32] and the switching signal flow method [33–36]. Such methods are relatively simple and they directly adopt the small ripple hypothesis and the averaging concept of the state space averaging method, so the inherent limitation of the method, such as the existence of direct current drift, the inaccuracy of steady state and stability analysis [37–39], makes itself disable to accurately analyze the ripple of the state variable and non-applicable in the resonance converters. The other is the large-signal analysis method based on the state variable ripple solution, such as extended ripple analysis method [37], n-order harmonic three-port model method [38], KBM method (or improved averaging method), general averaging method (or MFA analysis method), equivalent small parameter method [40–52], etc. These methods have the main advantage of more accuracy. Furthermore, there are Volterra series method [53–55] and description function method [56]. The Volterra series method has not attracted much attention because of the complex and incomprehensible calculation of its kernel and the mere application to weak nonlinear systems [40, 56]. The description function method is essentially the same as the general averaging method based on the Fourier series of state variables. Therefore, these two methods are not introduced here.

1.3.1 State-Space Averaging Method Based Large-Signal Averaging Method

1.3.1.1 Analytic Method and Phase Plane Method [28–30]

Take the CCM as an example. Small signal assumption should be no longer adopted when the disturbance of state space average equation of the system is made, that is, to retain the nonlinear term of the perturbation equation. The large-signal average model of the system can be found as

$$\frac{d\hat{x}(t)}{dt} = (D_0A_1 + D_0'A_2)\hat{x}(t) + [(A_1 - A_2)X_0 + (B_1 - B_2)V_s]\hat{d}(t) \\ + \hat{d}(t)(A_1 - A_2)\hat{x}(t) \tag{1.5}$$

where D_0 is the static duty ratio ($D_0' = 1 - D_0$) and X_0 represents the value of the respective state variable,

$$X_0 = -(D_0A_1 + D_0'A_2)^{-1} \cdot (D_0B_1 + D_0'B_2)V_s \tag{1.6}$$

The actual duty ratio of closed-loop system is

$$d(t) = D_0 + \hat{d}(t) \tag{1.7}$$

where $\hat{d}(t)$ is the linear combination of state variables for PWM feedback control,

$$\hat{d}(t) = -f^T\hat{x}(t) \tag{1.8}$$

where f^T is the vector of feedback gain.

Considering the saturation nonlinearity of $\hat{d}(t)$, i.e., $D_{\min} \le d(t) \le D_{\max}$, we can obtain that

$$d(t) = D_{\max} \text{ when } -f^T\hat{x}(t) \le D_{\max} - D_0 \tag{1.9}$$

$$d(t) = D_{\min} \text{ when } -f^T\hat{x}(t) \ge D_{\min} - D_0 \tag{1.10}$$

Thus, the continuous state differential equation of the regulator operating at the unsaturated mode can be attained as

$$\frac{d\hat{x}(t)}{dt} = A\hat{x}(t) - f^T\hat{x}(t) \cdot (A_1 - A_2) \cdot \hat{x}(t) \tag{1.11}$$

where $A = (D_0A_1 + D_0'A_2) - [(A_1 - A_2)x_0 + (B_1 - B_2)V_s]f^T$

When $d(t)$ is saturated, the state differential equation of the regulator can be described as

$$\frac{dx(t)}{dt} = (D_{sat}A_1 + D_{sat}'A_2)x(t) + (D_{sat}B_1 + D_{sat}'B_2)V_s \tag{1.12}$$

where $D_{sat} = D_{\max}$ or $D_{\min}$ and $D_{sat}'1 - D_{sat}$.

According to (1.11) and (1.12), the equilibrium point of the system in unsaturated and saturated regions can be determined. Directly solving the time-invariant differential equations of (1.11) and (1.12) leads to the time-domain analytic expressions of state variables at the given initial value, which is an analytic method [30]. Since (1.11) is nonlinear, the analytic method can merely adoptable for the simple two order system, the Buck regulator, for example. When it operates in the continuous inductor current mode (CCM), we have $A_1 = A_2$ and (1.11) can be linear. But for higher order systems, the solution of the analytical method becomes rather complicated, so a pure analytical method is seldom used currently.

Given $dx/dt = (x_{n+1} - x_n)/T_s$, the differential is replaced by the forward Euler difference. After discretization of (1.11) and (1.12), the discrete large-signal model of the closed loop system in the unsaturated and saturated state can be found as

$$\begin{aligned}\hat{x}_{n+1} = {} & [I + T_s(D_0A_1 + D_0'A_2)]\hat{x}_n + T_s[(A_1 - A_2)X_0 + (B_1 - B_2)V_s]\hat{d}_n \\ & + T_s\hat{d}_n(A_1 - A_2)\hat{x}_n\end{aligned} \tag{1.13}$$

and

$$x_{n+1} = [I + T_s(D_{sat}A_1 + D_{sat}'A_2)]x_n + T_s[D_{sat}B_1 + D_{sat}'B_2]V_s \tag{1.14}$$

where $\hat{d}_n = -f^T\hat{x}_n$

The trajectory diagram of the state variables can be made on the phase plane based on the above model and the stability of the system can be determined based on the equilibrium point, which is exactly the phase plane method [28]. Phase plane method is a graphical method with the feature of visualization, but limited to the analysis of two order systems and non-applicable for higher order systems.

1.3.1.2 Large-Signal Equivalent Circuit Method [31]

Similar to the derivation of the unified equivalent low-frequency small-signal circuit model from the state space averaging method, a specific average circuit model of the converter can be obtained on the basis of the nonlinear state space average equation, and a unified equivalent large-signal circuit model [31, 32] can be built with the duty ratio controlled variable $\hat{d}$. As a nonlinear time-domain circuit model with a clearer physical concept, it can be used for the numerical simulation calculation and for the analysis of the converters' operation performance under the large-signal condition. It is contemplated that the second order item $\hat{d}\hat{x}$ is retained and there is no more small disturbance assumption as the small-signal model. Despite the feasibility of modelling a unified large-signal equivalent circuit model of PWM DC-DC converter under the control of duty ratio, there are several disadvantages including: (1) disability to carry out the ripple analysis since its theoretical foundation is still state space averaging method, (2) inapplicability for the resonant converter, (3) negligence of the influence of control circuit on the whole system, for example, the feedback compensation network.

1.3.1.3 Switching Signal Flow Diagram Method [33–35]

Switching signal flow method is a nonlinear graphical tool for modeling. The switch converter as a whole is a nonlinear system, but its equivalent circuits during the on-state and the off-state of the switch are both linear and can be described by the signal flow diagram G_{on} and G_{off} respectively. By introducing two respective branches with the switching gain k and $\bar{k}$, the switch signal flow diagram of the whole converter can be represented as:

$$G = kG_{on} + \bar{k}_{off}G_{off} \tag{1.15}$$

where k and $\bar{k}$ are two complementary switching functions,

$$k = \begin{cases} 1 & 0 \le t \le T_{on} \\ 0 & T_{on} \le t \le T \end{cases} \text{ and } \bar{k} = \begin{cases} 0 & 0 \le t \le T_{on} \\ 1 & T_{on} \le t \le T \end{cases}$$

The switching branches and their steady-state, large-signal, small signal model are presented respectively in Fig. 1.4 where $d(t)$ and $d'(t)$ denote the average value of k and $\bar{k}$ respectively during a whole period. It can be seen that the large-signal model of switching signal flow is still an average model in essence (Fig. 1.4).

The dynamic linearization model and transfer function of Buck, Boost, Buck-Boost and Cuk converter, and the closed-loop dynamic model of the switching regulated power supply are obtained by using switching signal flow diagram method in [33–35].

The expression of the transfer functions is the same as that of the state space averaging method if the delay effect of controlled variables is ignored. It has been proved theoretically and experimentally that the frequency characteristic of state space averaging method demonstrates enough accuracy at low frequency but deviates from the exact value at high frequency.

A switching delay term $e^{-\frac{T}{2}s}$ is introduced in the PWM switching signal flow graph, which means an extra pole is added in the high frequency band for rectifying the deviation and obtaining accurate frequency characteristics similar to the discrete method or the discrete averaging method.

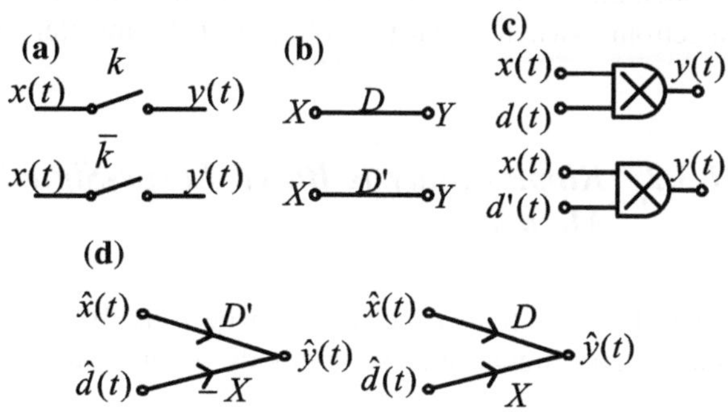

Fig. 1.4 **a** Switching branches; **b** steady-state model; **c** large-signal model; **d** small-signal model

Since the delay term can reflect the actual physical process of the converter, the PWM switch signal flow graph method has higher accuracy than the state space averaging method. However, no experimental or simulation proof is given. Afterwards, few literature had been published on modeling of converter by the use of such method. It was brought up again by K. Smedley and S. Cuk until 1994 [33], and further applied to the modeling and analysis of switching converters with polymorphic and multivariable control in 1997 [34].

The main advantage of the switching signal flow diagram is that it directly converts the switching converter into an integrated model that includes large signals, small signals, and steady state, clearly demonstrating the causality of dynamic characteristics in the system. However, the gain of the branches is still obtained by the averaging method. It can be said that the switching signal flow diagram is essentially a graphical state space averaging method, and its steady state and small signal models can be used for analysis and calculation. Since the large-signal model contains a multiplier, the graphical simulation program TUTSIM should also be used for obtaining the response characteristics [33]. Therefore, the method has certain limitations to the analysis of large-signal characteristics.

Another different method is proposed in [35] that obtains the signal flow diagram based on the small signal perturbation equation from state-space averaging method, and calculates the loop gain according to the Mason formula to analyze the stability of the entire system (including voltage mode feedback control or current mode feedback control). However, applying Mason's formula to calculate the determinant of the graph is really cumbersome due to the complexity of its signal flow diagram. Conversely, it is more intuitive and convenient if the state-space averaging method is applied to obtain the block diagram of the transfer function of the entire closed-loop system for stability analysis (see Sect. 1.2). Essentially this method is also a small signal analysis method, and its result is consistent with that of the small signal model using the switching signal flow diagram method. In addition, the method involves calculations of matrix inverse, trace, and adjoint matrix, so that extra computer processing is necessary when the order of the system is high.

The switching signal flow diagram method has obvious physical meaning, intuitive form and simple operation [36], and it can avoid the complicated matrix operations of the state space averaging method and can quickly obtain the dynamic model and transfer function of complex circuits. However, in the complex circuit model, the number of abstract signal switches is so larger than the actual number of electronic switches that the solution of the transfer function is quite complicated.

1.3.2 Ripple-Analysis Based Large-Signal Averaging Method

The afore mentioned methods can be roughly classified as state-space-averaging based large-signal analysis methods because all of them (such as the switching signal flow diagram method) directly or indirectly use the state-space averaging method to obtain the large-signal average model. The main drawback is that they cannot perform ripple analysis. Actually, ripple has a very important impact on the operation and performance of the converter. For example, the output voltage ripple is usually a design index of the power regulator, and the current ripple and its average value can determine whether the converter's operating mode has been changed, i.e., whether the converter operates at DCM mode or not [37]. Though the ripple performance can be evaluated by computer simulation with an accurate result, the relationship between ripple and circuit parameters is usually not obvious enough for further optimization of the converter.

1.3.2.1 Extended Ripple Analysis Method [37]

The proposed extended ripple analysis method in [37] can obtain the simplified expressions of the first and second ripples of the state variables at the steady state under CCM according to Taylor series expansion.

Suppose that $v(t)$ and $i(t)$ can be expanded as Eqs. (1.16a) and (1.16b).

$$v(t) = V + v^{(1)} \cdot (t - t_0) + v^{(2)} \cdot (t - t_0)^2 + \cdots \tag{1.16a}$$

$$i(t) = I + i^{(1)} \cdot (t - t_0) + i^{(2)} \cdot (t - t_0)^2 + \cdots \tag{1.16b}$$

where V or I is the average value of the waveform, $v^{(1)}, i^{(1)}; v^{(2)}, i^{(2)}; \ldots v^{(k)}, i^{(k)}; \ldots$ represent the 1-order (or linear), 2-order and k-order component of the waveform respectively. For PWM converters, V and I are can be determined by the volt-second balance principle of the inductor and the ampere-second balance principle of the capacitor respectively. The volt-ampere characteristics of the inductor and capacitor can be described respectively as

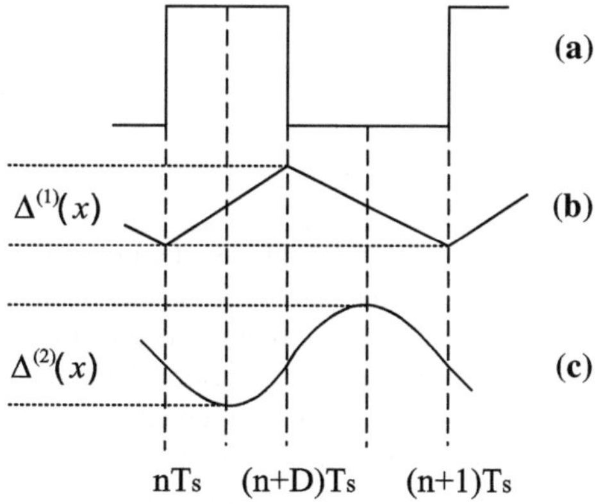

Fig. 1.5 **a** Piecewise 0-order ripple, **b** 1-order ripple and **c** 2-order ripple waveform

$$
\begin{aligned}
i_L(t) &= \frac{1}{L}\int_0^t v_L(\tau)d\tau \\
v_C(t) &= \frac{1}{C}\int_0^t i_C(\tau)d\tau
\end{aligned} \tag{1.17}
$$

The n-order component of the capacitor current (or inductor voltage) will produce the (n + 1)-order component. Thus, as shown in Fig. 1.5, the integral of (a) leads to (b) and the integral of (b) leads to (c). In other words, the piecewise zero-order waveform can generate the 1-order ripple waveform, while the piecewise linear ripple can generate the 2-order ripple. Taking CCM as an example, the piecewise state equation of the system is

$$
\dot{x}(t) = A_i x(t) + b_i V_s \tag{1.18}
$$

The exact solution of (1.18) can be written as

$$
x(t) = e^{A_i(t-t_i)}x(t_i) + A_i^{-1}[e^{A_i(t-t_i)} - I]b_i V_s, \quad i = 1, 2 \tag{1.19}
$$

We take $e^{At} = I + At$ for solving the 1-order ripple. According to Fig. 1.5a, the 1-order can be found as

$$
\Delta^{(1)}x = x[(n+D)T_s] - x(nT_s) \quad \text{or } \Delta^{(1)}x = x[(n+1)T_s] - x[(n+D)T_s] \tag{1.20}
$$

We take $e^{At} = I + At + (At)^2/2$ for getting the 2-order ripple. Substituting it into (1.19) leads to the secondary components $x^{(2)}(t)$ (detailed derivation can be found in [37]),

$$
x^{(2)}(t) = \frac{A_i}{2}[A_i x(t_i) + b_i V_s] \cdot (t - t_i)^2 \tag{1.21}
$$

According to Fig. 1.5c, the 2-order ripple can be derived as

$$
\Delta^{(2)}x = x^{(2)}[(n + D/2)T_s] - x^{(2)}[n + D + D'/2)T_s] \tag{1.22}
$$

From the above analysis, it can be seen that the method has a simple expression of the ripple, but it relies too much on the assumption that the waveform of the state-space averaging method is linear (see Fig. 1.5), and thus it is not applicable for analysis on resonant converters. This method only discusses the 1-order and 2-order ripples. It would be complicated to get the high-order ripples because multiples of the Taylor series expansion should be considered and re-derivation is needed every time we get a new higher order ripple. Besides, it must be assumed that the ripple is determined by the lowest-order non-zero component.

1.3.2.2 Three Port Model of *n*-Order Harmonic [38]

A so-called PWM switch model can be formed by extracting the transistors and diodes from the switching power converter, as shown in Fig. 1.6a, which represents the nonlinearity of the entire power stage circuit. Take CCM as an example, for the PWM switch, we can get the following transient relationship as shown in Eq. (1.23).

$$i_1(t) = \begin{cases} i_3(t) & 0 < t \le dT_s \\ 0 & dT_s < t \le T_s \end{cases}, \quad v_{32}(t) = \begin{cases} v_{12}(t) & 0 < t \le dT_s \\ 0 & dT_s < t \le T_s \end{cases} \tag{1.23}$$

Averaging (1.23) leads to the large-signal averaged model as shown in Fig. 1.6, which is

$$i_1 = di_3 \quad , v_{32} = dv_{12} \tag{1.24}$$

After perturbation to (1.24), we can get the low-frequency small-signal model, the same as the one that derived from the state-space averaging method [14, 15]. But Ref. [38] retains the nonlinear item of the perturbation model and assumes that

$$\begin{aligned} d &= D + \hat{d}, \\ i_3 &= I_3 + \hat{i}_3,\ v_{12} = V_{12} + \hat{v}_{12},\ i_1 = DI_3 + I_3\hat{d} + D\hat{i}_3 + \hat{i}_3\hat{d}, \\ v_{32} &= DV_{12} + V_{12}\hat{d} + D\hat{v}_{12} + \hat{v}_{12}\hat{d} \end{aligned} \tag{1.25}$$

Given $\hat{d}$ as the single frequency disturbance, i.e., $\hat{d} = d^{(1)} \cos \omega t, \quad \left|d^{(1)}\right| \ll 1$.

$\hat{i}_3$ and $\hat{v}_{12}$ can be expanded into the following Fourier Series form,

$$\hat{i}_3 = \sum_{n=0}^{\infty} i_3^{(n)} \cos(n\omega t + \phi_n), \quad \hat{v}_{12} = \sum_{n=0}^{\infty} v_{12}^{(n)} \cos(n\omega t + \theta_n) \tag{1.26}$$

Substituting (1.26) into (1.25) yields

$$i_1 = \sum_{n=0} i_1^{(n)}, \quad v_{32} = \sum_{n=0} v_{32}^{(n)} \tag{1.27}$$

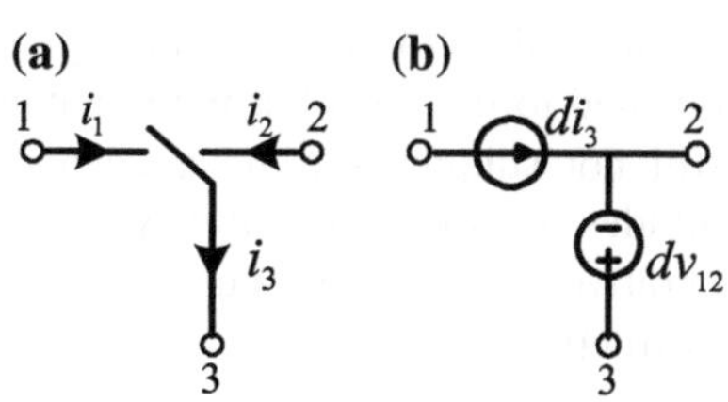

Fig. 1.6 **a** PWM switch and **b** its large-signal averaged model

where

$$i_1^{(0)} \approx DI_3 \quad \text{(0-order component or DC steady state value)}$$

$$i_1^{(1)} \approx I_3 d^{(1)} \cos \omega t + D i_3^{(1)} \cos(\omega t + \varphi_1) \quad \text{(1-order harmonic)}$$

$$i_1^{(n)} \approx (i_3^{(n-1)} d^{(1)}/2) \cos(n\omega t + \varphi_{n-1}) + D i_3^{(n)} \cos(\omega t + \varphi_n) \quad \text{(}n\text{-order harmonic)}$$

$$v_{32}^{(0)} \approx DV_{12} \quad \text{(0-order component or DC steady state value)}$$

$$v_{32}^{(1)} \approx V_{12} d^{(1)} \cos \omega t + D v_{12}^{(1)} \cos(\omega t + \theta_1) \quad \text{(1-order harmonic)}$$

$$v_{32}^{(n)} \approx \left(v_{12}^{(n-1)} d^{(1)}/2\right) \cos(n\omega t + \theta_{n-1}) + D v_{12}^{(n)} \cos(\omega t + \theta_n) \quad \text{(}n\text{-order harmonic).}$$

Thus, the switch model of Fig. 1.6b can be decomposed into multiple linear models such as DC model, fundamental frequency model, and other harmonics model.

This method is actually an analytical method by approximate decomposition of nonlinear systems. We consider a nonlinear system that responds to the input $x(t)$ as:

$$y(t) = f[x(t)] \tag{1.28}$$

When the nonlinear system is decomposed into linear subsystems $a_0, a_1, \ldots a_n$ and the input of each subsystem is $x(t), x^1(t), \ldots x^n(t)$, the response of the entire nonlinear system is approximately

$$y(t) \approx \sum a_n x^n(t) \tag{1.29}$$

And $y(t)$ can be expressed as the sum of the following series

$$y(t) = \sum B_n \cos(n\omega t + \theta_n) \tag{1.30}$$

Let (1.29) be equal to (1.30) and apply the harmonic balance principle for obtaining $B_0, B_1, \ldots B_n$, which is also the solution of the nonlinear system.

From the above, it can be seen that this method can get the large-signal averaged model by using the averaging concept, but it just takes into account the case where the duty cycle contains a single-frequency disturbance, and neglects the case when the system is closed-loop.

Since the closed-loop duty cycle is a function of the state variables, it is complicated to use such harmonic balance analysis method. In addition, this method is still not suitable for the analysis of resonant converters. However, this method divides the solution of the system into the sum of various harmonics and uses the iterative method to obtain the response of the entire system which is really instructive.

1.3.2.3 Development of Averaging Method—Asymptotic Method, Improved Averaging Method and General Averaging Method [21–26]

The state space averaging method is the first method to systematically analyze switching power converters, but its accuracy will decrease as the voltage or current ripple increases, and as the equivalent switching frequency increases [22]. That is because the time-continuous averaging method cannot preserve the information of the current or voltage ripple; moreover, the state-space time averaged model is proposed under the conditions that the waveform is linear and the ripple is small. In addition, Ref. [31] points out by an example that the original state space averaging method cannot accurately determine the stability of the system. Furthermore, many emerging converters (such as resonant or quasi-resonant type converters) do not satisfy waveform linearity or small-ripple assumptions, which urges researchers to study the mathematical basis and applicability of the state-space averaging method, and to propose new methods to make up for the deficiencies, such as the asymptotic method, the improved averaging method, and the general averaging method.

I. **Asymptotic method (KBM) [21] and improved averaging method** [22, 23]

Early literatures have discussed the rationality of the state space averaging method. For example, Ref. [21] rigorously demonstrates the approximation properties of the state-space averaging method, and proposes an asymptotic method (Krylov-Bogoliubov-Miltropolsky, KBM) for analyzing switching power conversion systems. Its solution can be approximated to any degree of accuracy by a power series represented by the small parameter. The small parameter is related to the switching period and the time constant of the system. For small parameter systems there are

$$\dot{x} = \varepsilon F(t,x), \quad \varepsilon \ll 1, \quad x(t_0) = x_0 \tag{1.31}$$

Given

$$G(\cdot) = \lim_{T\to\infty} \frac{1}{T} \int_0^T F(t,\cdot)dt \tag{1.32}$$

The averaged system can be written as

$$\dot{y} = \varepsilon G(y), \quad y(t_0) = y_0 \tag{1.33}$$

If the limit of (1.32) exists, it can be considered that the solution of the averaged system (1.33) can represent the solution of the original system (1.31). For a switching power converter, (1.31) is time-varying and discontinuous due to changes in the switching state and circuit topology. The small parameters ε is dependent on the switching frequency and the time constant of the system. The average of (1.31) over a switching period can lead to the classic large-signal state space averaged model.

Let

$$x(t) = y(t) + \varepsilon\psi_1(t, y) + \varepsilon^2\psi_2(t, y) + \varepsilon^3\psi_3(t, y) + \cdots \tag{1.34}$$

and

$$\dot{y}(t) = \varepsilon G_1(y) + \varepsilon^2 G_2(y) + \varepsilon^3 G_3(y) + \cdots \tag{1.35}$$

The steady-state periodic solution of (1.31) can be determined by solving $G_1(y), G_2(y)\ldots$ of (1.35) and $\psi_1(t, y), \psi_2(t, y)\ldots$ of (1.34).

The KBM method has the following deficiencies.

(1) The nonlinear system should be a small parameter system;
(2) $\psi_1(t, y), \psi_2(t, y)\ldots$ is a function of time whose average value is zero. Its physical meaning is not obvious and the solution of the system is not straightforward enough since it can be determined only after solving $G_1(y), G_2(y)\ldots$. Also the estimate of ripple is complicated.
(3) y, i.e., the average value of the state variable over one period, should be assumed to be constant when solving $\psi_i(t, y)$ and $G_i(y)$, which means that the steady-state DC solution of the system is assumed to be consistent with its state space average value. However, it has been confirmed in [22] that DC drift exists in the switching converter when the frequency is low.
(4) The method is still not suitable for analysis of resonant converters.

Reference [23] holds the view that the classical averaging method in [21] is only suitable for open-loop controlled converters and not suitable for closed-loop systems with feedback control. Because the feedback-controlled closed-loop system is state-discontinuous and does not satisfy the Lipschiz condition, the classical averaging theory that directly averages the system's state differential equations is not applicable. To solve this problem, it holds that as long as the power electronic switch only operates once in each period, the differential equation can be integrable and the corresponding integral equation can be used to replace the differential equation at discontinuous states. And it is applicable to average the integral model.

According to the theoretical proof and conclusion of [23], the frequency-dependent modeling method of the PWM feedback controlled DC-DC converter proposed by [22] considers only the 1-order ripple of the steady-state solution, so it can be regarded as an improved averaging method combined with KBM and 1-order ripple.

II. **General averaging method and MFA (Multi-frequency Averaging)** [24, 25]

A general averaging method is proposed in [24] based on description of the Fourier series with time-dependent sliding window of a given waveform. The MFA method proposed in [25] can be regarded as a further deepening of [24], and can also be regarded as the application of the general averaging method in the modeling of an actual closed-loop PWM converter system. Unlike all previous methods, the general averaging method does not average the system at first. Instead, a switching function is introduced to establish an exact nonlinear model of the switching power converter. Both the circuit state variables and nonlinear functions (such as switching functions) are expanded into Fourier series, whose coefficients are correlated with time. Thus an approximate model described by time-invariant state differential equation group can be obtained, and its state variables to be determined are the corresponding coefficient in the Fourier series expansion. The higher the order of the harmonics described in the model, the closer the result is to the exact solution of the circuit. The specific method can be briefly described as follows.

The periodic function $x(\tau)(\tau \in [t-T, t])$ can be described by the Fourier series as:

$$x(\tau) = \sum_{l=-\infty}^{+\infty} \langle x \rangle_l(t) e^{jl\omega_s \tau} = \langle x \rangle_0 + 2\sum_{l=1}^{\infty} \left[R\{\langle x \rangle_l\} \cos(l\omega_s \tau) + I\{\langle x \rangle_l\} \sin(l\omega_s \tau)\right] \tag{1.36}$$

where $\omega_s = 2\pi/T$, $\langle x_l(t) \rangle$ is the Fourier coefficients of complex form, $R(\cdot), I(\cdot)$ are the real part and the imaginary part of the complex respectively, T is the switching period.

$$\langle x \rangle_0 = \frac{1}{T}\int_{t-T}^{t} x(\tau)d\tau, \quad \langle x \rangle_l(t) = \frac{1}{T}\int_{t-T}^{t} x(\tau)e^{-jl\omega_s \tau}d\tau \tag{1.37}$$

$\langle x \rangle_0$ denotes the average value of the state variables over a switching period. If we only consider the DC term in (1.36), we can get the approximate index-0 model of the system, which is the same one as the state space averaging model. Therefore, state space averaging method is the approximation of the general averaging method at zero order. If both the 1-order harmonic and the DC term in (1.36) are taken into account, an index-1 model reflecting the influence of the 1-order ripple of the state variable can be obtained. And so on.

In the derivation of the approximate model, the following two important features are used.

(i) l-order coefficient differential to time:

$$d\langle x\rangle_l/dt = \langle dx/dt\rangle_l - j\omega_s\langle x\rangle_l(t) \tag{1.38}$$

(ii) The l-order coefficient of the product of functions $x(t)$ and $q(t)$:

$$\langle qx\rangle_l = \sum_{l=-\infty}^{\infty} \langle q\rangle_{l-i}\langle x\rangle_i \tag{1.39}$$

Since there is no waveform linearity and small ripple assumptions, this method can be applied to the analysis of resonant converters [24]. Compared with KBM, the general averaging method does not need to convert the system into a standard form like KBM, nor does it require strict average theory to support it. For KBM, the order of the state-differential equation is constant in each solution of the ripple, but for general averaging method, it is related to the order of the approximation.

As being pointed out in [26], for the 2-order open-loop circuit, using the 1-order approximation, the corresponding model matrix has six state variables. And if the 2-order approximation is used, the model matrix will contain ten state variables. For the 4-order Cuk circuit, the 1-order approximation yields a 12-order model matrix. In conclusion, if the order of a circuit is n, the order of the i-order approximation model matrix will be $n(2^i+1)$, which means that there will be $n(2^i+1)$ state variables to be solved.

The general averaging method also has some similarities with the equivalent small parameter (ESP) method that will be mentioned below. Both of them are based on the Fourier series expansion of the switching function $q(t)$ and state variables. But the ESP applies the concept from KBM of asymptotic iteration to solve the equations successively, and it is simpler and easier to understand than MFA.

1.3.2.4 Symbolic Analysis Method—Equivalent Small Parameter Method [40–52]

Compared with the numerical simulation method, the symbol analysis method can reflect the behavior of the circuit and obtain an analytical model for circuit optimization and control design [57–60]. Reference [57] uses MATLAB and its symbol analysis toolbox to obtain a new design curve for the conversion topology. Reference [58] can obtain the symbolic transfer function of the converter in each mode and calculate its time domain response by using Laplace's transformation. In contrast to the frequency-domain method [57], Ref. [58] directly expresses the circuit equation in the time domain. The above methods are limited to hard-switching PWM converters. Based on general averaging method, a symbolic analysis method is proposed in [59], and it can be used for hard-switching and quasi-resonant soft-switching converters. But the intrinsic limitation of the averaging method will also be inherited. The equivalent small parametric method is a relatively simple symbolic algorithm in combination with the advantages of the perturbation method and the harmonic balance method [40–42]. It has been used in the analysis of open-loop systems of PWM switching converters, Class E amplifiers, single-phase inverters and fractional-order DC-DC converters [43–52]. This book further apply it to the analysis of closed-loop systems of PWM and quasi-resonant switching power converter. The theoretical analysis and examples indicate that the equivalent small parameter method provide a simple and relatively precise way to the analysis of PWM and the closed-loop system of the resonant converter in the steady-state. All the obtain results are symbolic expressions. The method can reflect the mechanism of the circuit, and can obtain the output ripple easily. Therefore, it has more applications in engineering design and computer symbol analysis. When the equivalent small parameter method is applied to closed-loop system analysis, the main oscillation component equation can be transformed into a univariate high-order equation, which can be used for multiple solution problems and system stability analysis [43].

The equivalent small parameter method can easily adopt computer symbolic algebraic systems to create symbolic analysis programs, and make an important impact on the practical design where large amount of repeated calculation is required. Symbolic analysis opens up a new field for the application of computers in power electronic systems, and the hybrid simulation method which combines it with numerical analysis method will be a direction for the development of the switching power converter analysis method [52, 57].

The basic principle, solving process and application of the equivalent small parameter method are introduced in Chap. 3 of this book.

1.4 Summary

This chapter gives a comprehensive review on the existing analysis methods of the switching converters (especially the large-signal analysis method), and compares the advantages and disadvantages of these methods. The large-signal modeling analysis methods for the closed-loop system of switching power converters can be roughly divided into two categories. One is state space averaging based average large-signal modeling method. The solution procedure can be summarized as: Average–Perturbation–Large-signal averaged model. The obtained model cannot analyze the system ripple accurately, but it can adopt the phase plane method or characteristic equation root analysis method to judge the stability of the system. These methods have more or less the following disadvantages: (1) Averaging processing makes the system unable to carry out ripple analysis effectively. (2) The stability analysis cannot be performed accurately. (3) Only applicable to non-resonant converters. The other is ripple-analysis based large-signal averaging method, which can obtain the ripple of the state variables and analyze the stability of the system based on some kind of mathematical analysis methods of nonlinear system. Among these methods, the equivalent small parameter method combines the advantages of the perturbation method and the harmonic balance method, and is proved to be an effective method for analyzing strong nonlinear systems such as switching power converters.

References

1. Bose BK (1989) Power electronics—an emerging technology. IEEE Trans Ind Electron 36 (3):403–412
2. Bose BK (1992) Recent advances in power electronics. IEEE Trans PE 1:2–16
3. Wilson TG (1999) The evolution of power electronics. In: Proceedings APEC, pp 3–9
4. Adler MS, Owyang KW, Baliga BJ, Kokosa RA (1984) The evolution of power device technology. IEEE Trans Electron Devices ED-31(11):1570–1591
5. Smith S (2002) Developments in power electronics, machines and drives. IEEE Power Eng J 16(1):13–17
6. Baliga BJ (2001) The future of power semiconductor device technology. Proc IEEE 89:822–832
7. Bose BK (2013) Global energy scenario and impact of power electronics in 21st century. IEEE Trans Ind Electron 60(7):2638–2651
8. Bose BK (2000) Energy environment and power electronics. IEEE Trans Power Electron 15 (4):688–701
9. Middlebrook RD, Cuk S (1976) A general unified approach to modelling switching-converters power stages. IEEE PESC Rec pp 18–34
10. Middlebrook RD, Cuk S (1983) Advances in switched-mode power conversion. Published by TESLAco
11. Chetty PRK (1981) Current injected equivalent circuit approach to modeling of switching DC-DC converters in continuous inductor conduction mode. IEEE Trans AES 17(6):802–808
12. Ben-Yaakov S, Edry D (1994) Averaged models and tools for studying the dynamics of switch mode dc-dc converters. In: Proceedings IEEE power electronics specialists conference PESC'94, pp 1218–1225
13. Pejovic P, Maksimovic D (1995) A new algorithm for simulation of power electronic systems using piecewise-linear device models. IEEE Trans Power Electron 10(3):340–348
14. Vorperian V (1990) Simplified analysis of PWM converters using the model of the PWM switch : part I continuous conduction mode. IEEE Trans AES 26(3):490–496
15. Vorperian V (1990) Simplified analysis of PWM converters using the model of the PWM switch : part II discontinuous conduction mode. IEEE Trans AES 26(3):497–505
16. van Dijk E, Spruijt HJN, O'Sullivan DM et al (1995) PWM-switch modeling of DC-DC converters. IEEE Trans PE 10(6):659–664
17. Brown AR, Middlebrook RD (1981) Sampled-data modeling of switching regulators. IEEE PESC Rec 349–369
18. Verghese G, Bruzos C, Mahabir K (1989) Averaged and sampled-data modes for current-Mode control: a reexamination. IEEE PESC Rec 484–491
19. Short DJ, Lee FC (1982) An improved switching converter model using discrete and average techniques. IEEE PESC Rec 192–212
20. Short DJ, Lee FC (1983) Extension of the discrete-average models for converter power Stages. IEEE PESC Rec 23–37

21. Krein PT, Bentsman J, Bass RM et al (1990) On the use of averaging for the analysis of power electronic systems. IEEE Trans PE 5(2):182–190
22. Lehman B, Bass RM (1996) Switching frequency dependent averaged models for PWM DC-DC converters. IEEE Trans PE 11(1):89–98
23. Lehman B, Bass RM (1996) Extensions of averaging theory for power electronic systems. IEEE Trans PE 11(4):542–553
24. Sanders SR, Noworolski JM, Liu XZ et al (1991) Generalized averaging method for power conversion circuits. IEEE Trans PE 6(2):251–258
25. Caliskan VA, Verghese GC, Stankovic AM (1999) Multifrequency averaging of DC/DC converters. IEEE Trans PE 14(1):124–133
26. Mahdavi J, Emaddi A, Bellar MD et al (1997) Analysis of power electronic converters using the generalized state-space averaging approach. IEEE Trans CAS-I 44(8):767–770
27. Tse KK, Chung HS et al (1999) Stepwise quadratic state-space modeling technique for simulation of power electronics circuits. IEEE Trans IEEE 46(1):91–99
28. Erickson RW, Cuk S, Middlebrook RD (1982) Large-signal modelling and analysis of switching regulators. IEEE PESC Rec 240–250
29. Bass RM, Sun J (1998) Large-signal averaging methods under large ripple conditions. IEEE PESC Rec 630–632
30. Takashi Nabeshima, Kovsuke H (1982) Large-signal transient response of switching regulators. IEEE Trans AES 18(5):545–551
31. Ilic M, Leeb S, Liebezeit K (1988) Exact nonlinear analysis of switched mode power converters. IEEE PESC Rec 977–986
32. Fen Chen, Xuan San Cai (1989) Design of feedback control laws for switching regulators based on the bilinear large signal model. In: Power electronics specialists conference 1989, PESC'89 record, 20th annual IEEE, vol 1, pp 468–475
33. Smedley K, Cuk S (1994) Switching flow-graph nonlinear modeling technique. IEEE Trans PE 9(4):405–413
34. Ma Yunhong, Ma Keyue (1997) Switching flow-graph nonlinear modeling method for multistate-switching converters. IEEE Trans PE 12(5):854–861
35. Ki WH (1998) Signal flow graph in loop gain analysis of DC-DC PWM CCM switching converters. IEEE Trans CAS-I 45(6):644–655
36. Smedley KM, Cuk S (1995) Dynamics of one-cycle controlled Cuk converters. IEEE Trans Power Electron 10:634–639. ISSN 0885-8993
37. Tymerski R, Li D (1993) Extended ripple analysis of PWM DC-to-DC converters. IEEE Trans PE 8(4):588–595
38. Tymerski R, Vorperian V, Lee FC et al (1988) Nonlinear modelling of the PWM switch. IEEE PESC Rec 968–976
39. Griffio A, Wang Jiabin (2012) Large signal stability analysis of 'more electric' aircraft power systems with constant power loads. IEEE Trans Aerosp Electron Syst 48(1):477–489
40. Qiu SS, Filanovsky IM, Stromsmoe KA (1985) An equivalent small parameter method for strongly nonlinear systems. In: Proceedings 28th Midwest symposium on circuits and systems, Louisille, pp 61–64
41. Qiu SS, Filanovsky IM, Stromsmoe KA (1987) On one combination of the harmonic balance method and perturbation techniques. Control-Theory Adv Technol 3:189–195
42. Qiu SS, Filanovsky IM (1990) A method of calculation of steady state oscillations in autonomous nonlinear systems. J Sound Vibrat 136(1):35–44
43. Qiu SS, Filanovsky IM, Lin BT (1999) A new method of analysis for PWM switching power converters. Int J Electron 6(11):1395–1410
44. Qiu SS, Filanovsky IM (2000) Harmonic analysis of PWM converters. IEEE Trans Circuits systems I: Fundam Theory Appl 47:1340–1349
45. Chen W, Floberg H, Qiu S (1999) A new analytical method for analysis and design of Class E power amplifiers taking into account the switching device on resistance. Int J Circuit Theory Appl 27(4):421–436

46. Wu Y, Hu AP, Budgett D, Malpas S, Dissanayake T (2010) Harmonic analysis of voltage-fed full-bridge converter for TET application. In: 2010 the 5th IEEE conference on industrial electronics and applications (ICIEA), pp 1040–1045
47. Wu Y, Hu AP, Budgett D, Malpas SC, Dissanayake T (2011) Efficient power-transfer capability analysis of the TET system using the equivalent small parameter method. IEEE Trans Biomed Circuits Syst 5(3):272–282
48. Chen Y, Chen X, Zhang B, Qiu D (2015) A new analyzing scheme for non-integer order DC/DC converters. In: IECON 2015—41st annual conference of the ieee industrial electronics society, pp 001633–001638
49. Cheng X, Chen Y, Chen X, Zhang B, Qiu D (2017) An extended analytical approach for obtaining the steady-state periodic solutions of SPWM single-phase inverters. In: Energy conversion congress and exposition (ECCE) 2017 IEEE, pp 1311–1316
50. Chen X, Chen Y, Zhang B, Qiu D (2017) A method of modeling and analysis for fractional-order DC-DC converters. IEEE Trans Power Electron 32(9):1–10
51. Chen Y, Chen X, Hu J, Zhang B, Qiu D, (2017) A symbolic analysis method for fractional-order boost converter in discontinuous conduction mode. In: Industrial electronics society IECON 2017—43rd annual conference of the IEEE, pp 8738–8743
52. Li X, Chen Y, Chen X, Zhang B, Qiu D, (2017) An analytical approach for obtaining the transient solution of the fractional-order buck converter in CCM. In: Industrial electronics society IECON 2017—43rd annual conference of the IEEE, pp 6859–6863
53. Tymerski R (1991) Volterra series modeling of power conversion systems. IEEE Trans PE 6 (4):712–718
54. Rugh WJ (1981) Nonlinear system theory: the volterra/wiener approach. The John Hopkins University Press
55. Schetzen M (1981) Nonlinear system modeling based on the Wiener theory. Proc IEEE 69 (12):1557–1573
56. Sanders SR (1993) On limit cycles and the describing function method in periodically switched circuits. IEEE Trans CAS-I 40(9):572
57. Sun J, Grotstollen H (1997) Symbolic analysis methods for averaged modeling of switching power converters. IEEE Trans PE 12(3):537–546
58. A. Liberatore et al (1991) A new symbolic approach to the analysis of switch power converters. In: Proceedings EPE'91, pp 489–494
59. Webster R, Ngo KDT (1992) Computer-based symbolic circuit analysis and simulation. In: APEC, pp 772–779
60. Wambacq P et al (1998) Symbolic network analysis method for practical analog integrated circuits: a survey. IEEE Trans Circuits Syst-II 45(10):1331–1341

Chapter 2
A General Nonlinear Mathematical Model of DC/DC Converter

Switching converter system consists of two parts: power stage main circuit and control circuit. The power stage main circuit contains nonlinear components such as semiconductor switches, and the control circuit is usually composed of linear components. Therefore, the modeling of the power stage circuit is the key to the modeling of the entire system.

2.1 Basic Topology and Constraints of DC/DC Converters

The traditional PWM switching converters includes the two basic topologies, i.e., Buck and Boost, and other four derived topologies such as Buck-boost, Cuk, Zeta, and Sepic, as shown in Fig. 2.1. The corresponding equivalent operation model of PWM switch is shown in a dashed box in Fig. 2.2, where V_b, I_b denote the voltage and current buffers respectively, and V_o, I_o represent the voltage-type and current-type load respectively.

It is clear that the basic converters have the following topological characteristics [1].

(1) Inclusion of an active switch and a passive switch.
(2) Planar trapezoidal structure.
(3) No coupled inductor.

Each basic topology can be divided into two parts or three parts, that is, the input, intermediate and output parts, as shown in Fig. 2.3. It should be noted that Buck and Boost are two most basic converters, and both of them only consist of the input part and output part.

The input section is composed of a voltage/current source and an active switch and it has the following constraints [1]. (1) It contains only a source (voltage-type or current-type), and the current source can be realize by connecting the voltage source with a large inductor in series. (2) According to the characteristics of the

Y. Chen and B. Zhang, *Equivalent-Small-Parameter Analysis of DC/DC Switched-Mode Converter*, CPSS Power Electronics Series,
https://doi.org/10.1007/978-981-13-2574-8_2

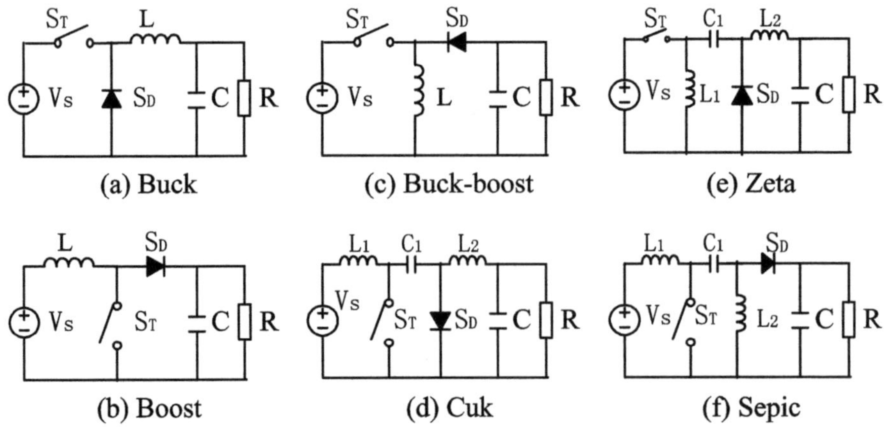

Fig. 2.1 Basic PWM DC/DC converters

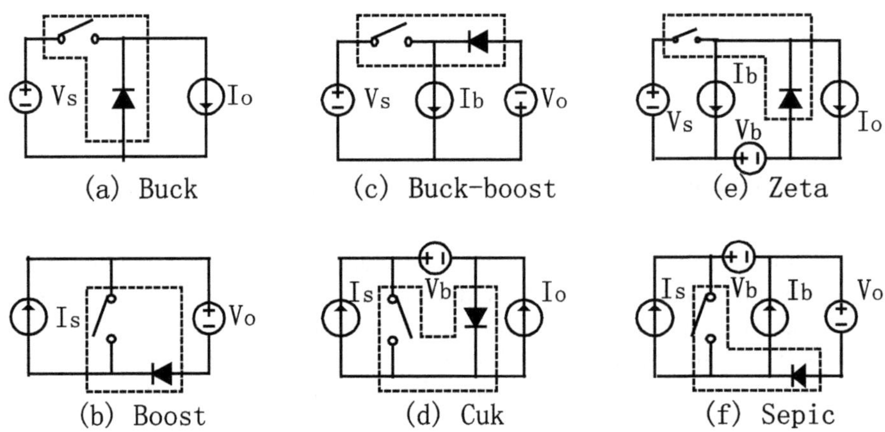

Fig. 2.2 Basic PWM DC/DC converters' operation models

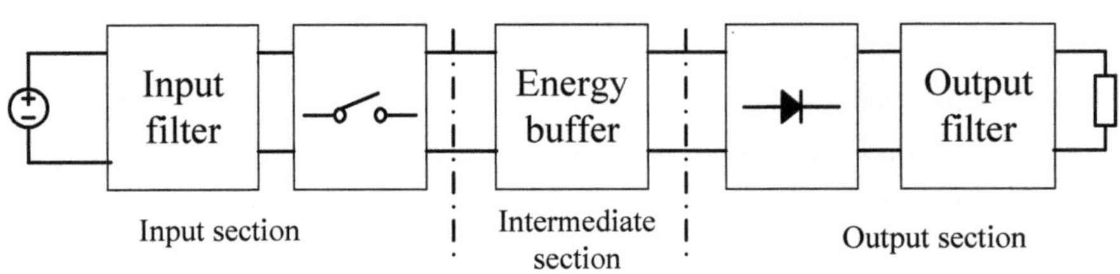

Fig. 2.3 Block diagram of basic converters

ideal voltage source and current source, the input part has only two forms, namely, the voltage source is connected in series with the switch, or the current source is connected in parallel with the switch.

The intermediate section is usually a combination of two basic linear energy storage elements, i.e., L and C (dissipative components (R) are generally omitted). The inductor is considered as a current buffer and the capacitor is regarded as a voltage buffer. According to the principle that the inductor current and the capacitor voltage cannot be mutated, the constraint on the intermediate part is that each branch contains only one current buffer or one voltage buffer, and the intermediate part can only contain up to two different types of buffers, namely a current buffer and a voltage buffer.

The output part consists of a passive switch (such as rectifier diode) and the load. Basically, the load can be divided into two types: voltage-type and current-type, which can be regarded as an ideal voltage source and ideal current source respectively. The rectifier diode is considered as the ideal switch. According to the characteristics of the voltage source current source, similar to the aforementioned input part, only two structures of the output part are effective, namely, the rectifier diode is connected in series with the voltage-type load, or the rectifier diode is connected in parallel with the current-type load.

The structure of the basic converter is connected in the following two ways:

(1) Voltage Source–Current Buffer–Voltage Buffer–Current Type Load.
(2) Current Source–Voltage Buffer–Current Buffer–Voltage Type Load.

According to the above constraints of the basic topologies, it can easily be derived that the basic structure can only have six forms as shown in Fig. 2.2. In other words, Fig. 2.1 can be seen as the circuit implementation of Fig. 2.2. Under the premise of ideal components, regarding both the buffer and the load as the ideal source, some topology structural principles are proposed in [1] according to the characteristics of the ideal source and the simplified rules of series and parallel connection, which is,

(1) A topology should include at least two parts: the input and the output; (2) According to the form of the input, the basic topology can be divided into two categories. One is that the input is voltage source type, including Buck, Buck-boost and Zeta. The other is that the input is current source type, including Boost, Cuk and Sepic. The current source in the actual circuit is always achieved by the connection of voltage source and the inductor in series. In [2] they are called Buck families and Boost families respectively. Buck and Boost are considered to be the basic converter units (BCU), while Buck-boost, Zeta, and Cuk Sepic can be constructed by its corresponding BCU and suitable linear feedback network. (3) The relationship between the two families is duality [3]. According to the principle of duality, the voltage source is dual to current source, and the branches in series connection are dual to the branches in parallel connection, etc. It can be seen easily from Fig. 2.2 that Buck and Boost, Buck-boost and Cuk, Zeta and Sepic are mutually dual.

2.2 Operating Modes of DC/DC Converters

The operating modes of PWM DC/DC converters can be divided into continuous conduction mode (CCM) and discontinuous conduction mode (DCM) [4]. The converter has two sub-topologies in each switching period at CCM, and three or more linear sub-topologies at DCM. The steady-state and transient characteristics of the converter are completely different when operating at different modes [5].

The premise that the converter can work in the CCM is that the diode must be gated on during off-state of the transistor and it must be reversed during the on-state of the transistor. In other words, at the steady state all the inductor current and capacitor voltage must meet the small ripple assumption [6], namely,

Assumption 1: small inductor current ripple, that is, the AC ripple of the inductor current is sufficiently small compared to its DC component.
Assumption 2: small capacitor voltage ripple, that is, the AC ripple of the capacitor voltage is sufficiently small compared to its DC component.

DCM can be further divided into the following three categories [6].

(1) Discontinuous Inductor Current Mode (DICM): The converter enters the DICM when the assumption of a small inductor current ripple is not satisfied. The DCM that we usually talk of is actually this case that the inductor current is discontinuous.
(2) Discontinuous Capacitor Voltage Mode (DCVM): When the assumption of a small capacitor voltage ripple is not satisfied, the converter enters DCVM.
(3) Discontinuous Quasi-Resonant Mode (DQRM): When both the small inductor current ripple and the small capacitor voltage ripple are not satisfied, the converter enters the DQRM.

Obviously, the circuit topology must meet certain conditions before entering the DCM mode [6]. For DICM, a cut set must be formed by the PWM switch and a group of inductors (at least one inductor). For DCVM, the condition is that a loop must be formed by the PWM switch and the capacitor and the possible voltage source (the output filter capacitor is excluded). In other words, the order of circuit must be higher than two-order, such as Buck converter with input filter and Cuk converter [6]. From another point of view, if the circuit enters the DVCM or DQRM, two switches must be gated on at the same time. It can be seen from Fig. 2.2 that for the basic second-order PWM converters, such as Buck, Boost, and Buck-Boost, their voltage sources will connected in parallel or short-circuited. Therefore, their possible operation modes of the converters are CCM and DICM.

Each basic converter contains only one passive switch and one active switch. If we use “1” to indicate that the switch is on and “0” to indicate that the switch is off, there can be only four combination results of the active switch S_T and passive switch S_D, which are denoted by A, B, C, and D, respectively. Table 2.1 shows the possible operating modes of the basic DC/DC switching converters and the operating sequences of each mode.

Table 2.1 Operation modes of DC/DC converters

Combination results	Switching function	S_T	S_D	Operating mode	Operating sequences
A	$\delta^{(1)}$	1	0	CCM	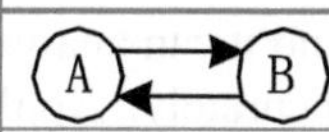
B	$\delta^{(2)}$	0	1	DICM	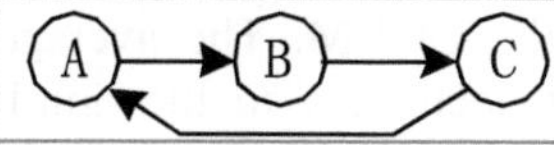
C	$\delta^{(3)}$	0	0	DCVM	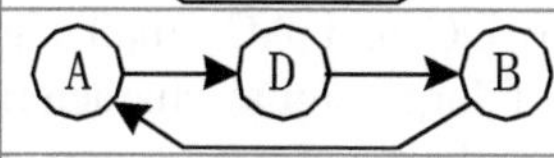
D	$\delta^{(4)}$	1	1	DQRM	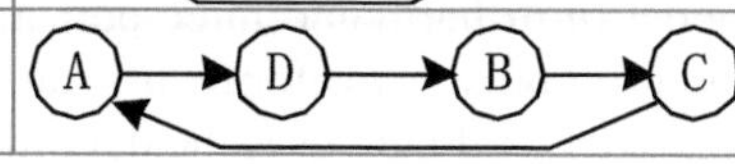

Four binary (1,0) periodic switching functions $\delta^{(i)}(i = 1, 2, 3, 4)$ is defined to represent the operating state of the switch. For $\delta^{(1)} = S_T\overline{S_D}$, it means that only when the active switch S_T is gated on and the passive switch S_D is gated off, we have $\delta^{(1)} = 1$, and in other time of switching periods we have $\delta^{(1')} = 0$. Similarly, $\delta^{(2)} = \overline{S_T}S_D$ means that its value is 1 only when S_T is turned off and S_D is on, $\delta^{(3)} = \overline{S_T S_D}$ means that its value can be 1 only when both S_T and S_D are off, and $\delta^{(4)} = S_T S_D$ means that its value can be 1 only when both S_T and S_D are on. The specific sequence and definition of $\delta^{(i)}$ are determined by the actual operation of the circuit. Apparently within one switching period, there is $\sum \delta^{(i)} = 1$.

2.3 General Nonlinear Modeling of DC/DC Converters

In this section, a unified nonlinear large-signal model of the converter operating in CCM and DCM is proposed based on the switching function defined in Sect. 2.2. For the convenience of analysis, it is assumed that all the switches and components in the circuit are ideal.

DCM usually occurs when the DC/DC converter is operating at light load. For low power applications, the converter is required to operate in DCM, even at full load [7]. In the PFC circuit, the converters are specially designed to work in DCM [8–10]. Hence, it is of great importance to modeling the converter under DCM. Currently, the method of analyzing PWM converter in DCM mainly focus on the state space averaging method [4]. Taking DICM as an example, the inductor current drops to zero before the end of the switching period and remains at zero until the start of the next cycle, so the inductor current is always zero at the beginning and end of each cycle. The traditional state space averaging method can use it as a constraint to obtain a reduced-order model that no longer contains the inductor current [4]. But the reduced-order model is inaccurate enough to analyze the dynamic characteristics of the system [7]. And in the PFC circuit, the inductor current is the ultimate control target, so the disappearance of the inductor current in this model is not what we want. A full-order equivalent duty cycle model of DICM is proposed in [6], and it can analyze the AC small-signal dynamic characteristics of the system more accurately. Applying the principle of duality can further easily analyze the DCVM. The average switched model proposed in [11] applies the averaging technique to the non-linear part of the converter circuit and directly obtain the DC and AC small signal model of the PWM converter, but for the analysis of large signal characteristics, it must resort to a more complex circuit model (n-th harmonic three-port model) [12]. In [13], a unified modeling method based on pulse waveforms first analyzes the operation of each sub-topology of the converter, and then uses the definition of the pulse function to obtain the expression of the equivalent source of the active and passive switches, without taking into account the cases in DCVM and DQRM. From the following analysis, it can be found that the method proposed in this chapter can directly obtain a unified equivalent large-signal non-linear circuit model that characterizes the converter working in CCM, DICM, DCVM, and DQRM. This model shows convenience to analyze closed-loop systems in DCM.

Figure 2.4a is a PWM switched model with the inductor L where 1 is the active terminal, usually connected with the source, 2 is the passive terminal, often connected with the load, 3 is the common terminal, often connected with the inductor. According to Fig. 2.2, for Cuk, Zeta, and Sepic converters, the inductor L in the figure represents the equivalent inductance in the circuit, that is, $L = L_1/L_2$. If $i_3 = 0$ before the end of a switching cycle, the circuit enters DICM where both switches are off. If $v_{12} = 0$ at certain time in a switching cycle, the circuit enters DVCM or DQRM where both switches are on. In order to characterize the discontinuous state of the inductor current, an imaginary diode D' is connected in series between the common terminal and the inductor, and it is assumed that $V_{D'} = 0$. Similarly, to characterize the discontinuity of the capacitor voltage, an imaginary switch is connected in parallel between 1 and 2, and it is assumed that $I_{S'} = 0$. According to the actual operation of the circuit, such assumptions are reasonable. Since the inductor current (or capacitor voltage) does not satisfy the piecewise linear characteristic at the discontinuous state, it can be essentially regarded as a non-linear element. By introducing an imaginary diode (switch) element, we can assume that the inductor current (or capacitor voltage) still satisfies piecewise linearity in a switching cycle and the non-linearity is caused by the switching element. A full-order model of the circuit can be built at discontinuous conducting state where S' is replaced with an equivalent controlled voltage source, D' is represented by an equivalent controlled current source, S_T is replaced with a controlled current source or voltage source, and S_D is replaced with a controlled voltage source or current source. The value of the controlled source is determined by the product of the switching function and the value of the current flowing through the switch (or the voltage across the switch). Note that S_T and S_D cannot be replaced by the same kind of controlled source because if they were replaced both by current source or voltage source, there would be a cut-set consisted of pure current sources (or current

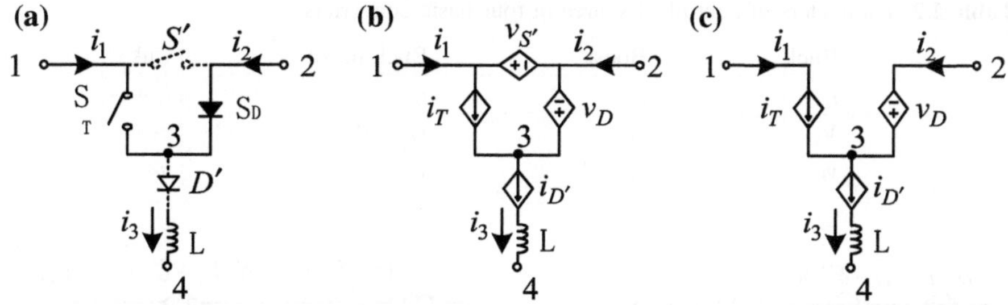

Fig. 2.4 **a** PWM switched model with inductor; **b** unified equivalent model for CCM and DCM; **c** unified equivalent model for CCM and DICM

sources and the inductors) or a loop consisted of pure voltage sources (or voltage sources and the capacitors). Here,we choose a controlled current source to represent S_T and a controlled voltage source to represent S_D. Thus, the equivalent controlled source model can be obtained as shown in Fig. 2.4b, which is exactly the unified nonlinear large-signal equivalent circuit model that can characterize the converter in CCM, DICM, DCVM, DQRM. It retains all the nonlinearity of the original circuit, wherein the value of each nonlinear time-varying controlled source is determined by (2.1) according to the definition of the switching function previously.

$$
\begin{aligned}
i_T &= \delta^{(1)} \cdot i_3 + \delta^{(2)} \cdot 0 + \delta^{(3)} \cdot 0 + \delta^{(4)} \cdot i_1 = \delta^{(1)} \cdot i_3 + \delta^{(4)} \cdot i_1 \\
v_D &= \delta^{(1)} \cdot v_{12} + \delta^{(2)} \cdot 0 + \delta^{(3)} \cdot v_{42} + \delta^{(4)} \cdot 0 = \delta^{(1)} \cdot v_{12} + \delta^{(3)} \cdot v_{42} \\
v_{S'} &= \delta^{(1)} \cdot v_{12} + \delta^{(2)} \cdot v_{12} + \delta^{(3)} \cdot v_{12} + \delta^{(4)} \cdot 0 = (\delta^{(1)} + \delta^{(2)} + \delta^{(3)}) \cdot v_{12} \\
i_{D'} &= \delta^{(1)} \cdot i_3 + \delta^{(2)} \cdot i_3 + \delta^{(3)} \cdot 0 + \delta^{(4)} \cdot i_3 = (\delta^{(1)} + \delta^{(2)} + \delta^{(4)}) \cdot i_3
\end{aligned}
\tag{2.1}
$$

where i_3 is the value of the inductor current when it is non-zero, and v_{12} is the value of the voltage between 1 and 2 when it is non-zero. According to the discussion of 2.2, it is impossible for the basic second-order converters to have two switches turned on at the same time. If we only consider CCM and DICM, that is, $\delta^{(4)} = 0$ and $\delta^{(1)} + \delta^{(2)} + \delta^{(3)} = 1$, we can get the unified nonlinear model of basic PWM DC/DC converters in CCM and DICM, as shown in Fig. 2.4c. Note that the value of the controlled source is determined by (2.2). For 4-order circuits such as Cuk, when the energy-transferring capacitor is connected between 1 and 2, the current flowing through the capacitor can be determined by i_{12} in (2.2).

$$
\begin{aligned}
i_T &= \delta^{(1)} \cdot i_3 \\
v_D &= \delta^{(1)} \cdot v_{12} + \delta^{(3)} \cdot v_{42} \\
i_{D'} &= (1 - \delta^{(3)}) \cdot i_3 \\
i_{12} &= i_1 - \delta^{(1)}(i_1 + i_2) - \delta^{(3)}(i_1 + i_2)
\end{aligned}
\tag{2.2}
$$

Table 2.2 Parameters of controlled source of four basic converters

	Buck	Boost	Buck-boost	Cuk
i_3	i_L	i_L	i_L	$i_1 + i_2$
v_{12}	V_s	v_C	$v_C + V_s$	v_1
v_{42}	v_C	$v_C - V_s$	v_C	v_C

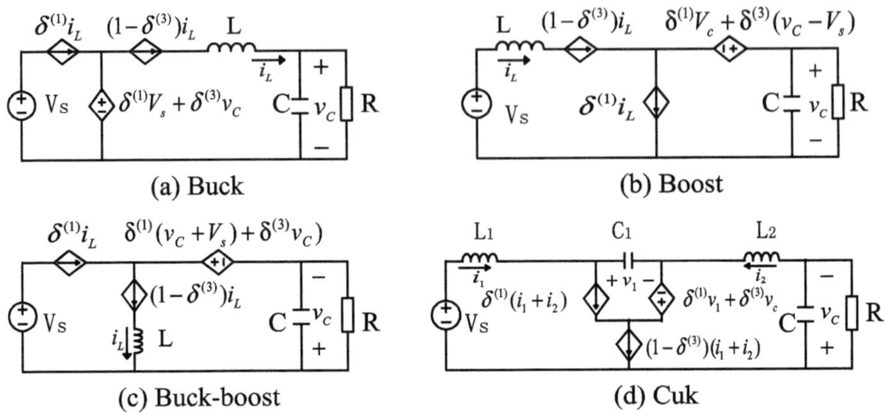

Fig. 2.5 Equivalent circuit models of four basic converters

Considering that DICM is the most common discontinuous mode that occurs in actual circuits, the DCM we refer to below denotes the case of DICM unless otherwise specified. Since there are not many practical applications of Zeta and Sepic converters, they won't be discussed below. Table 2.2 shows the relationship between the variables in (2.2) and the parameters of the four basic converters (Buck, Boost, Buck-boost, and Cuk).

The equivalent circuit model of the four basic converters can be derived as shown in Fig. 2.5. Applying KVL and KCL to the equivalent circuit leads to the unified time-varying mathematical model of PWM DC/DC converter,

$$G_0(p)\mathbf{x} + G_1(p)\delta^{(1)}(F_1x + \mathbf{e}_1) + G_3(p)\delta^{(3)}(x + \mathbf{e}_3) = \mathbf{u} \tag{2.3}$$

where x denote the state variables (inductor current and capacitor voltage), e_i is an constant vector, u is the input power vector, $G_i(p)$ is the coefficient matrix related to the circuit structure, F_1 indicates the strength of nonlinearity. Among the four converters, only the Buck converter has weak nonlinearity, so we have $F_1 = 0$ for Buck while $F_1 = 1$ for other three converters. Table 2.3 gives the mathematical description of the four basic converters. The results is the same as that in [13] when given $\delta^{(3)} = 1 - \delta^{(1)} - \delta^{(2)}$. It should be noted that for Cuk, the current flowing through the energy-transferring capacitor C_1 is determined by (2.2). That is because according to the definition of the controlled source in the model of Fig. 2.4c, i_1, i_2

Table 2.3 Mathematical description of four basic converters

	Mathematical description
Buck	$\begin{bmatrix} p & \frac{1}{L} \\ \frac{-1}{C} & p+\frac{1}{RC} \end{bmatrix} \cdot \begin{bmatrix} i_L \\ v_c \end{bmatrix} + \begin{bmatrix} 0 & \frac{-1}{L} \\ 0 & 0 \end{bmatrix} \cdot \delta^{(1)} \cdot \begin{bmatrix} 0 \\ V_s \end{bmatrix} + \begin{bmatrix} 0 & \frac{-1}{L} \\ \frac{1}{C} & 0 \end{bmatrix} \cdot \delta^{(3)} \cdot \begin{bmatrix} i_L \\ v_c \end{bmatrix} = \begin{bmatrix} 0 \\ 0 \end{bmatrix}$
Boost	$\begin{bmatrix} p & \frac{1}{L} \\ \frac{-1}{C} & p+\frac{1}{RC} \end{bmatrix} \cdot \begin{bmatrix} i_L \\ v_c \end{bmatrix} + \begin{bmatrix} 0 & \frac{-1}{L} \\ \frac{1}{C} & 0 \end{bmatrix} \cdot \delta^{(1)} \cdot \begin{bmatrix} i_L \\ v_c \end{bmatrix} + \begin{bmatrix} 0 & \frac{-1}{L} \\ \frac{1}{C} & 0 \end{bmatrix} \cdot \delta^{(3)} \cdot \left(\begin{bmatrix} i_L \\ v_c \end{bmatrix} + \begin{bmatrix} 0 \\ -V_s \end{bmatrix} \right) = \begin{bmatrix} \frac{1}{L}V_s \\ 0 \end{bmatrix}$
Buck-boost	$\begin{bmatrix} p & \frac{1}{L} \\ -\frac{1}{C} & p+\frac{1}{RC} \end{bmatrix} \cdot \begin{bmatrix} i_L \\ v_c \end{bmatrix} + \begin{bmatrix} 0 & \frac{-1}{L} \\ \frac{1}{C} & 0 \end{bmatrix} \cdot \delta^{(1)} \cdot \left(\begin{bmatrix} i_L \\ v_c \end{bmatrix} + \begin{bmatrix} 0 \\ V_s \end{bmatrix} \right) + \begin{bmatrix} 0 & \frac{-1}{L} \\ \frac{1}{C} & 0 \end{bmatrix} \cdot \delta^{(3)} \cdot \begin{bmatrix} i_L \\ v_c \end{bmatrix} = \begin{bmatrix} 0 \\ 0 \end{bmatrix}$
Cuk	$\begin{bmatrix} p & 0 & \frac{1}{L_1} & 0 \\ 0 & p & 0 & \frac{1}{L_2} \\ \frac{-1}{C_1} & 0 & p & 0 \\ 0 & \frac{-1}{C} & 0 & p+\frac{1}{RC} \end{bmatrix} \cdot \begin{bmatrix} i_1 \\ i_2 \\ v_1 \\ v_c \end{bmatrix} + \begin{bmatrix} 0 & 0 & \frac{-1}{L_1} & 0 \\ 0 & 0 & \frac{-1}{L_2} & 0 \\ \frac{1}{C_1} & \frac{1}{C_1} & 0 & 0 \\ 0 & 0 & 0 & 0 \end{bmatrix} \delta^{(1)} \cdot \begin{bmatrix} i_1 \\ i_2 \\ v_1 \\ v_c \end{bmatrix}$ $+ \begin{bmatrix} 0 & 0 & 0 & \frac{-1}{L_1} \\ 0 & 0 & 0 & \frac{-1}{L_2} \\ \frac{1}{C_1} & \frac{1}{C_1} & 0 & 0 \\ 0 & 0 & 0 & 0 \end{bmatrix} \delta^{(3)} \cdot \begin{bmatrix} i_1 \\ i_2 \\ v_1 \\ v_c \end{bmatrix} = \begin{bmatrix} \frac{1}{L_1}V_s \\ 0 \\ 0 \\ 0 \end{bmatrix}$

only represent the value of the inductor current during the time of continuous current, and does not represent the value during the entire switching cycle. Therefore, the current through the capacitor C_1 cannot be directly obtained by applying the KVL to Fig. 2.5d.

It can be observed from the follow-up analysis of this chapter that the equivalent circuit model represented by $\delta^{(1)}, \delta^{(3)}$ manifests convenience to solve the time-varying equation of the closed loop systems of the PWM converters in DCM, since $\delta^{(1)}, \delta^{(3)}$ can be expanded into the same series form.

2.4 Summary

This chapter introduces a unified nonlinear large signal model for PWM DC/DC converters in CCM and DCM, after defining the switching functions. The model can preserve all the nonlinearity of the original circuit. Such modeling method based on switching functions can be easily applied to the analysis and construction of resonant converters.

References

1. Liu KH, Lee FC (1988) Topological constraints on basic PWM converters. IEEE PESC Rec 164–172
2. Wu TF, Chen YK (1998) Modeling PWM DC/DC converters out of basic converter units. IEEE Trans PE 13(5):370–381
3. Tymerski R, Vorperian V (1986) Generation classification and analysis of switched—mode. Dc-to-dc converters by the use of converter cells, IEEE International telecommunications energy conference

4. Cuk S, Middlebrook RD (1977) A general unified approach to modelling switching DC-to-DC converters in discontinuous conduction mode. IEEE PESC Rec 36–57
5. Femia N, Tucci V (1994) On the modeling of PWM converters for large signal analysis in discontinuous conduction mode. IEEE Trans PE 9(5):487–496
6. Maksimovic D, Cuk S (1991) A unified analysis of PWM converters in discontinuous modes. IEEE Trans PE 6(3):476–490
7. Sun J, Mitchell DM, Grenel MF et al (1998) Modeling of PWM converters in discontinuous conduction mode—a reexamination. APEC Rec 615–622
8. Sebastian J, Cobos JA et al (1995) The determination of the boundaries between continuous and discontinuous conduction modes in PWM DC-to-DC converters used as power factor preregulators. IEEE Trans PE 10(5):574–582
9. Wang J, Dunford WD, Mauch K (1997) Analysis of a ripple-free input-current boost converter with discontinuous conduction characteristics. IEEE Trans PE 12(4):684–694
10. Lee YS, Wang SJ, Hui SYR (1997) Modeling, analysis, and application of buck converters in discontinuous-input-voltage mode operation. IEEE Trans PE 12(2):350–360
11. van Dijk E, Spruijt HJN, O'Sullivan DM et al (1995) PWM-switch modeling of DC/DC converters. IEEE Trans PE 10(6):659–664
12. Tymerski R, Vorperian V, Lee FC et al (1988) Nonlinear modelling of the PWM switch. IEEE PESC Rec 968–976
13. Wu Y, Hu AP, Budgett D, Malpas S, Dissanayake T (2010) Harmonic analysis of voltage-fed full-bridge converter for TET application. In: The 5th IEEE conference on industrial electronics and applications (ICIEA), pp 1040–1045

Chapter 3
Equivalent-Small-Parameters Method (ESPM)

3.1 Fundamental Principles of ESPM

Assuming a strongly nonlinear non-autonomous high-order system [2–6] can be described by the following formula:

$$a_n x^{(n)} + \cdots + a_2 x^{(2)} + a_1 x^{(1)} + a_0 x + f(x) = E \cos\tau \tag{3.1}$$

the right hand of which is an additional excitation signal, and $\tau = \omega t$. On the left hand of the equation, x is the state variable to be solved, $f(x)$ is a strong nonlinear function of state variable and its derivatives, which are expressed as:

$$x^{(\mathrm{n})} = \frac{d^{\mathrm{n}} x}{d\tau^{\mathrm{n}}} \tag{3.2}$$

in which n is a positive integer (n = 1, 2, 3, …). Let p represent the differential operator, i.e., $p = \frac{d}{d\tau}$, then the n-th order differential operator is $p^n = \frac{d^n}{d\tau^n}$, and Eq. (3.1) can be rewritten as:

$$(a_n p^n + \cdots + a_2 p^2 + a_1 p + a_0)x + f(x) = E \cos\tau \tag{3.3}$$

or simply written as:

$$g(p)x + f(x) = E \cos\tau \tag{3.4}$$

in which $g(p)$ is a polynomial function of p:

$$g(p) = a_n p^n + \cdots + a_2 p^2 + a_1 p + a_0 \tag{3.5}$$

Y. Chen and B. Zhang, *Equivalent-Small-Parameter Analysis of DC/DC Switched-Mode Converter*, CPSS Power Electronics Series,
https://doi.org/10.1007/978-981-13-2574-8_3

To find the periodic solution of the high-order strong nonlinear non-autonomous system (3.4), the solution is expanded into the following series in form of the sum of the main wave x_0 and correction x_i:

$$x = x_0 + \varepsilon^1 x_1 + \varepsilon^2 x_2 + \cdots + \varepsilon^i x_i + \cdots \tag{3.6}$$

where $\varepsilon = 1$, and it is only a symbolic mark introduced temporarily to indicate that the correction $\boldsymbol{x}_\mathbf{i}$ is much smaller than the term x_{i-1}, that is, $|x_{i-1}| < |x_i| \ll |x_0|$. Then the introduction of Eq. (3.6) into the function $f(x)$ leads to:

$$f(x) = f_0 + \varepsilon^1 f_1 + \varepsilon^2 f_2 + \cdots + \varepsilon^i f_i + \cdots \tag{3.7}$$

in which i is the order of corrections, and

$$f_0 = f(x_0) \tag{3.8a}$$

$$f_i = f(x_0, x_1, \ldots x_i), \quad (i = 1, 2 \ldots) \tag{3.8b}$$

The terms x_0 and f_0 are the main waves of x and $f(x)$, respectively. The terms x_i and f_i are the i-th order corrections of x and $f(x)$, respectively.

Assume that the main wave x_0 of the solution includes harmonic contents like:

$$x_0 = \sum_{n \in E_0} (A_{n0} \cos n\tau + B_{n0} \sin n\tau) \tag{3.9}$$

where the capital word E_0 represents the frequency set, for example, if n = 0, then the set $E_0 = \{0\}$, and it means the main wave x_0 has only DC component. If n = 0, 1, 3, then the set $E_0 = \{0, 1, 3\}$, and it means that the main wave x_0 has DC component, basic harmonic and 3rd harmonic. Actually in many physical models, the main wave x_0 has only a limited number of harmonics.

Introduce Eq. (3.9) into Eq. (3.8a), then one can obtain the expression for $f_0 = f(x_0)$. Along with the introduction process, new kinds of harmonics appear, sine the existence of the nonlinear function $f(.)$. So the term $\boldsymbol{f}_0$ is written as:

$$f_0 = f_{0m} + \varepsilon R_1 \tag{3.10}$$

where f_{0m} represents the main wave which has the same frequency set with $\boldsymbol{x}_0$, and it is called as the main term, whereas the term R_1 represents the newly generated harmonics in the above introduction procedures, that is, $R_1 = f_0 - f_{0m}$, and it is called as the remaining term. Note that, the term R_1 is usually much smaller than the term $\boldsymbol{f}_{0m}$, so it is also marked with the symbol ε. Generally, the main term $\boldsymbol{f}_{0m}$ and the remaining term R_1 are composed by different kinds of harmonics, so they can be expressed as:

$$f_{0m} = \sum_{n \in E_0} (C_{k0} \cos n\tau + D_{k0} \sin n\tau) \tag{3.11a}$$

$$R_1 = f_0 - f_{0m} = \sum_{k \in E_1} \left(C'_{k1} \cos k\tau + D'_{k1} \sin k\tau \right) \tag{3.11b}$$

where the capital words E_0 and E_1 represent the frequency set of $\boldsymbol{x}_0$ and $\boldsymbol{R}_1$, respectively. Additionally, the coefficient matrix C_{k0} and D_{k0} are linear function of A_{k0} and B_{k0}, respectively, yet C'_{k1} and D'_{k1} are nonlinear function of A_{k0} and B_{k0}, respectively.

Further, assume that the 1st correction is:

$$x_1 = \sum_{k \in E_1} (A_{k1} \cos k\tau + B_{k1} \sin k\tau) \tag{3.12}$$

which has the same frequency set E_1 with the remaining term R_1. Once the harmonic content of the main wave x_0 is obtained, one can introduce x_0 and x_1 into the expression for f_1:

$$f_1 = f_1(x_0, x_1) = f_{1m} + \varepsilon R_2 \tag{3.13}$$

Similarly, $f_{1\mathrm{m}}$ is the main component of f_1, which has the same frequency set with x_1. The remaining term $R_2 = f_1 - f_{1m}$, it is also marked with the symbol ε, which means the harmonic amplitudes of R_2 are much smaller than those of $\boldsymbol{f}_{1\mathrm{m}}$. Generally, the term R_2 includes those harmonic contents which are not in the set E_1, and the harmonic contents of which can be used to determine the harmonic set of x_2.

$$f_{1m} = \sum_{k \in E_1} (C_{k1} \cos k\tau + D_{k1} \sin k\tau) \tag{3.14a}$$

$$R_2 = f_1 - f_{1m} = \sum_{k \in E_2} \left(C'_{k2} \cos k\tau + D'_{k2} \sin k\tau \right) \tag{3.14b}$$

where C_{k1} and D_{k1} are linear functions of A_{k1} and B_{k1}, respectively, whereas C'_{k2} and D'_{k2} are nonlinear functions of A_{k1} and B_{k1} respectively.

Similarly, Let

$$x_i = \sum_{k \in E_i} (A_{ki} \cos k\tau + B_{ki} \sin k\tau) \tag{3.15}$$

And one can obtain:

$$f_i = f(x_0, x_1, \ldots, x_i) = f_{im} + \varepsilon R_{i+1} \tag{3.16}$$

$$f_{im} = \sum_{k \in E_i} (C_{ki} \cos k\tau + D_{k1} \sin k\tau) \tag{3.17a}$$

$$R_{i+1} = f_i - f_{im} = \sum_{k \in E_{i+1}} \left(C'_{k(i+1)} \cos k\tau + D'_{k(i+1)} \sin k\tau \right) \tag{3.17b}$$

where E_i and E_{i+1} are the harmonic sets of x_i and R_{i+1}, respectively. Additionally, the set E_{i+1} can be used to determine the harmonic contents of $\boldsymbol{x}_{i+1}$. The term f_{im} represents the main components of f_i, which has the same harmonic contents with $\boldsymbol{x}_i$. The term R_{i+1} represents the remaining components of f_i, which includes the harmonics different with x_i, and the amplitudes of which are usually much smaller than those of f_{im}.

Introduce Eq. (3.16), into Eq. (3.7), one obtains:

$$f = (f_{0m} + \varepsilon f_{1m} + \varepsilon^2 f_{2m} + \cdots) + (\varepsilon R_1 + \varepsilon^2 R_2 + \cdots) \tag{3.18}$$

According to the above equation, one can find that the nonlinear function can be divided into two parts. The first one includes the main waves of the nonlinear function and the main components of corrections, which have absolutely the same frequency set with the frequency set of each correction. The second part is the remaining part, which contains the newly generated harmonics. By substituting Eqs. (3.18) and (3.6) into Eq. (3.4), and by equaling those terms on both sides of equations which have the same indicator parameter ε^i, one can acquire the following equation set:

$$g(p)x_0 + f_{0m} = E \cos \tau \tag{3.19a}$$

$$g(p)x_1 + f_{1m} = -R_1 \tag{3.19b}$$

$$g(p)x_2 + f_{2m} = -R_2 \quad \cdots \tag{3.19c}$$

Note that each one of Eqs. (3.19) has a fixed frequency set. According to Eqs. (3.9)–(3.17), one can conclude that, the term f_{0m} is the main term of f_0, which has the same harmonic contents of x_0, whereas the term R_1 is the remaining part of f_0, which includes higher order harmonics than those in x_0, and thus the amplitudes of which are generally smaller than those in f_{0m}. The harmonic contents of R_1 are used to determine the harmonic contents of the first order correction x_1 and the harmonic set E_1 of f_{1m}. So on and so forth, the harmonic sets of x_i and f_{im} can be determined by the harmonic set E_i of the remaining term R_i. Accordingly, once the harmonic contents of the main wave x_0 is determined, then the harmonic contents of x_i can be calculated step by step. Besides, the harmonic set E_O of x_0 is heavily dependent on the physical properties associated with Eq. (3.1). For instance, for DC/DC switching converters, when the converter is steady-state, the main components of the output voltage and the inductor current are DC values, so the set $E_0 = \{0\}$, and the following set E_i can be determined step by step in the solving procedures.

By adopting the principle of harmonic balance [1], Eq. (3.19) can be solved one by one, in which (3.19a) is related with the main solution x_0, and the others are correction equation, which can be used to get the higher order harmonics of the system and the corresponding corrections. As a result, the solution of the system (3.1) has the form as follows:

$$x = x_0 + x_1 + x_2 + \cdots + x_i + \cdots \tag{3.20}$$

For those circuits with low-pass properties, such as switching power converters, the higher harmonics are generally quite small, thus the accuracy can be ensured by only considering two times of corrections.

3.2 A Simplified Edition

According to the results of the last section, it can be deduced that the approximate periodic solution of a high-order strong nonlinear system is in the form Eq. (3.20), the main wave and corrections of which are expressed by Fourier series in the form of trigonometric functions:

$$x_0 = \sum_{n \in E_0} (A_{n0} \cos n\tau + B_{n0} \sin n\tau) \tag{3.21a}$$

$$x_i = \sum_{k \in E_i} (A_{ki} \cos k\tau + B_{ki} \sin k\tau) \quad (\mathrm{i} = 1, 2, 3\ldots) \tag{3.21b}$$

Actually, to simplify the analyzing procedure, one can expand sinusoidal and cosine functions by Euler equation:

$$\cos k\tau = \frac{1}{2}(e^{jk\tau} + e^{-jk\tau}) \tag{3.22a}$$

$$\sin k\tau = -\frac{1}{2}j(e^{jk\tau} - e^{-jk\tau}) \tag{3.22b}$$

Then the above equations are transformed into the following complex form of Fourier series:

$$x_0 = \sum_{n \in E_0} \left(a_{n0} e^{jn\tau} + \bar{a}_{k0} e^{-jn\tau}\right) \tag{3.23a}$$

$$x_i = \sum_{k \in E_i} \left(a_{ki} e^{jk\tau} + \bar{a}_{ki} e^{-jk\tau} \right) \tag{3.23b}$$

where the coefficients are:

$$a_{k0} = \frac{1}{2}(A_{k0} - jB_{k0}), \quad \bar{a}_{k0} = \frac{1}{2}(A_{k0} + jB_{k0})$$
$$a_{ki} = \frac{1}{2}(A_{ki} - jB_{ki}), \quad \bar{a}_{ki} = \frac{1}{2}(A_{ki} + jB_{ki})$$

According to the property of exponential function, that is,

$$p^n(e^{jk\tau}) = \frac{d^n(e^{jk\tau})}{d\tau^n} = (jk)^n e^{jk\tau} \quad (n = 1, 2, 3, \ldots) \tag{3.24}$$

in which the term p is the differential operator, and the term $p^{\mathrm{n}}(.)$ is the n-th order derivative of p. The above formula can be used to simplify the solution of Eq. (3.19). Moreover, $\bar{a}_{ki} e^{-jk\tau}$ is the conjugate term of $a_{ki} e^{jk\tau}$, as a result, the calculating process is greatly simplified.

3.3 Error Estimation for the Approximate Periodic Solutions

Rewrite the strong nonlinear non-autonomous system shown in Eq. (3.4) as follows:

$$g(p)x + f(x) = E \cos \tau \tag{3.24}$$

in which the linear and nonlinear parts in Eq. (3.24) are defined as:

$$f_L = g(p)x - E \cos \tau \tag{3.25a}$$

$$f_N = f(x) \tag{3.25b}$$

Assume that the approximate periodic expression of the system has the following form:

$$x \simeq \sum_{k=0}^{K} (A_k \cos k\tau + B_k \sin k\tau) \tag{3.26}$$

in which $x \neq x_0$. Then the introduction of the above expression into Eq. (3.25a) will lead to:

$$f_L = \sum_{k=0}^{K} (A_{kL} \cos k\tau + B_{kL} \sin k\tau) \tag{3.27}$$

Substitute the Eq. (3.26) into the Eq. (3.25b) and select those having the same frequency component as the approximate periodic solution x, then one gets:

$$f_{Nm} = \sum_{k=0}^{K} (A_{kN} \cos k\tau + B_{kN} \sin k\tau) \tag{3.28}$$

Define

$$\delta_k \triangleq \sqrt{\frac{(A_{kL} - A_{kN})^2 + (B_{kL} - B_{kN})^2}{A_{kL}^2 + B_{kl}^2}} \tag{3.29}$$

where k = 0, 1, 2, … K. If the following condition is fulfilled,

$$\delta_k < \gamma_k \tag{3.30}$$

Then Eq. (3.26) is a valid periodic solution. Note that the term γ_k in the above equation is a constant number, and the value of which can be selected according to the relevant requirements. If the formula (3.30) is not satisfied, it means that the error of the periodic solution is too large, and the k-th harmonic should be considered in the correcting process [3].

If only the two main harmonic components of the periodic solution are examined, then the concept of mean error is adopted, that is,

$$\delta_{AV}^{(2)} \triangleq \sqrt{\frac{\delta_1^2 + \delta_H^2}{2}} \tag{3.31}$$

where δ_H represents which has the maximum values of all δ_k $(k \neq 1)$. Similarly, one can define the average error that counts more harmonics. By using the above criterion, one can estimate the overall error level of the approximate periodic solution. Meanwhile, one can also compare the results of different analytical methods. It should be noted that, δ_H should be selected from those have the same max value of k when comparing the error obtained by different kinds of methods.

3.4 General Solutions for Strong Nonlinear High-Order Systems Described by Matrix Equations

Rewrite the high-order strong nonlinear non-autonomous system shown in Eq. (3.1) as follows:

$$a_n x^{(n)} + \cdots + a_2 x^{(2)} + a_1 x^{(1)} + a_0 x + f(x) = E \cos\tau \tag{3.32}$$

in which the nonlinear function $f(x)$ is related to x and its derivatives $x^{(n)}$ ($n = 1, 2, \ldots$), and it can be described by:

$$f = f\left(x, \dot{x}, \ldots, x^{(n)}\right) \tag{3.33}$$

Let $x_1 = x$, $x^{(1)} = \frac{dx_1}{d\tau} = x_2$, $x^{(2)} = \frac{d^2x}{d\tau^2} = \frac{dx_2}{d\tau} = x_3$, …, $x^{(n-1)} = \frac{d^{n-1}x}{d\tau^{n-1}} = \frac{dx_{n-1}}{d\tau} = x_n$, and $x^{(n)} = \frac{d^n x}{d\tau^n} = \frac{dx_n}{d\tau}$ then one can deduce the following equation set by combining Eq. (3.32)

$$\begin{cases} \frac{dx_1}{d\tau} - x_2 = 0 \\ \frac{dx_2}{d\tau} - x_3 = 0 \\ \cdots \\ \frac{dx_{n-1}}{d\tau} - x_n = 0 \\ a_n \frac{dx_n}{d\tau} + \cdots + a_2 \frac{dx_2}{d\tau} + a_1 \frac{dx_1}{d\tau} + a_0 x + f(x_1, x_2, \ldots x_n) = E \cos\tau \end{cases} \tag{3.34}$$

in which the term $p = d/dt$, represent the differential operator, and $\mathbf{x} = [x_1, x_2, \ldots x_n]^{\mathrm{Tr}}$(The superscript "Tr" represents the transpose of the matrix) is a $n \times 1$ state vector. Then Eq. (3.34) can be described by the following matrix equation [7–10]:

$$\mathbf{G}_0(p)\mathbf{x} + \mathbf{G}_1(p) \cdot \mathbf{f}(\mathbf{x}) = \mathbf{u}(\tau) \tag{3.35}$$

where both $G_0(p)$ and $G_1(p)$ are $n \times n$ coefficient matrices related with the differential operator p and the highest number of p equals to 1. $\mathbf{u}(\tau)$ is the input signal vector. Additionally, $\mathbf{f}(\mathbf{x})$ is a nonlinear function of the vector $\mathbf{x}$, also, it is a $n \times 1$ vector. Equation (3.35) is equivalent to Eq. (3.1), which can be used to describe general high-order strong nonlinear non-autonomous systems represented by matrix functions.

Similarly, the Eq. (3.35) can be solved by the ESP method presented in Sect. 3.1, in which the steady-state periodic solution $\mathbf{x}$ is expanded to the series form of the sum of the main components and the corrections:

$$\mathbf{x} = \mathbf{x}_0 + \sum_{i=1}^{\infty} \varepsilon^i \mathbf{x}_i \tag{3.36}$$

where the term $\varepsilon = 1$ is the indicator parameter to indicate that the correction $\mathbf{x_i}$ is much smaller than the term $\mathbf{x_{i-1}}$, that is, $|\mathbf{x}_{i-1}| < |\mathbf{x}_i| \ll |\mathbf{x}_0|$.

The introduction of Eq. (3.36) into $\mathbf{f}(\mathbf{x})$ leads to:

$$\mathbf{f}(\mathbf{x}) = \mathbf{f}_0 + \sum_{i=1}^{\infty} \varepsilon^i \mathbf{f}_i \tag{3.37}$$

where

$$\mathbf{f}_0 = \mathbf{f}(\mathbf{x}_0) \tag{3.38a}$$

$$\mathbf{f}_i = \mathbf{f}(\mathbf{x}_0, \mathbf{x}_1, \ldots \mathbf{x}_i), \quad (i = 1, 2\ldots) \tag{3.38b}$$

and in which the term i is the Approximate order, $\mathbf{x}_0$ and $\mathbf{f}_0$ are the main components of $\mathbf{x}$ and $\mathbf{f}(\mathbf{x})$, respectively. Additionally, $\mathbf{x}_i$ and $\mathbf{f}_i$ are the i-th corrections of $\mathbf{x}$ and $\mathbf{f}(\mathbf{x})$, respectively. Here $\mathbf{x}_0$ and $\mathbf{x}_i$ can be expressed as the complex form Fourier series:

$$\mathbf{x}_0 = \sum_{n \in E_0} \left(\mathbf{a_{n0}} e^{jn\tau} + \bar{\mathbf{a}}_{n0} e^{-jn\tau}\right) \tag{3.39a}$$

$$\mathbf{x}_i = \sum_{k \in E_i} \left(\mathbf{a_{ki}} e^{jk\tau} + \bar{\mathbf{a}}_{\mathbf{ki}} e^{-jk\tau}\right) \tag{3.39b}$$

In the above equation, the term $\tau = \omega t = 2\pi t/T$, the coefficient $\mathbf{a_{n0}}$ can be used to calculate the amplitude of the n-th harmonic in the main wave, and the coefficient $\mathbf{a_{ki}}$ can be used to calculate the amplitude of the k-th harmonic in the i-th correction.

The substitution of Eq. (3.39) into Eq. (3.38) will bring the expressions for the main component and each order of corrections of the nonlinear vector function $\mathbf{f}_i (i = 0, 1, 2\ldots)$, which can be divided into the summation form of main components and remaining terms:

$$\mathbf{f}_i = \mathbf{f}(\mathbf{x}_0, \mathbf{x}_1, \ldots, \mathbf{x}_i) = \mathbf{f}_{im} + \varepsilon \mathbf{R}_{i+1} \tag{3.40}$$

where the terms $\mathbf{f}_{im}$ and $\mathbf{R}_{i+1}$ can be expression by the following complex form Fourier series:

$$\mathbf{f}_{0m} = \sum_{k \in E_0} \left(\mathbf{h}_{n0} e^{jn\tau} + \bar{\mathbf{h}}_{n0} e^{-jn\tau}\right) \tag{3.41a}$$

$$\mathbf{f}_{im} = \sum_{k \in E_i} \left(\mathbf{h}_{ki} e^{jk\tau} + \bar{\mathbf{h}}_{ki} e^{-jk\tau}\right) \tag{3.41b}$$

$$\mathbf{R}_{i+1} = \sum_{k \in E_{i+1}} \left(\mathbf{h}'_{k(i+1)} e^{jk\tau} + \bar{\mathbf{h}}'_{k(i+1)} e^{-jk\tau}\right) \tag{3.41c}$$

Substitute Eq. (3.40) into Eq. (3.37), and it can be deduced:

$$\mathbf{f} = (\mathbf{f}_{0m} + \varepsilon\mathbf{f}_{1m} + \varepsilon^2\mathbf{f}_{2m} + \cdots) + (\varepsilon\mathbf{R}_1 + \varepsilon^2\mathbf{R}_2 + \cdots) \tag{3.42}$$

Then introduce Eqs. (3.36) and (3.42) into Eq. (3.35), and collect the terms with the same order of indicator parameter ε^i on the left- and right-hand sides. Then one obtains the equivalent linear differential equations:

$$\begin{cases} \mathbf{G}_0(p)\mathbf{x}_0 + \mathbf{G}_1(p)\mathbf{f}_{0m} = \mathbf{u} \\ \mathbf{G}_0(p)\mathbf{x}_1 + \mathbf{G}_1(p)\mathbf{f}_{1m} = -\mathbf{G}_1(p)\mathbf{R}_1 \\ \mathbf{G}_0(p)\mathbf{x}_2 + \mathbf{G}_1(p)\mathbf{f}_{2m} = -\mathbf{G}_1(p)\mathbf{R}_2 \\ \qquad \cdots \end{cases} \tag{3.43}$$

Similarly, based on the principle of harmonic balance, the above equations can also be calculated like Eq. (3.19).

At last, one can get the steady-state periodic solution:

$$\mathbf{x} = \mathbf{x}_0 + \mathbf{x}_1 + \mathbf{x}_2 + \cdots + \mathbf{x}_i + \cdots \tag{3.44}$$

Consider a more general situation to extend the above method to the analysis of general nonlinear time-varying systems, and then one can write the system which has N nonlinear vector functions as:

$$G_0(p)\mathbf{x} + \sum_{n=1}^{N} G_n(p) \cdot \mathbf{f}^{(n)}(\mathbf{x}) = \mathbf{u} \tag{3.45}$$

in which N is a positive integer. As a result of the method presented above, one obtains the iterative equations for solving the steady-state periodic solution:

$$G_0(p)\mathbf{x}_0 + \sum_{n=1}^{N} G_n(p)\mathbf{f}_{0m}^{(n)} = \mathbf{u} \tag{3.46a}$$

$$G_0(p)\mathbf{x}_1 + \sum_{n=1}^{N} G_n(p)\mathbf{f}_{1m}^{(n)} = -\sum_{n=1}^{N} G_n(p)\mathbf{R}_1^{(n)} \tag{3.46b}$$

$$G_0(p)\mathbf{x}_2 + \sum_{n=1}^{N} G_n(p)\mathbf{f}_{2m}^{(n)} = -\sum_{n=1}^{N} G_n(p)\mathbf{R}_2^{(n)} \tag{3.46c}$$

$$\cdots$$

Obviously, the above equations can be calculated like Eq. (3.43) based on the principle of harmonic balance, since each one of them has a fixed frequency set.

3.5 Summary

The Equivalent-Small-Parameter method [2–6] is somewhat a kind of approximate analytical method, which can be adopted in the analysis of high-order nonlinear system, and the characteristics of which are:

(1) The advantages of both harmonic balance and perturbation method are absorbed.
(2) In the calculating processes, an equivalent symbol ε is introduced temporarily, and the i-th power of which is used to indicate the position of the i-th order correcting term.
(3) An iterative procedure is adopted, and the principle of harmonic balance is applied in the calculation. As a result, the initial conditions are used and there is no need to eliminate the long-term item that proportional to time.
(4) The procedures of high-order derivatives are simplified by adopting a combined form of differential operators and complex form Fourier series.

References

1. Atadan AS, Huseyin K (1984) An intrinsic method of harmonic analysis for nonlinear oscillations. J Sound Vibr 95
2. Qiu SS, Filanovsky IM, Stromsmoe KA (1985) An equivalent small parameter method for strongly nonlinear systems. In: Proceedings of 28th Midwest symposium on circuits and systems, Louisille, pp 61–64
3. Qiu S-S, Filanovsky IM (1987) Periodic solutions of the Van der Pol equation with moderate values of damping coefficient. IEEE Trans Circuits Syst CAS-34:913–918
4. Qiu SS, Filanovsky IM, Stromsmoe KA (1987) On one combination of the harmonic balance method and perturbation techniques. Control-Theory Adv Technol 3:189–195
5. Qiu SS, Filanovsky IM (1989) Calculation of steady-state oscillation in nonlinear circuits. Int J Electron 67(3):403–414
6. Qiu SS, Filanovsky IM (1990) A method of calculation of steady state oscillations in autonomous nonlinear systems. J Sound Vibr 136(1):35–44
7. Qiu SS, Filanovsky IM, Lin BT (1999) A new method of analysis for PWM switching power converters. Int J Electron 86(11):1395–1410
8. Qiu S-S, Filanovsky IM (2000) Harmonic analysis of PWM converters. IEEE Trans Circuits Syst I: Fundam Theory Appl 47:1340–1349 (ISSN 1057-7122)
9. Wu Y, Hu AP, Budgett D, Malpas S, Dissanayake T (2010) Harmonic analysis of voltage-fed full-bridge converter for TET application. In: The 5th IEEE conference on industrial electronics and applications (ICIEA), pp 1040–1045
10. Wu Y, Hu AP, Budgett D, Malpas SC, Dissanayake T (2011) Efficient power-transfer capability analysis of the TET system using the equivalent small parameter method. IEEE Trans Biomed Circuits Syst 5(3):272–282

Chapter 4
Analysis of Open-Loop PWM DC/DC Converters Based on ESPM

4.1 Introduction

Harmonic balance method [1] is one of the commonly used methods for analyzing the periodic solution of nonlinear systems. It transforms the problem of solving nonlinear differential equations into solving nonlinear algebraic equations, which greatly simplifies the analysis. However, the harmonic balance method has the following disadvantages: (1) It is only suitable for obtaining a steady-state solution but cannot be used for transient analysis; (2) If a periodic solution is approximated by harmonics containing orders from 0 (DC component) to N, then there will be (2N + 1) nonlinear algebraic equations to be solved. Hence, the amount of calculation using this method will increase sharply as N increases.

The general averaging method [2] proposed in [3] generalizes the harmonic balance method to transient analysis, and successfully analyzes the resonant converter and PWM (Pulse Width Modulation) switching converter. For a PWM switching converter with a duty cycle D close to 0 (or 1), it is not enough to estimate the waveform with only one harmonic. Therefore, it is often necessary to solve the higher harmonics. The general average method is not easy to solve higher harmonics. The reason is similar to the harmonic balance method, that is, if a solution needs to be approximated by harmonics containing orders from 0 (DC component) to N, then there will be (2N + 1) nonlinear algebraic equations to be solved. Thus, it is quite difficult to have the solutions, which usually need to be obtained numerically. For an open-loop PWM switching converter, although the problem is shown in the form of the linear differential equations, the number of orders is higher bringing about large amount of calculation. It is obvious that this method fails to overcome the second shortcoming of the harmonic balance method.

The equivalent-small-parameter method [4] introduces the perturbation technique into the harmonic balance method, and approximately converts the periodic solution into an expanded triangular series according to the (equivalent) small parameter. According to the order of the small parameters and the type of

Y. Chen and B. Zhang, *Equivalent-Small-Parameter Analysis of DC/DC Switched-Mode Converter*, CPSS Power Electronics Series,
https://doi.org/10.1007/978-981-13-2574-8_4

harmonics, the corresponding algebraic equations can be obtained. Hence, there is no need to solve nonlinear algebraic equations with many variables. In fact, the equivalent-small-parameter method is only applied to solve the nonlinear equations of the main oscillation. The other higher harmonics and corresponding correction terms can be obtained by solving the linear equations. The main oscillation only contains one or two harmonics, and the corresponding algebraic equation has fewer variables. It can be seen that the calculation amount is greatly reduced.

In this chapter, the equivalent-small-parameter method is used to analyze the PWM switching converters operating in CCM (continuous current mode) and DCM (discontinuous current mode), and the analytical expression of the steady-state periodic solution is obtained.

Furthermore, applying the idea in [2, 3], the equivalent-small-parameter method is generalized to the transient analysis of the PWM converter to obtain the equivalent differential (rather than algebraic) equations. Moreover, for open-loop PWM switching converters, only linear differential equations need to be solved. In [5], the progressive method is applied to the switching converter for analyzing transient process and ripple. However, the asymptotic method combines the transient and steady-state solutions, which results in complicated solution. The equivalent-small-parameter method clearly distinguishes the transient and steady-state solutions. Hence, the solution can be easily obtained.

4.2 General Method for Analysis of PWM Switching Power Converter by ESPM

According to the analysis in Chap. 2, the PWM switching power converter can be uniformly represented by the following time-varying differential equations as

$$\mathbf{G}_0(p)\mathbf{x} + \sum_{n=1}^{N} \mathbf{G}_n(p) \cdot \mathbf{f}^{(n)}(\mathbf{x}) = \mathbf{u} \tag{4.1}$$

In which $p = d/dt$ represents the differential operator, $\mathbf{G}_0(p)$ and $\mathbf{G}_1(p)$ are coefficient matrices related to the circuit parameters and the operator p, $\mathbf{u}$ is the input vector, and the nonlinear vector function can be expressed as

$$\mathbf{f}^{(n)}(\mathbf{x}) = \delta^{(n)}(\mathbf{x} + \mathbf{e}^{(n)}) \tag{4.2}$$

Here $\mathbf{e}^{(n)}$ is a constant vector, and the switching function is defined as

$$\delta^{(n)} = \begin{cases} 1 & (t_{n-1}, t_n) \\ 0 & others \end{cases} \quad (n = 1,\, 2,\, \ldots N) \tag{4.3}$$

where $t_0 = kT$, $t_N = (k+1)T$ (k is an integer), which means the converter occupies (N + 1) sub-topologies during one switching cycle. The duty cycle of each sub-topology is defined as $d^{(n)} = (t_n - t_{n-1})/T$, $(n = 1, 2, \ldots N)$, and $\sum_n^N d^{(n)} = 1$. The state variable vector **x** is still expressed as the sum of the following main components and the order corrections:

$$\mathbf{x} = \mathbf{x}_0 + \sum_{i=1}^{\infty} \varepsilon^i \mathbf{x}_i \tag{4.4}$$

Here $\mathbf{x}_0$ and $\mathbf{x}_i$ is the main component and the *i*th order correction term respectively. The parameter $\varepsilon = 1$ is introduced temporarily to indicate that different corrections would have different amplitudes, for example, the indicator ε^k means the $\mathbf{x}_k$ is a small amount higher than $\mathbf{x}_{k-1}$, that is $|\mathbf{x}_k| < |\mathbf{x}_{k-1}| \ll |\mathbf{x}_0|$. Substituting (4.4) into (4.2) gives

$$\mathbf{f}^{(n)}(\mathbf{x}) = \mathbf{f}_0^{(n)} + \sum_{i=1}^{\infty} \varepsilon^i \mathbf{f}_i^{(n)} \tag{4.5}$$

where

$$\begin{cases} \mathbf{f}_0^{(n)} = \mathbf{f}^{(n)}(x_0) \\ \mathbf{f}_i^{(n)} = \mathbf{f}^{(n)}(\mathbf{x}_0, \mathbf{x}_1, \ldots \mathbf{x}_i), \quad (i = 1, 2 \ldots) \end{cases} \tag{4.6}$$

The switching function $\delta^{(n)}$ characterizes the periodic time-varying properties of the converter system. Correct processing of $\delta^{(n)}$ is the key to extend the equivalent small-parameter method to the analysis of switching converter systems. Since $\delta^{(n)}$ is a periodic function, it is usually expanded into a Fourier series, as follows.

$$\delta^{(n)} = b_0^{(n)} + \sum_{m=1}^{\infty} \left[b_m^{(n)} \exp(jm\tau) + \bar{b}_m^{(n)} \exp(-jm\tau) \right] \tag{4.7}$$

where

$$\begin{aligned} b_0^{(n)} &= d^{(n)}, \quad b_m^{(n)} = \tfrac{1}{2}(\alpha_m^{(n)} - j\beta_m^{(n)}), \quad m = 1, 2 \ldots \\ \alpha_m^{(n)} &= \left[\sin 2(\sum_{k=0}^{n} d^{(k)}) m\pi - \sin 2(\sum_{k=0}^{n-1} d^{(k)}) m\pi \right] / m\pi \\ \beta_m^{(n)} &= \left[\cos 2(\sum_{k=0}^{n-1} d^{(k)}) m\pi - \cos 2(\sum_{k=0}^{n} d^{(k)}) m\pi \right] / m\pi \end{aligned} \tag{4.8}$$

Similarly, the switch function is expanded to the sum of the main terms and the correction terms as

$$\delta^{(n)} = \delta_{.0}^{(n)} + \varepsilon\delta_1^{(n)} + \varepsilon^2\delta_2^{(n)} + \cdots \tag{4.9}$$

Since the magnitude of the coefficient $b_m^{(n)}$ decreases as m increases, it is generally assumed that $\delta_{.0}^{(n)}$ contains a DC component and a fundamental component, and $\delta_i^{(n)}$ contains two adjacent higher harmonic components, namely:

$$\begin{aligned} \delta_0^{(n)} &= d^{(n)} + b_1^{(n)} e^{j\tau} + c.c \\ \delta_i^{(n)} &= b_{2i}^{(n)} e^{j2i\tau} + b_{(2i+1)}^{(n)} e^{j(2i+1)\tau} + c.c \end{aligned} \tag{4.10}$$

In which $\tau = \omega t$, the symbol $c.c$ represents the conjugate complex term (the meaning is the same below). Substituting (4.4) and (4.9) into (4.2) gives

$$\begin{aligned} \mathbf{f}_0^{(n)} &= \delta_0^{(n)}(\mathbf{x}_0 + \mathbf{e}^{(n)}) \\ \mathbf{f}_1^{(n)} &= \delta_0^{(n)}\mathbf{x}_1 + \delta_1^{(n)}(\mathbf{x}_0 + \mathbf{e}^{(n)}) \\ \mathbf{f}_2^{(n)} &= \delta_0^{(n)}\mathbf{x}_2 + \delta_1^{(n)}\mathbf{x}_1 + \delta_2^{(n)}(\mathbf{x}_0 + \mathbf{e}^{(n)}) \\ &\cdots \end{aligned} \tag{4.11}$$

Using the same method as in Chap. 3, the main term $\mathbf{f}_{im}^{(n)}$ and the remainder $\mathbf{R}_i^{(n)}$ of each nonlinear function vector $\mathbf{f}_i^{(n)}$ are got according to Eq. (4.11), and then the steady-state periodic solution of the system is obtained according to Eq. (3.46).

4.3 Analysis of the Open-Loop Boost Converter Under CCM Operation

4.3.1 Modeling of the CCM-Boost Converter

The principle circuit of Boost converter is shown in Fig. 4.1a. It's assumed that the converter operates in continuous-current-mode (CCM), that is, the converter has two equivalent topologies during one switching cycle T, which correspond to the switch S_T is turned on and the S_D is turned off (Fig. 4.1b), the switch S_T is turned off, and the S_D is turned on (Fig. 4.1c). Before modeling, the following assumptions are made:

(1) The input power supply is ideal, that is, its internal resistance is ignored;
(2) Inductors and capacitors are considered as ideal components, that is, their parasitic parameters are not taken into account;
(3) Both the controllable switch S_T and the diode S_D are ideal switches, their on-resistance is zero, and the resistance is infinite when disconnected.

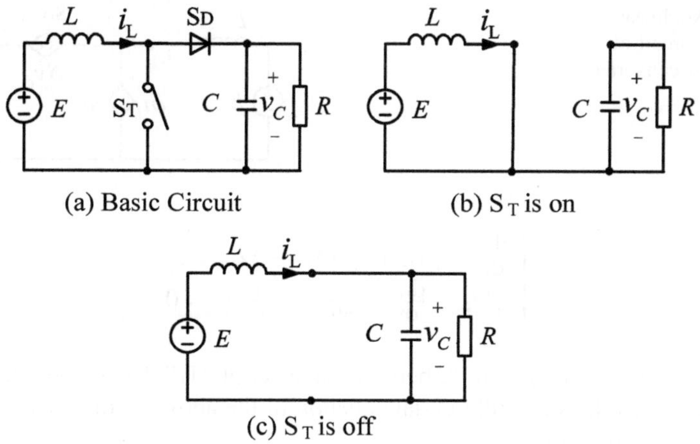

Fig. 4.1 CCM-Boost converter with its two equivalent topologies during one cycle

Let S_T have an on time of t_{on} in one switching cycle T and d as a duty cycle, which is defined as $d = t_{on}/T$. According to the analysis in Chap. 2, a switching function $\delta(t)$ (abbreviated as δ) is defined to describe the on/off state of the switch S_T, that is, $\delta(t) = 1$ when S_T is on, and $\delta(t) = 0$ when S_T is off. For the Boost converter operating in CCM, it is assumed that the controllable switch S_T is turned on at the beginning of each switching cycle. After t_{on} time, S_T is turned off, and remains off state until the end of this cycle, then the definition of the switching function can be described by the following formula:

$$\delta(t) = \begin{cases} 1 & t \in [nT,\ (n+d)T] \\ 0 & t \in [(n+d)T,\ (n+1)T] \end{cases} \tag{4.12}$$

Here n = 0, 1, 2 …, is an integer.

As can be seen from Fig. 4.1b and c, when S_T is on, S_D is naturally turned off due to the reverse bias voltage. During this period, the current flowing through the switch S_T is equal to the inductor current i_L, and the reverse bias voltage that the S_D is subjected to is v_C. When the S_T is turned off, the S_D is turned on due to the positive bias voltage. And during this period, the current flowing through the switch S_T is 0, and the voltage drop across the S_D is also 0 under the assumption of the ideal switch. Thus, in one switching cycle, the current flowing through S_T is δi_L, and the reverse bias voltage that S_D is subjected to is δv_C. Using a controlled current source with a value of δi_L to replace the switch S_T, and a controlled voltage source with a value of δv_C to replace the diode S_D, then the nonlinear equivalent circuit of the Boost converter with CCM operation can be obtained, as shown in Fig. 4.2.

Taking the inductor current i_L and the capacitor voltage v_C as the state variables, according to Fig. 4.2, the state differential equation describing the CCM-Boost converter can be obtained as follows:

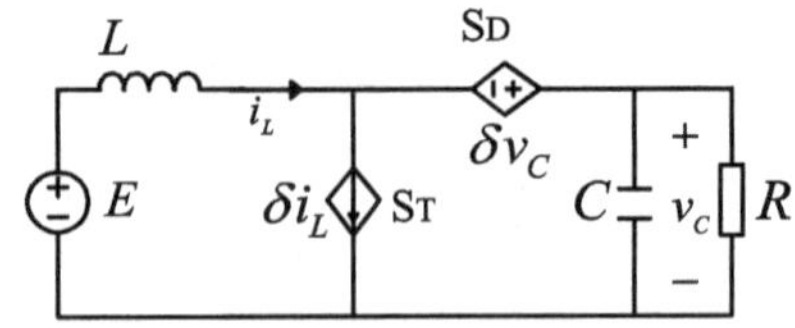

Fig. 4.2 Nonlinear equivalent circuit of CCM-Boost converter

$$\begin{cases} \frac{\mathrm{d}i_L}{\mathrm{d}t} + \frac{1}{L}v_C + \frac{-1}{L}\delta v_C = \frac{1}{L}E \\ \frac{-1}{C}i_L + \frac{\mathrm{d}v_C}{\mathrm{d}t} + \frac{1}{RC}v_C + \frac{1}{C}\delta i_L = 0 \end{cases} \tag{4.13}$$

Let $\mathbf{x} = \begin{bmatrix} i_L & v_C \end{bmatrix}^{Tr}$ and p = d/dt, here the superscript "Tr" denote the transposition of a matrix, then the state differential equation of the above formula can be written in matrix form as follows

$$G_0(p)\mathbf{x} + G_1(p)\mathbf{f} = \mathbf{u} \tag{4.14}$$

where

$$G_0(p) = \begin{bmatrix} p & \frac{1}{L} \\ \frac{-1}{C} & p + \frac{1}{RC} \end{bmatrix}, \quad G_1(p) = \begin{bmatrix} 0 & \frac{-1}{L} \\ \frac{1}{C} & 0 \end{bmatrix}, \quad \mathbf{u} = \begin{bmatrix} E/L \\ 0 \end{bmatrix} \tag{4.15}$$

and the nonlinear vector is

$$\mathbf{f} = \delta\mathbf{x} \tag{4.16}$$

4.3.2 The Equivalent Mathematical Model Based on ESPM

According to the method introduced in Chap. 3 and Sect. 4.1, we assume that the state vector **x** and the switching function δ(t) in Eq. (4.12) could be expanded into the following series:

$$\mathbf{x} = \mathbf{x}_0 + \varepsilon\mathbf{x}_1 + \varepsilon^2\mathbf{x}_2 + \cdots \tag{4.17}$$

and

$$\delta(t) = \delta_0 + \varepsilon\delta_1 + \varepsilon^2\delta_2 + \cdots \tag{4.18}$$

where the zero order approximation $\mathbf{x}_0$ and the *i*th order approximation $\mathbf{x}_i$ are called as the main wave and corrections, and represent the main components and harmonics of state vector **x** respectively. The parameter ε^i, which can also be called as "intrinsic small parameter", is introduced temporarily to indicate that the terms $\mathbf{x}_i$ have different order of magnitude. Similarly, δ(t) is expressed as the sum of main wave δ_0 and the corrections δ_i too, as shown in Eq. (4.18). Substituting Eqs. (4.17) and (4.18) into $\mathbf{f}(\mathbf{x}) = \delta(t)\mathbf{x}$, and combining terms which have the same order of indicator parameter ε^i, we find that the nonlinear function $\mathbf{f}(\mathbf{x})$ can be expressed in the same way with state vector $\mathbf{x}$:

$$\mathbf{f}(\mathbf{x}) = \mathbf{f}_0 + \varepsilon^1 \mathbf{f}_i + \varepsilon^2 \mathbf{f}_2 + \cdots \tag{4.19}$$

in which:

$$\begin{cases} \mathbf{f}_0 = \delta_0 \mathbf{x}_0 \\ \mathbf{f}_1 = \delta_0 \mathbf{x}_1 + \delta_1 \mathbf{x}_0 \\ \mathbf{f}_2 = \delta_0 \mathbf{x}_2 + \delta_1 \mathbf{x}_1 + \delta_2 \mathbf{x}_0. \\ \cdots \end{cases} \tag{4.20}$$

Similarly, $\mathbf{f}_0$ and $\mathbf{f}_i$ represent the main components and corrections of nonlinear vector function $\mathbf{f}(\mathbf{x})$ respectively.

Generally the terms $\mathbf{x}_0$ and $\mathbf{x}_i$ in Eq. (4.17) for state variables $\mathbf{x}$ can be expressed in the Fourier series as follows

$$\begin{aligned} \mathbf{x}_0 &= \sum_{n \in E_0} \mathbf{x}_{00} = \mathbf{a}_{00} + \sum (\mathbf{a}_{n0} e^{jn\tau} + \bar{\mathbf{a}}_{n0} e^{-jn\tau}) \\ \mathbf{x}_i &= \sum_{k \in E_i} \mathbf{x}_{ki} = \mathbf{a}_{0i} + \sum (\mathbf{a}_{ki} e^{jk\tau} + \bar{\mathbf{a}}_{ki} e^{-jk\tau}) \end{aligned} \tag{4.21}$$

in which n and k are integers, and $\tau = \omega t$ (where $\omega = 2\pi/T$) is the normalized time. The terms $\mathbf{a}_{00}$ and $\mathbf{a}_{0i}$ are the DC components of $\mathbf{x}_0$ and $\mathbf{x}_i$ respectively, and the terms $\mathbf{a}_{n0}$ and $\mathbf{a}_{ki}$ are used to determine the amplitudes of the nth and the kth harmonics that belong to $\mathbf{x}_0$ and $\mathbf{x}_i$ separately. The spectral content E_0 of vector $\mathbf{x}_0$, which consists of a set of numbers representing relative frequencies of harmonics, is dictated by the physical sense of the study object. For example, as DC/DC converters have low-pass filtering properties, we usually assume that the main wave $\mathbf{x}_0$ includes DC components only, i.e. $\mathbf{x}_0 = \mathbf{a}_{00} = [I_{00}\ V_{00}]^{Tr}$, thus the set $\{E_0\}$ is $\{0\}$. Similarly, the set E_i contains the spectral contents for each $\mathbf{x}_i$, yet it is not known in advance and is determined in the process of iterated calculations, i.e. it is determined by $\mathbf{x}_0$, $\mathbf{x}_1$, …, and $\mathbf{x}_{i-1}$ during the iteration. The term $\mathbf{a}_{i0}$ and $\mathbf{a}_{ik}$ represent the DC components and the kth harmonics that belong to x_i respectively.

The periodic switch function δ defined by (4.12) can be expanded into Fourier series as

$$\delta = b_0 + \sum_{m=1}^{\infty} [b_m \exp(jm\tau) + \bar{b}_m \exp(-jm\tau)] \tag{4.22}$$

where

$$
\begin{aligned}
b_0 &= d \\
b_m &= \frac{1}{2}(\alpha_m - j\beta_m), \quad m = 1, 2 \ldots \\
\alpha_m &= \frac{2}{T}\int_0^T \delta(t)\cos(m\omega t)dt = \frac{\sin 2dm\pi}{m\pi} \\
\beta_m &= \frac{2}{T}\int_0^T \delta(t)\sin(m\omega t)dt = \frac{1-\cos 2dm\pi}{m\pi}
\end{aligned}
\tag{4.23}
$$

Generally for DC/DC converters, we can choose that δ_0 contains only the DC component and the fundamental component, and δ_i contains the (2*i*)th and (2*i* + 1) th harmonic components, here *i* = 1, 2, 3 …, is a positive integer, that is,

$$
\begin{aligned}
\delta_0 &= d + b_1 e^{j\tau} + c.c \\
\delta_i &= b_{2i} e^{j2i\tau} + b_{(2i+1)} e^{j(2i+1)\tau} + c.c
\end{aligned}
\tag{4.24}
$$

in which *c.c* denotes conjugate items.

The nonlinear function vector is further expressed as a two-level form of the sum of the main item series and the remainder series, as follows.

$$
\mathbf{f} = (\mathbf{f}_{0m} + \varepsilon\mathbf{f}_{1m} + \varepsilon^2\mathbf{f}_{2m} + \cdots) + (\varepsilon\mathbf{R}_1 + \varepsilon^2\mathbf{R}_2 + \ldots) \tag{4.25}
$$

The method for determining the main terms $\mathbf{f_{0m}}$ and $\mathbf{f_{im}}$, and the residual R_i is the same as described in Chap. 3. We can also refer to the specific solution process in the next section.

Substituting (4.17) and (4.25) into (4.14), one can obtain the equivalent mathematical model describing the system i.e. the iterative equations shown as follows.

$$
\begin{cases}
\mathbf{G}_0(p)\mathbf{x}_0 + \mathbf{G}_1(p)\mathbf{f}_{0m} = \mathbf{u} \\
\mathbf{G}_0(p)\mathbf{x}_1 + \mathbf{G}_1(p)\mathbf{f}_{1m} = -\mathbf{G}_1(p)\mathbf{R}_1 \\
\mathbf{G}_0(p)\mathbf{x}_2 + \mathbf{G}_1(p)\mathbf{f}_{2m} = -\mathbf{G}_1(p)\mathbf{R}_2 \\
\quad\cdots\cdots
\end{cases}
\tag{4.26}
$$

Each equation in (4.26) can be solved step by step by using the method of harmonic balance, where the first equation is used to obtain the main wave $\mathbf{x}_0$, and following equations are used to solve the corrections $\mathbf{x}_1$, $\mathbf{x}_2$, …, etc. If the harmonic amplitude in the *k*th correction $\mathbf{x}_\mathrm{k}$ is much smaller than those in the (*k* − 1)th correction $\mathbf{x}_{\mathrm{k}-1}$, the calculation process is terminated. Hence, according to discussions above, the steady-state solution for the vector state variable Eq. (4.14) can be approximated by $\mathbf{x} \approx \mathbf{x}_0 + \mathbf{x}_1 + \mathbf{x}_2 + \ldots$. Usually, as the low-pass filtering property of DC/DC converters, the magnitudes of harmonics with much higher frequencies are small, so they are neglected, and only the first three equations in Eq. (4.26) need to be solved, these solutions suffice for most technical applications.

4.3.3 *The Steady-State Periodic Solution of the Boost Converter Based on ESPM*

4.3.3.1 Solution of the Main Term

As it is known that DC/DC converters have low-pass properties, the main wave $\mathbf{x}_0$ to the steady-state value of state variable $\mathbf{x}$ could be chosen as

$$\mathbf{x}_0 = \mathbf{a}_{00} = [I_{00} \quad V_{00}]^{Tr} \tag{4.27}$$

here I_{00} and V_{00} are dc values. Substituting it into (4.20) gives

$$\mathbf{f}_0 = \delta_0 \mathbf{x}_0 = b_0 \mathbf{a}_{00} + b_1 \mathbf{a}_{00} e^{j\tau} + c.c \tag{4.28}$$

According to assumptions in Chap. 3, $\mathbf{f}_{0m}$ should include the same harmonics as $\mathbf{x}_0$, and the rest harmonics in $\mathbf{f}_0$ belong to $\mathbf{R}_1$, hence we can get

$$\begin{cases} \mathbf{f}_{0m} = b_0 \mathbf{a}_{00} \\ \mathbf{R}_1 = b_1 \mathbf{a}_{00} e^{j\tau} + c.c \end{cases} \tag{4.29}$$

Introducing $\mathbf{x}_0$ and $\mathbf{f}_{0m}$ into the first equation of Eq. (4.26), it can be attained

$$[\mathbf{G}_0(0) + \mathbf{G}_1(0) b_0] \mathbf{a}_{00} = \mathbf{u} \tag{4.30}$$

Since the derivative of the dc component is zero, $\mathbf{G}_0(0)$ and $\mathbf{G}_1(0)$ can be obtained by setting the differential operator $p = 0$ in Eq. (4.15). For the converter with open loop operation, the duty ratio d is a known amount, i.e. $d = D$, then Eq. (4.30) can be rewritten in the matrix form as

$$\left(\begin{bmatrix} 0 & \frac{1}{L} \\ -\frac{1}{C} & \frac{1}{RC} \end{bmatrix} + \begin{bmatrix} 0 & -\frac{D}{L} \\ \frac{D}{C} & 0 \end{bmatrix} \right) \begin{bmatrix} I_{00} \\ V_{00} \end{bmatrix} = \begin{bmatrix} E/L \\ 0 \end{bmatrix} \tag{4.31}$$

Thereby the term $\mathbf{a}_{00}$ could be got from (4.30) or (4.31).

$$V_{00} = \frac{E}{1-D} \qquad I_{00} = \frac{V_{00}}{(1-D)R} \tag{4.32}$$

4.3.3.2 Solution of the First Correction Term

According to the content of harmonics in the term $\mathbf{R}_1$ (there only exits the first-harmonic), we can deduce that the spectral content of vector $\mathbf{x}_1$ is $\{E_1\} = \{1\}$, and then the first order correction $\mathbf{x}_1$ can be assumed to be:

$$\mathbf{x}_1 = \mathbf{a}_{11}e^{j\tau} + \bar{\mathbf{a}}_{11}e^{-j\tau} \tag{4.33}$$

in which $\bar{\mathbf{a}}_{11}$ means complex-conjugate, and $\mathbf{a}_{11} = [I_{11}\ V_{11}]^{Tr}$. Substituting $\mathbf{x}_0$, $\mathbf{x}_1$, δ_0 and δ_1 into the equation $\mathbf{f}_1 = \delta_0\mathbf{x}_1 + \delta_1\mathbf{x}_0$, we can get

$$\mathbf{f}_1 = (b_0\mathbf{a}_{11}e^{j\tau} + b_0\bar{\mathbf{a}}_{11}e^{-j\tau}) + (\bar{b}_1\mathbf{a}_{11} + b_1\bar{\mathbf{a}}_{11}) + (b_1\mathbf{a}_{11} + b_2\mathbf{a}_{00})e^{j2\tau} + b_3\mathbf{a}_{00}e^{j3\tau} + c.c$$

And the following expressions for $\mathbf{f}_{1m}$ and $\mathbf{R}_2$ can be got as

$$\begin{cases} \mathbf{f}_{1m} = b_0\mathbf{a}_{11}e^{j\tau} + c.c \\ \mathbf{R}_2 = (b_1\bar{\mathbf{a}}_{11} + \bar{b}_1\mathbf{a}_{11}) + (b_1\mathbf{a}_{11} + b_2\mathbf{a}_{00})e^{j2\tau} + b_3\mathbf{a}_{00}e^{j3\tau} + c.c \end{cases} \tag{4.34}$$

where $\mathbf{f}_{1m}$ should consist of all the terms in $\mathbf{f}_1$ with the same harmonics as $\mathbf{x}_1$, and $\mathbf{R}_2$ includes the rest. Introducing $\mathbf{x}_1$, $\mathbf{R}_1$ and $\mathbf{f}_{1m}$ into the second equation in Eq. (4.26), and using the method of harmonic balance, one can get:

$$[\mathbf{G}_0(j\omega) + \mathbf{G}_1(j\omega)b_0]\mathbf{a}_{11} = -\mathbf{G}_1(j\omega)\cdot b_1\mathbf{a}_{00} \tag{4.35}$$

where the coefficient b_1 can be obtained according to (4.23). $\mathbf{G}_0(j\omega)$ and $\mathbf{G}_1(j\omega)$ can be obtained by setting the differential operator $p = j\omega$ in Eq. (4.15), as for the derivative of the exponential function, there is $pe^{jk\omega t} = (jk\omega)e^{jk\omega t}$ (k is an integer, here k = 1). For the Boost converter under study, Eq. (4.35) can be rewritten in the matrix form as:

$$\left(\begin{bmatrix} j\omega & \frac{1}{L} \\ -\frac{1}{C} & j\omega + \frac{1}{RC} \end{bmatrix} + \begin{bmatrix} 0 & -\frac{D}{L} \\ \frac{D}{C} & 0 \end{bmatrix}\right)\begin{bmatrix} I_{11} \\ V_{11} \end{bmatrix} = \begin{bmatrix} 0 & \frac{1}{L} \\ \frac{-1}{C} & 0 \end{bmatrix}\begin{bmatrix} b_1 I_{00} \\ b_1 V_{00} \end{bmatrix} \tag{4.36}$$

Thus the vector $\mathbf{a}_{11}$ could be found from (4.36)

$$\begin{cases} I_{11} = \frac{b_1 V_{00} - (1-D)V_{11}}{j\omega L} \\ V_{11} = \frac{b_1[(1-D)V_{00} - j\omega L I_{00}]}{(1-D)^2 + (j\omega)^2 LC + j\omega L/R} \end{cases} \tag{4.37}$$

4.3.3.3 Solution of the Second Correction Term

It can be seen from Eq. (4.34) that there exist DC components, second- and third-harmonics in the term $\mathbf{R}_2$, thus the spectral content of $\mathbf{x}_2$ is determined, that is

$\{E_2\} = \{0, 2, 3\}$. Following the same procedure, we can choose the second order correction $\mathbf{x}_2$ as:

$$\mathbf{x}_2 = \mathbf{a}_{20} + \mathbf{a}_{22}e^{j2\tau} + \bar{\mathbf{a}}_{22}e^{-j2\tau} + \mathbf{a}_{23}e^{j3\tau} + \bar{\mathbf{a}}_{23}e^{-j3\tau} \tag{4.38}$$

Here $\mathbf{a}_{20} = [I_{20}\ V_{20}]^{Tr}$ is the vector of dc values, it will give the corrections of dc components in $\mathbf{a}_{00}$, and $\mathbf{a}_{22} = [I_{22}\ V_{22}]^{Tr}$ and $\mathbf{a}_{23} = [I_{23}\ V_{23}]^{Tr}$ correspond to the amplitudes second- and third-harmonics of state variable $\mathbf{x}$ respectively.

Similarly, substituting $\mathbf{x}_0$, $\mathbf{x}_1$, $\mathbf{x}_2$, and δ_0, δ_1, δ_2 into $\mathbf{f}_2 = \delta_0\mathbf{x}_2 + \delta_1\mathbf{x}_1 + \delta_2\mathbf{x}_0$, we can get

$$\begin{aligned}\mathbf{f}_2 = {} & b_0\mathbf{a}_{20} + (b_1\mathbf{a}_{00} + \bar{b}_1\mathbf{a}_{22} + b_2\bar{\mathbf{a}}_{11})e^{j\tau} + (b_0\mathbf{a}_{22} + \bar{b}_1\mathbf{a}_{23} + b_3\bar{\mathbf{a}}_{11})e^{j2\tau} \\ & + (b_0\mathbf{a}_{23} + b_1\mathbf{a}_{22} + b_2\mathbf{a}_{11})e^{j3\tau} + (b_1\mathbf{a}_{23} + b_3\mathbf{a}_{11} + b_4\mathbf{a}_{00})e^{j4\tau} + b_5\mathbf{a}_{00}e^{j5\tau} + c.c\end{aligned}$$

and the expressions for $\mathbf{f}_{2m}$, it should include the same harmonics as $\mathbf{x}_2$.

$$\mathbf{f}_{2m} = b_0\mathbf{a}_{20} + (b_0\mathbf{a}_{22} + \bar{b}_1\mathbf{a}_{23} + b_3\bar{\mathbf{a}}_{11})e^{j2\tau} + (b_0\mathbf{a}_{23} + b_1\mathbf{a}_{22} + b_2\mathbf{a}_{11})e^{j3\tau} + c.c \tag{4.39}$$

Next, substituting $\mathbf{x}_2$, $\mathbf{R}_2$ and $\mathbf{f}_{2m}$ into the third equation of Eq. (4.26), and still using the method of harmonic balance, one can obtain the following three equations:

$$[G_0(0) + G_1(0)b_0]\mathbf{a}_{20} = -G_1(0)(b_1\bar{\mathbf{a}}_{11} + \bar{b}_1\mathbf{a}_{11}) \tag{4.40a}$$

$$[G_0(j2\omega) + G_1(j2\omega)b_0]\mathbf{a}_{22} = -G_1(j2\omega)\cdot(b_2\mathbf{a}_{00} + b_1\mathbf{a}_{11} + b_3\bar{\mathbf{a}}_{11}) \tag{4.40b}$$

and

$$[G_0(j3\omega) + G_1(j3\omega)b_0]\mathbf{a}_{23} = -G_1(j3\omega)\cdot(b_3\mathbf{a}_{00} + b_2\mathbf{a}_{11} + b_1\mathbf{a}_{22}) \tag{4.40c}$$

Likewise, the coefficient matrix $\mathbf{G}_0(jk\omega)$ and $\mathbf{G}_1(jk\omega)$ can be obtained by setting the differential operator $p = jk\omega$ in Eq. (4.15), as for the derivative of the exponential function, there is $pe^{jk\omega t} = (jk\omega)e^{jk\omega t}$ (here $k = 0, 2, 3$).

It should be noticed that during the derivation process, the term $\bar{b}_1\mathbf{a}_{23}$ in Eq. (4.40b) can be omitted with respect to the term $(b_0\mathbf{a}_{22} + b_3\bar{\mathbf{a}}_{11})$, as the magnitudes of harmonics with higher frequencies are smaller than those with lower frequencies.

So Eq. (4.40a) can be rewritten in matrix form as:

$$\left(\begin{bmatrix} 0 & \frac{1}{L} \\ -\frac{1}{C} & \frac{1}{RC} \end{bmatrix} + \begin{bmatrix} 0 & -\frac{D}{L} \\ \frac{D}{C} & 0 \end{bmatrix}\right)\begin{bmatrix} I_{20} \\ V_{20} \end{bmatrix} = \begin{bmatrix} 0 & \frac{1}{L} \\ \frac{-1}{C} & 0 \end{bmatrix}\begin{bmatrix} \bar{b}_1 I_{11} + b_1\bar{I}_{11} \\ \bar{b}_1 V_{11} + b_1\bar{V}_{11} \end{bmatrix} \tag{4.41}$$

where $b_0 = D$ and $b_1 = [\sin 2D\pi - j(1 - \cos 2D\pi)]/2\pi$, the vector $\mathbf{a}_{02}$ could be found from (4.41)

$$\begin{cases} I_{20} = \frac{R(\bar{b}_1 I_{11} + b_1 \bar{I}_{11}) + V_{20}}{R(1-D)} \\ V_{20} = \frac{\bar{b}_1 V_{11} + b_1 \bar{V}_{11}}{1-D} \end{cases} \tag{4.42}$$

Then Eq. (4.40b) is rewritten in matrix form as:

$$\left(\begin{bmatrix} j2\omega & \frac{1}{L} \\ -\frac{1}{C} & j2\omega + \frac{1}{RC} \end{bmatrix} + \begin{bmatrix} 0 & -\frac{D}{L} \\ \frac{D}{C} & 0 \end{bmatrix}\right)\begin{bmatrix} I_{22} \\ V_{22} \end{bmatrix} = \begin{bmatrix} 0 & \frac{1}{L} \\ \frac{-1}{C} & 0 \end{bmatrix}\begin{bmatrix} b_1 I_{11} + b_2 I_{00} + b_3 \bar{I}_{11} \\ b_1 V_{11} + b_2 V_{00} + b_3 \bar{V}_{11} \end{bmatrix} \tag{4.43}$$

And Eq. (4.40c) can be rewritten in matrix form as:

$$\left(\begin{bmatrix} j3\omega & \frac{1}{L} \\ -\frac{1}{C} & j3\omega + \frac{1}{RC} \end{bmatrix} + \begin{bmatrix} 0 & -\frac{D}{L} \\ \frac{D}{C} & 0 \end{bmatrix}\right)\begin{bmatrix} I_{23} \\ V_{23} \end{bmatrix} = \begin{bmatrix} 0 & \frac{1}{L} \\ \frac{-1}{C} & 0 \end{bmatrix}\begin{bmatrix} b_1 I_{22} + b_2 I_{11} + b_3 I_{00} \\ b_1 V_{22} + b_2 V_{11} + b_3 V_{00} \end{bmatrix} \tag{4.44}$$

where $b_k = [\sin 2kD\pi - j(1 - \cos 2kD\pi)]/2k\pi$ $(k = 1, 2, 3)$. Then the vectors $\mathbf{a}_{22}$ and $\mathbf{a}_{23}$ could be found from (4.43) and (4.44), as shown in (4.45) and (4.46) respectively.

$$\begin{cases} I_{22} = \frac{b_1 V_{11} + b_2 V_{00} + b_3 \bar{V}_{11} - (1-D) V_{22}}{j2\omega L} \\ V_{22} = \frac{(1-D)(b_1 V_{11} + b_2 V_{00} + b_3 \bar{V}_{11}) - j2\omega L(b_1 I_{11} + b_2 I_{00} + b_3 \bar{I}_{11})}{(1-D)^2 + (j2\omega)^2 LC + j2\omega L/R} \end{cases} \tag{4.45}$$

$$\begin{cases} I_{23} = \frac{b_1 V_{22} + b_2 V_{11} + b_3 V_{00} - (1-D) V_{23}}{j3\omega L} \\ V_{23} = \frac{(1-D)(b_1 V_{22} + b_2 V_{11} + b_3 V_{00}) - j3\omega L(b_1 I_{22} + b_2 I_{11} + b_3 I_{00})}{(1-D)^2 + (j3\omega)^2 LC + j3\omega L/R} \end{cases} \tag{4.46}$$

Following this procedure, the term $\mathbf{a}_{ik}$ could be solved step by step, and detailed information is listed in Table 4.1.

Table 4.1 Values of $\mathbf{a}_{ik}$ for CCM-Boost converter

i	k	a_{ik}	
		V_{ik}	I_{ik}
0	0	$\frac{E}{(1-D)}$	$\frac{E}{R(1-D)^2}$
1	1	$\frac{(1-D) b_1 V_{00} - j\omega L b_1 I_{00}}{\Delta(j\omega)}$	$\frac{b_1 V_{00} - (1-D) V_{11}}{(j\omega) L}$
2	0	$\frac{\bar{b}_1 V_{11} + b_1 \bar{V}_{11}}{(1-D)}$	$\frac{R(\bar{b}_1 I_{11} + b_1 \bar{I}_{11}) + V_{20}}{R(1-D)}$
	2	$\frac{(1-D)(b_1 V_{11} + b_2 V_{00} + b_3 \bar{V}_{11}) - j2\omega L(b_1 I_{11} + b_2 I_{00} + b_3 \bar{I}_{11})}{\Delta(j2\omega)}$	$\frac{(b_1 V_{11} + b_2 V_{00} + b_3 \bar{V}_{11}) - (1-D) V_{22}}{j2\omega L}$
	3	$\frac{(1-D)(b_1 V_{22} + b_2 V_{11} + b_3 V_{00}) - (j3\omega L(b_1 I_{22} + b_2 I_{11} + b_3 I_{00})}{\Delta(j3\omega)}$	$\frac{(b_1 V_{22} + b_2 V_{11} + b_3 V_{00}) - (1-D) V_{23}}{j3\omega L}$

In Table 4.1, the Δ(jkω) is defined as:

$$\Delta(jk\omega) = (1 - D)^2 + (jk\omega)^2 LC + jk\omega L/R \tag{4.47}$$

And the coefficient b_i in Eq. (4.23) is rewritten as follows

$$\begin{cases} b_0 = d \\ b_m = \frac{\sin 2dm\pi - j(1-\cos 2dm\pi)}{2m\pi}, \quad m = 1, 2\ldots \end{cases} \tag{4.48}$$

$\bar{b}_m$ is the conjugate complex of b_m.

Hence, according to the discussions above, the approximate periodic solutions for CCM-Boost converter could be expressed as

$$\begin{aligned} \mathbf{x} &= [\, i_L \quad v_C \,]^{Tr} \\ &= (\mathbf{a}_{00} + \mathbf{a}_{02}) + \left(\mathbf{a}_{11}e^{j\tau} + \mathbf{a}_{22}e^{j2\tau} + \mathbf{a}_{32}e^{j3\tau} + c.c\right) \end{aligned} \tag{4.49}$$

The components of this vector are

$$\begin{aligned} i_L \approx\ & (I_{00} + I_{02}) + 2(\mathrm{Re}\, I_{11} \cos \omega t - \mathrm{Im}\, I_{11} \sin \omega t \\ & + \mathrm{Re}\, I_{22} \cos 2\omega t - \mathrm{Im}\, I_{22} \sin 2\omega t + \mathrm{Re}\, I_{32} \cos 3\omega t - \mathrm{Im}\, I_{32} \sin 3\omega t) \end{aligned} \tag{4.50a}$$

in amperes, and

$$\begin{aligned} v_C \approx\ & (V_{00} + V_{02}) + 2(\mathrm{Re}\, V_{11} \cos \omega t - \mathrm{Im}\, V_{11} \sin \omega t \\ & + \mathrm{Re}\, V_{22} \cos 2\omega t - \mathrm{Im}\, V_{22} \sin 2\omega t + \mathrm{Re}\, V_{32} \cos 3\omega t - \mathrm{Im}\, V_{32} \sin 3\omega t) \end{aligned} \tag{4.50b}$$

in volts. In Eqs. (4.50a) and (4.50b) **Re** $\mathbf{x}_{ik}$ and **Im** $\mathbf{x}_{ik}$ represent the real- and imaginary-part of the complex term $\mathbf{x}_{ik}$ separately. Obviously, the DC values and harmonic components of state variables can be obtained from Eqs. (4.49) or (4.50a).

4.3.4 Simulations

The circuit parameters of the open-loop Boost converter are list in Table 4.2.

According to (4.48), when the duty ratio is known, the coefficients b_m can be obtained. Then according to Table 4.1, the approximate solution of the state variable $\mathbf{x} = [\, i_L \quad v_C \,]^{Tr}$ can be obtained as follows, here i_L is the inductor current and v_C is the voltage over the output capacitor.

Table 4.2 Circuit parameters of the open-loop Boost converter

Parameters	Values
Input voltage E	37.5 V
Switching frequency f	1 kHz
Inductance L	6 mH
Capacitance C	45 μF
Load resistance R	30 Ω
Duty ratio D	0.25

$$\begin{aligned} i_L &= 2.11 - 0.487\cos\tau + 0.392\sin\tau - 0.205\cos 2\tau - 0.002\sin 2\tau \\ &\quad - 0.0476\cos 3\tau - 0.0408\sin 3\tau \qquad \text{(A)} \\ v_C &= 48.08 + 0.942\cos\tau - 3.945\sin\tau + 1.24\cos 2\tau - 0.026\sin 2\tau \\ &\quad + 0.1878\cos 3\tau + 0.3523\sin 3\tau \qquad \text{(V)} \end{aligned} \tag{4.51}$$

For the analysis of the steady state solution of the state variable, the ESP analysis method (dashed line) and the simulation results (solid line) are compared in Figs. 4.3 and 4.4. Figure 4.3 shows the comparison of the inductor current waveform and the capacitor voltage ripple waveform with $f_s = 1\,\text{kHz}$, and Fig. 4.4 shows the comparison waveforms with $f_s = 10\,\text{kHz}$. It can be seen that the results from the ESPM are in good coincidence with those from simulations, indicating that the ESPM can analyze the steady-state solution of the open-loop converter well. Even when the operating frequency of the converter is small, such as $f_s = 1\,\text{kHz}$, the state variable ripple is relatively large, the ESPM can still effectively obtain the analytical expression of the periodic solution of the converter, and only three iterations are needed to obtain enough precision. It can be seen that the equivalent small-parameter method is effective for analyzing the strong nonlinear system of the switching converter.

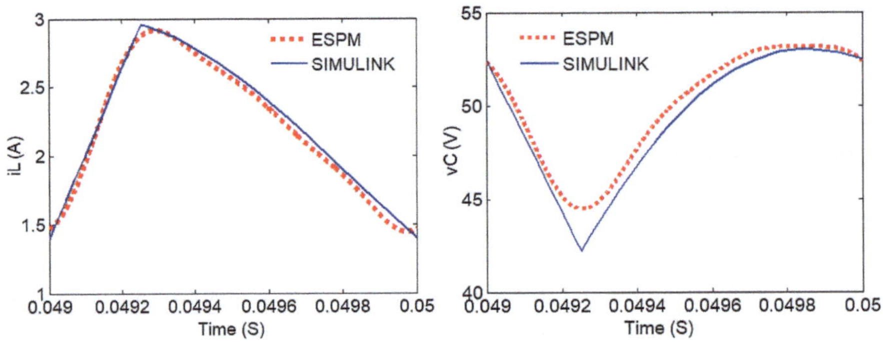

Fig. 4.3 Steady-state ripple waveforms of state variables for CCM-Boost converter with $f_s = 1$ kHz

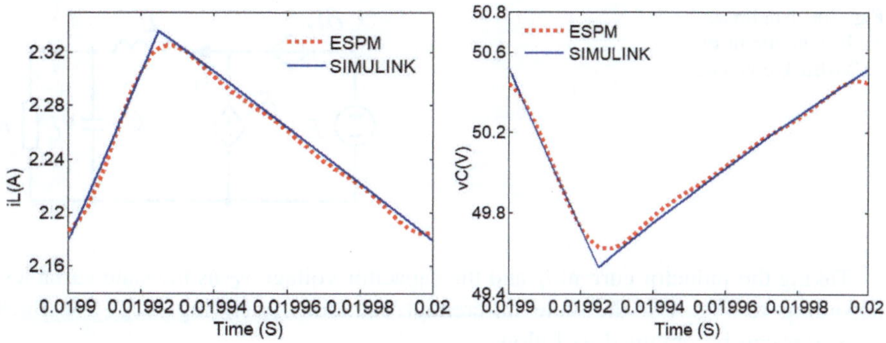

Fig. 4.4 Steady-state ripple waveforms of state variables for CCM-Boost converter with $f_s = 10$ kHz

4.4 Analysis of the Open-Loop Buck Converter Under CCM Operation

4.4.1 Modeling of the CCM-Buck Converter

The principle circuit of Buck converter is shown in Fig. 4.5a. It's assumed that the converter operates in continuous-current-mode (CCM), that is, the converter has two equivalent topologies during one switching cycle T, which correspond to the switch S_T is turned on and the S_D is turned off (Fig. 4.5b), the switch S_T is turned off, and the S_D is turned on (Fig. 4.5c).

Introduce a switching function to describe the on/off state of the switch S_T, which is defined as (4.12). Then according to the method in Chap. 2, the Buck converter with CCM operation can be replaced by a nonlinear equivalent circuit, as shown in Fig. 4.6.

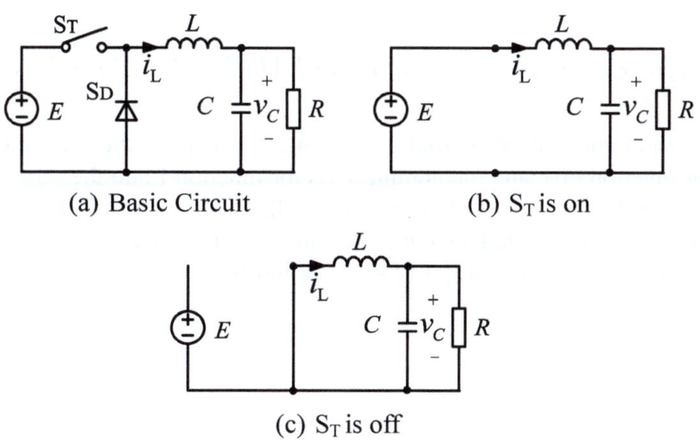

Fig. 4.5 CCM-Buck converter with its two equivalent topologies during one cycle

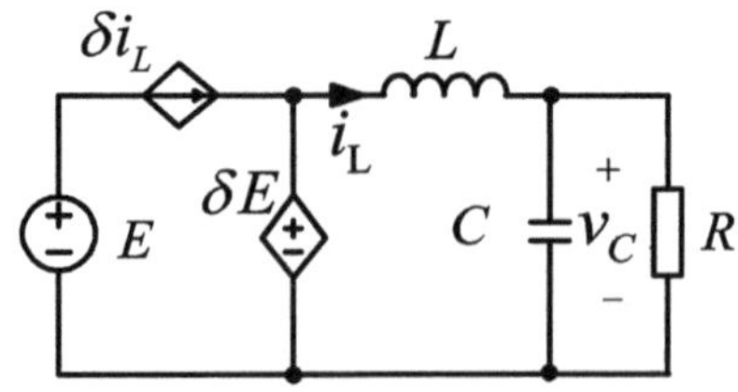

Fig. 4.6 Nonlinear equivalent circuit of CCM-Buck converter

Taking the inductor current i_L and the capacitor voltage v_C as the state variables, according to Fig. 4.6, the state differential equation describing the CCM-Buck converter can be obtained as follows:

$$\begin{cases} \frac{di_L}{dt} + \frac{1}{L} v_C + \frac{-1}{L} \delta E = 0 \\ \frac{-1}{C} i_L + \frac{dv_C}{dt} + \frac{1}{RC} v_C = 0 \end{cases} \quad (4.52)$$

Let $\mathbf{x} = \begin{bmatrix} i_L & v_C \end{bmatrix}^{Tr}$ and p = d/dt, then the state differential equation of the above formula can be written in matrix form as follows

$$G_0(p)\mathbf{x} + G_1(p)\mathbf{f} = 0 \quad (4.53)$$

Here the nonlinear vector is

$$\mathbf{f} = \delta \mathbf{e} \quad (4.54)$$

The constant vector **e** and the coefficient matrices are given as follows

$$G_0(p) = \begin{bmatrix} p & \frac{1}{L} \\ \frac{-1}{C} & p + \frac{1}{RC} \end{bmatrix}, \quad G_1(p) = \begin{bmatrix} 0 & \frac{-1}{L} \\ 0 & 0 \end{bmatrix}, \quad \mathbf{e} = \begin{bmatrix} 0 \\ E \end{bmatrix} \quad (4.55)$$

4.4.2 The Equivalent Mathematical Model Based on ESPM

Using the same method introduced in Sect. 4.2, we expand the state vector **x**, the switching function δ(t), and the nonlinear vector function **f** into the series as shown in Eqs. (4.17), (4.18) and (4.19), respectively. It should be noted that for the Buck converter, the main term and the correction amount of the nonlinear vector function are determined according to Eq. (4.54), it should be

$$\mathbf{f}_i = \delta_i \mathbf{e} \quad (4.56)$$

Using a similar processing method to solve Eq. (4.53), one can obtain the equivalent mathematical model describing the Buck converter, the iterative equations shown as follows, i.e.

$$\begin{cases} \mathbf{G}_0(p)\mathbf{x}_0 + \mathbf{G}_1(p)\mathbf{f}_{0m} = 0 \\ \mathbf{G}_0(p)\mathbf{x}_1 + \mathbf{G}_1(p)\mathbf{f}_{1m} = -\mathbf{G}_1(p)\mathbf{R}_1 \\ \mathbf{G}_0(p)\mathbf{x}_2 + \mathbf{G}_1(p)\mathbf{f}_{2m} = -\mathbf{G}_1(p)\mathbf{R}_2 \\ \cdots\cdots \end{cases} \tag{4.57}$$

4.4.3 *The Steady-State Periodic Solution of the Buck Converter Based on ESPM*

Similar to Sect. 4.3.3, we can solve Eq. (4.57) step by step. First, the main wave $\mathbf{x}_0$ to the steady-state value of state variable $\mathbf{x}$ is chosen as

$$\mathbf{x}_0 = \mathbf{a}_{00} = \begin{bmatrix} I_{00} & V_{00} \end{bmatrix}^{Tr} \tag{4.58}$$

The main wave of nonlinear vector is $\mathbf{f}_0 = \delta_0\mathbf{e} = b_0\mathbf{e} + b_1\mathbf{e}\cdot e^{j\tau} + c.c$, and hence we can get

$$\begin{cases} \mathbf{f}_{0m} = b_0\mathbf{e} \\ \mathbf{R}_1 = b_1\mathbf{e}\cdot e^{j\tau} + c.c \end{cases} \tag{4.59}$$

Introducing $\mathbf{x}_0$ and $\mathbf{f}_{0\mathrm{m}}$ into the first equation of Eq. (4.57), it can be attained

$$\mathbf{G}_0(0)\mathrm{a}_{00} = -\mathbf{G}_1(0)\mathrm{b}_0\mathbf{e} \tag{4.60}$$

And Eq. (4.60) can be rewritten in the matrix form as

$$\begin{bmatrix} 0 & \frac{1}{L} \\ -\frac{1}{C} & \frac{1}{RC} \end{bmatrix} \begin{bmatrix} I_{00} \\ V_{00} \end{bmatrix} = \begin{bmatrix} 0 & \frac{b_0}{L} \\ 0 & 0 \end{bmatrix} \begin{bmatrix} 0 \\ E \end{bmatrix} \tag{4.61}$$

Thereby the term $\mathbf{a}_{00}$ could be got from (4.61).

$$V_{00} = b_0E \qquad I_{00} = \frac{V_{00}}{R}$$

Then according to the content of harmonics in the term $\mathbf{R}_1$, the first order correction $\mathbf{x}_1$ can be assumed to be:

$$\mathbf{x}_1 = \mathbf{a}_{11}e^{j\tau} + \bar{\mathbf{a}}_{11}e^{-j\tau} \tag{4.62}$$

The first-order correction term of the nonlinear vector is $\mathbf{f}_1 = \delta_1\mathbf{e} = b_2\mathbf{e}\cdot e^{j2\tau} + b_3\mathbf{e}\cdot e^{j3\tau} + c.c$, and the following expressions for $\mathbf{f}_{1\mathrm{m}}$ and $\mathbf{R}_2$ can be got as

$$\begin{cases} \mathbf{f}_{1m} = 0 \\ \mathbf{R}_2 = b_2\mathbf{e}\cdot e^{j2\tau} + b_3\mathbf{e}\cdot e^{j3\tau} + c.c \end{cases} \tag{4.63}$$

Introducing $\mathbf{x}_1$, $\mathbf{R}_1$ and $\mathbf{f}_{1\mathrm{m}}$ into the second equation in Eq. (4.57), and using the method of harmonic balance, one can get

$$\mathbf{G}_0(j\omega)\mathbf{a}_{11} = -\mathbf{G}_1(j\omega)b_1\mathbf{e} \tag{4.64}$$

And Eq. (4.64) can be rewritten in the matrix form as:

$$\begin{bmatrix} j\omega & \frac{1}{L} \\ -\frac{1}{C} & j\omega + \frac{1}{RC} \end{bmatrix}\begin{bmatrix} I_{11} \\ V_{11} \end{bmatrix} = \begin{bmatrix} 0 & \frac{b_1}{L} \\ 0 & 0 \end{bmatrix}\begin{bmatrix} 0 \\ E \end{bmatrix} \tag{4.65}$$

Thus the vector $\mathbf{a}_{11}$ could be found from (4.65).

Next according to the harmonics in the term $\mathbf{R}_2$, we can choose the second order correction $\mathbf{x}_2$ as:

$$\mathbf{x}_2 = \mathbf{a}_{22}e^{j2\tau} + \bar{\mathbf{a}}_{22}e^{-j2\tau} + \mathbf{a}_{23}e^{j3\tau} + \bar{\mathbf{a}}_{23}e^{-j3\tau} \tag{4.66}$$

Similarly, from $\mathbf{f}_2 = \delta_2\mathbf{e}$, there is $\mathbf{f}_{2m} = 0$, thus one can obtain the following two equations

$$\mathbf{G}_0(\mathrm{j}2\omega)\mathbf{a}_{22} = -\mathbf{G}_1(\mathrm{j}2\omega)b_2\mathbf{e} \tag{4.67a}$$

$$G_0(j3\omega)\mathbf{a}_{23} = -G_1(j3\omega)b_3\mathbf{e} \tag{4.67b}$$

And they can be rewritten in matrix form as:

$$\begin{bmatrix} j2\omega & \frac{1}{L} \\ -\frac{1}{C} & j2\omega + \frac{1}{RC} \end{bmatrix}\begin{bmatrix} I_{22} \\ V_{22} \end{bmatrix} = \begin{bmatrix} 0 & \frac{b_2}{L} \\ 0 & 0 \end{bmatrix}\begin{bmatrix} 0 \\ E \end{bmatrix} \tag{4.68a}$$

And

$$\begin{bmatrix} j3\omega & \frac{1}{L} \\ -\frac{1}{C} & j3\omega + \frac{1}{RC} \end{bmatrix}\begin{bmatrix} I_{23} \\ V_{23} \end{bmatrix} = \begin{bmatrix} 0 & \frac{b_3}{L} \\ 0 & 0 \end{bmatrix}\begin{bmatrix} 0 \\ E \end{bmatrix} \tag{4.68b}$$

Then the vector $\mathbf{a}_{22}$ and $\mathbf{a}_{23}$ be found.

Following this process, the term $\mathbf{a}_{\mathrm{ik}}$ could be solved step by step, and detailed information is listed in Table 4.3.

And the coefficient b_{i} are still determined by Eq. (4.48).

Table 4.3 Values of $\mathbf{a}_{ik}$ for CCM-Buck converter

i	k	a_{ik}	
		V_{ik}	I_{ik}
0	0	$V_{00} = b_0 E$	$I_{00} = \frac{V_{00}}{R}$
1	1	$V_{11} = \frac{b_1 E}{1+(j\omega)^2 LC + j\omega L/R}$	$I_{11} = \frac{b_1 E - V_{11}}{j\omega L}$
	2	$V_{22} = \frac{b_2 E}{1+(j2\omega)^2 LC + j2\omega L/R}$	$I_{22} = \frac{b_2 E - V_{11}}{j2\omega L}$
	3	$V_{23} = \frac{b_3 E}{1+(j3\omega)^2 LC + j3\omega L/R}$	$I_{23} = \frac{b_3 E - V_{11}}{j3\omega L}$

4.4.4 Simulations

The circuit parameters of the open-loop Buck converter is list in Table 4.5.

According to (4.48), when the duty ratio is known, the coefficients b_m can be obtained. Then according to Table 4.4, the approximate solution of the state variable $\mathbf{x} = \begin{bmatrix} i_L & v_C \end{bmatrix}^{Tr}$ can be obtained as follows, here i_L is the inductor current and v_C is the voltage over the output capacitor.

$$\begin{aligned} i_L &= 0.5267 - 0.1631\cos\tau + 0.0833\sin\tau - 0.0333\cos 2\tau - 0.0242\sin 2\tau\ (\text{A}) \\ v_C &= 5.25 - 0.0636\cos\tau - 0.1062\sin\tau + 0.0078\cos 2\tau - 0.0308\sin 2\tau \quad (\text{V}) \end{aligned} \tag{4.69}$$

It should be noted that, as the amplitude of the 3rd harmonic is quite small, which is ignored in (4.69).

The ESP analysis method (solid line) and the simulation results (dashed line) are compared in Fig. 4.7, which shows the comparison of the inductor current ripple and the capacitor voltage ripple waveform during one cycle. It can be seen that the results from ESPM are in good coincidence with those from simulations, indicating that the ESPM can analyze the steady-state solution of the open-loop converter well (Fig. 4.7).

Table 4.4 Circuit parameters of the open-loop Buck converter

Parameters	Values
Input voltage E	15 V
Switching frequency f	50 kHz
Inductance L	150 μH
Capacitance C	4.7 μF
Load resistance R	10 Ω
Duty ratio D	0.35

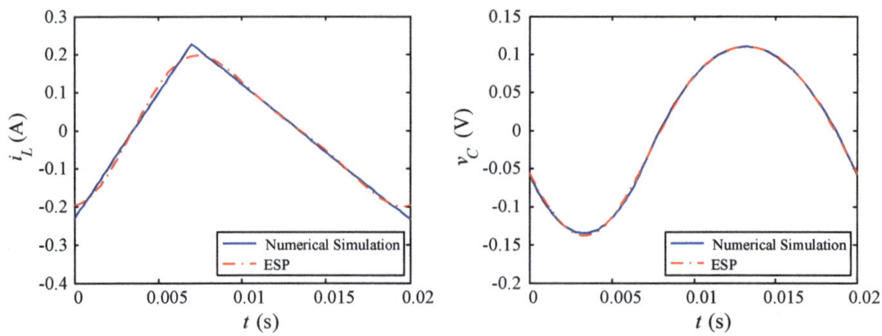

Fig. 4.7 Steady-state ripple waveforms of state variables for CCM-Buck converter

4.5 Analysis of the Open-Loop Cuk Converter Under CCM Operation

4.5.1 Modeling of the CCM-Cuk Converter

The principle circuit of Cuk converter is shown in Fig. 4.8a. It's assumed that the converter operates in continuous-current-mode (CCM), that is, the converter has two equivalent topologies during one switching cycle T, which correspond to the switch S_T is turned on and the S_D is turned off (Fig. 4.8b), the switch S_T is turned off and the S_D is turned on (Fig. 4.8c).

Introduce a switching function to describe the on/off state of the switch S_T, which is defined as (4.12). Then according to the method in Chap. 2, the Cuk converter with CCM operation can be replaced by a nonlinear equivalent circuit, as shown in Fig. 4.9.

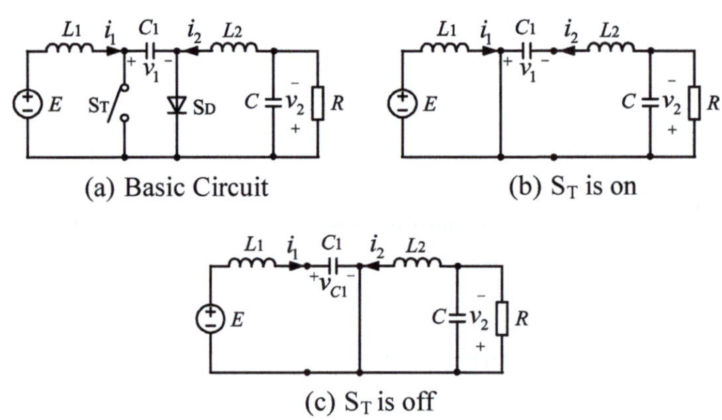

Fig. 4.8 CCM-Cuk converter with its two equivalent topologies during one cycle

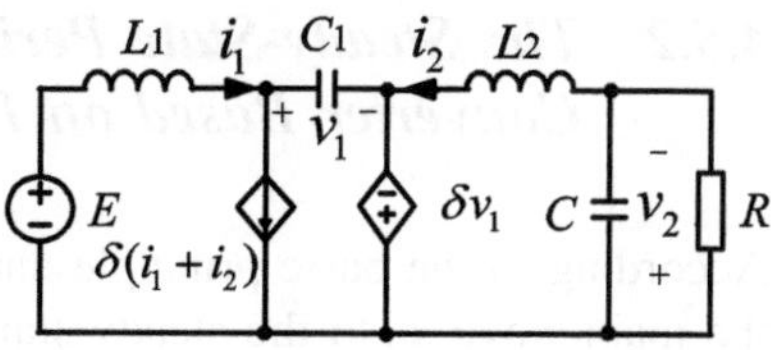

Fig. 4.9 Nonlinear equivalent circuit of CCM-Cuk converter

Taking the inductor currents i_1, i_2 and the capacitor voltages v_1,v_2 as the state variables, according to Fig. 4.9, the state differential equation describing the CCM-Cuk converter can be obtained as follows:

$$\begin{cases} \frac{\mathrm{d}i_1}{\mathrm{d}t} + \frac{1}{L_1}v_1 + \frac{-1}{L_1}\delta v_1 = \frac{1}{L_1}E \\ \frac{\mathrm{d}i_2}{\mathrm{d}t} + \frac{1}{L_2}v_2 + \frac{-1}{L_2}\delta v_1 = 0 \\ \frac{-1}{C_1}i_1 + \frac{\mathrm{d}v_1}{\mathrm{d}t} + \frac{1}{C_1}\delta(i_1+i_2) = 0 \\ \frac{-1}{C}i_2 + \frac{\mathrm{d}v_2}{\mathrm{d}t} + \frac{1}{RC}v_2 = 0 \end{cases} \tag{4.70}$$

Let $\mathbf{x} = \begin{bmatrix} i_1 & i_2 & v_1 & v_2 \end{bmatrix}^{Tr}$ and p = d/dt, then the state differential equation of the above formula can be written in matrix form as follows

$$G_0(p)\mathbf{x} + G_1(p)\mathbf{f} = \mathbf{u} \tag{4.71}$$

Here the nonlinear vector is

$$\mathbf{f} = \delta\mathbf{x} \tag{4.72}$$

The constant vector **u** and the coefficient matrices are given as follows

$$G_0(p) = \begin{bmatrix} p & 0 & \frac{1}{L_1} & 0 \\ 0 & p & 0 & \frac{1}{L_2} \\ \frac{-1}{C_1} & 0 & p & 0 \\ 0 & \frac{-1}{C} & 0 & p+\frac{1}{RC} \end{bmatrix}, \quad G_1(p) = \begin{bmatrix} 0 & 0 & \frac{-1}{L_1} & 0 \\ 0 & 0 & \frac{-1}{L_1} & 0 \\ \frac{1}{C_1} & \frac{1}{C_1} & 0 & 0 \\ 0 & 0 & 0 & 0 \end{bmatrix}, \quad \mathbf{u} = \begin{bmatrix} E/L_1 \\ 0 \\ 0 \\ 0 \end{bmatrix} \tag{4.73}$$

Equations (4.71) and (4.72) are identical in form to Eqs. (4.14) and (4.16) respectively. And the equivalent mathematical model of the CCM-operated Cuk converter is also the same as Eq. (4.26), which can be solved in exactly the same way as in Sect. 4.3.

4.5.2 The Steady-State Periodic Solution of the Cuk Converter Based on ESPM

According to the basic principle and the low-pass properties of DC/DC converter, the main wave $\mathbf{x}_0$ to the steady-state value of state variable $\mathbf{x}$ could be chosen as

$$\mathbf{x}_0 = \mathbf{a}_{00} = \begin{bmatrix} I_{100} & I_{200} & V_{100} & V_{200} \end{bmatrix}^{Tr}$$

And $\mathbf{a}_{00}$ can be got by solving the following matrix equation:

$$[\mathbf{G}_0(0) + \mathbf{G}_1(0)b_0]\mathbf{a}_{00} = \mathbf{u} \tag{4.74}$$

Here $\mathbf{G}_0(0)$ and $\mathbf{G}_1(0)$ can be obtained by setting the differential operator $p = 0$ in Eq. (4.73). For the converter with open loop operation, the duty ratio d is a known amount, i.e. $d = D$, then Eq. (4.74) can be rewritten in the matrix form as

$$\left(\begin{bmatrix} 0 & 0 & \frac{1}{L_1} & 0 \\ 0 & 0 & 0 & \frac{1}{L_2} \\ \frac{-1}{C_1} & 0 & 0 & 0 \\ 0 & \frac{-1}{C} & 0 & \frac{1}{RC} \end{bmatrix} + \begin{bmatrix} 0 & 0 & \frac{-D}{L_1} & 0 \\ 0 & 0 & \frac{-D}{L_1} & 0 \\ \frac{D}{C_1} & \frac{D}{C_1} & 0 & 0 \\ 0 & 0 & 0 & 0 \end{bmatrix}\right) \cdot \begin{bmatrix} I_{100} \\ I_{200} \\ V_{100} \\ V_{200} \end{bmatrix} = \begin{bmatrix} \frac{1}{L_1}E \\ 0 \\ 0 \\ 0 \end{bmatrix} \tag{4.75}$$

From (4.75), the solutions of DC components of the state variables can be obtained as

$$V_{100} = \frac{E}{1-D}, \quad V_{200} = \frac{L_2}{L_1}\frac{DE}{(1-D)}, \quad I_{200} = \frac{V_{200}}{R}, \quad I_{100} = \frac{DI_{200}}{1-D}$$

The first order correction $\mathbf{x}_1$ can be assumed to be:

$$\mathbf{x}_1 = \mathbf{a}_{11}e^{j\tau} + \bar{\mathbf{a}}_{11}e^{-j\tau}$$

in which $\bar{\mathbf{a}}_{11}$ means complex-conjugate, and $\mathbf{a}_{11} = [\mathrm{I}_{111}\ \mathrm{I}_{211}\ \mathrm{V}_{111}\ \mathrm{V}_{211}]^{\mathrm{Tr}}$. And $\mathbf{a}_{11}$ can be got from the following equation as

$$[\mathbf{G}_0(j\omega) + \mathbf{G}_1(j\omega)b_0]\mathbf{a}_{11} = -\mathbf{G}_1(j\omega) \cdot b_1\mathbf{a}_{00} \tag{4.76}$$

where the coefficient b_1 can be obtained according to (4.23). $\mathbf{G}_0(\mathrm{j}\omega)$ and $\mathbf{G}_1(\mathrm{j}\omega)$ can be obtained by setting the differential operator $\mathrm{p} = \mathrm{j}\omega$ in Eq. (4.73). For the Cuk converter under study, Eq. (4.76) can be rewritten in the matrix form as:

$$\left(\begin{bmatrix} j\omega & 0 & \frac{1}{L_1} & 0 \\ 0 & j\omega & 0 & \frac{1}{L_2} \\ \frac{-1}{C_1} & 0 & j\omega & 0 \\ 0 & \frac{-1}{C} & 0 & j\omega+\frac{1}{RC} \end{bmatrix} + \begin{bmatrix} 0 & 0 & \frac{-D}{L_1} & 0 \\ 0 & 0 & \frac{-D}{L_1} & 0 \\ \frac{D}{C_1} & \frac{D}{C_1} & 0 & 0 \\ 0 & 0 & 0 & 0 \end{bmatrix}\right) \cdot \begin{bmatrix} I_{111} \\ I_{211} \\ V_{111} \\ V_{211} \end{bmatrix} = \begin{bmatrix} 0 & 0 & \frac{-1}{L_1} & 0 \\ 0 & 0 & \frac{-1}{L_1} & 0 \\ \frac{1}{C_1} & \frac{1}{C_1} & 0 & 0 \\ 0 & 0 & 0 & 0 \end{bmatrix} \begin{bmatrix} b_1 I_{100} \\ b_1 I_{200} \\ b_1 V_{100} \\ b_1 V_{200} \end{bmatrix} \tag{4.77}$$

Following the same procedure, we can choose the second order correction $\mathbf{x}_2$ as:

$$\mathbf{x}_2 = \mathbf{a}_{20} + \mathbf{a}_{22}e^{j2\tau} + \bar{\mathbf{a}}_{22}e^{-j2\tau} + \mathbf{a}_{23}e^{j3\tau} + \bar{\mathbf{a}}_{23}e^{-j3\tau}$$

Here $\mathbf{a}_{20} = [\mathrm{I}_{120}\ \mathrm{I}_{220}\ \mathrm{V}_{120}\ \mathrm{V}_{220}]^{\mathrm{Tr}}$ is the vector of DC values, it will give the corrections of DC components in $\mathbf{a}_{00}$, and $\mathbf{a}_{22} = [\mathrm{I}_{122}\ \mathrm{I}_{222}\ \mathrm{V}_{122}\ \mathrm{V}_{222}]^{\mathrm{Tr}}$ and $\mathbf{a}_{23} = [\mathrm{I}_{123}\ \mathrm{I}_{223}\ \mathrm{V}_{123}\ \mathrm{V}_{223}]^{\mathrm{Tr}}$ correspond to the amplitudes of second- and third-harmonics of state variable $\mathbf{x}$ respectively.

Similarly, $\mathbf{a}_{20}$, $\mathbf{a}_{22}$, $\mathbf{a}_{23}$ can be obtained from the following equations.

$$[G_0(0) + G_1(0)b_0]\mathbf{a}_{20} = -G_1(0)(b_1\bar{\mathbf{a}}_{11} + \bar{b}_1\mathbf{a}_{11}) \tag{4.78a}$$

$$[G_0(j2\omega) + G_1(j2\omega)b_0]\mathbf{a}_{22} = -G_1(j2\omega)\cdot(b_2\mathbf{a}_{00} + b_1\mathbf{a}_{11} + b_3\bar{\mathbf{a}}_{11}) \tag{4.78b}$$

$$[G_0(j3\omega) + G_1(j3\omega)b_0]\mathbf{a}_{23} = -G_1(j3\omega)\cdot(b_3\mathbf{a}_{00} + b_2\mathbf{a}_{11} + b_1\mathbf{a}_{22}) \tag{4.78c}$$

The coefficient matrices $\mathbf{G}_0(\mathrm{jk}\,\omega)$ and $\mathbf{G}_1(\mathrm{jk}\,\omega)$ can be obtained by setting the differential operator $\mathrm{p} = \mathrm{jk}\,\omega$ in Eq. (4.73) (here k = 0, 2, 3). Equation (4.78) can be rewritten in matrix form as:

$$\left(\begin{bmatrix} 0 & 0 & \frac{1}{L_1} & 0 \\ 0 & 0 & 0 & \frac{1}{L_2} \\ \frac{-1}{C_1} & 0 & 0 & 0 \\ 0 & \frac{-1}{C} & 0 & \frac{1}{RC} \end{bmatrix} + \begin{bmatrix} 0 & 0 & \frac{-D}{L_1} & 0 \\ 0 & 0 & \frac{-D}{L_1} & 0 \\ \frac{D}{C_1} & \frac{D}{C_1} & 0 & 0 \\ 0 & 0 & 0 & 0 \end{bmatrix}\right) \begin{bmatrix} I_{120} \\ I_{220} \\ V_{120} \\ V_{220} \end{bmatrix} = \begin{bmatrix} 0 & 0 & \frac{-1}{L_1} & 0 \\ 0 & 0 & \frac{-1}{L_1} & 0 \\ \frac{1}{C_1} & \frac{1}{C_1} & 0 & 0 \\ 0 & 0 & 0 & 0 \end{bmatrix} \cdot \begin{bmatrix} \bar{b}_1 I_{111} + b_1\bar{I}_{111} \\ \bar{b}_1 I_{211} + b_1\bar{I}_{211} \\ \bar{b}_1 V_{111} + b_1\bar{V}_{111} \\ \bar{b}_1 V_{211} + b_1\bar{V}_{211} \end{bmatrix} \tag{4.79a}$$

$$\left(\begin{bmatrix} j2\omega & 0 & \frac{1}{L_1} & 0 \\ 0 & j2\omega & 0 & \frac{1}{L_2} \\ \frac{-1}{C_1} & 0 & j2\omega & 0 \\ 0 & \frac{-1}{C} & 0 & j2\omega + \frac{1}{RC} \end{bmatrix} + \begin{bmatrix} 0 & 0 & \frac{-D}{L_1} & 0 \\ 0 & 0 & \frac{-D}{L_1} & 0 \\ \frac{D}{C_1} & \frac{D}{C_1} & 0 & 0 \\ 0 & 0 & 0 & 0 \end{bmatrix}\right) \cdot \begin{bmatrix} I_{122} \\ I_{222} \\ V_{122} \\ V_{222} \end{bmatrix}$$
$$= \begin{bmatrix} 0 & 0 & \frac{-1}{L_1} & 0 \\ 0 & 0 & \frac{-1}{L_1} & 0 \\ \frac{1}{C_1} & \frac{1}{C_1} & 0 & 0 \\ 0 & 0 & 0 & 0 \end{bmatrix} \cdot \begin{bmatrix} b_1 I_{111} + b_2 I_{100} + b_3 \bar{I}_{111} \\ b_1 I_{211} + b_2 I_{200} + b_3 \bar{I}_{211} \\ b_1 V_{111} + b_2 V_{100} + b_3 \bar{V}_{111} \\ b_1 V_{211} + b_2 V_{200} + b_3 \bar{V}_{211} \end{bmatrix} \tag{4.79b}$$

$$\left(\begin{bmatrix} j3\omega & 0 & \frac{1}{L_1} & 0 \\ 0 & j3\omega & 0 & \frac{1}{L_2} \\ \frac{-1}{C_1} & 0 & j3\omega & 0 \\ 0 & \frac{-1}{C} & 0 & j3\omega + \frac{1}{RC} \end{bmatrix} + \begin{bmatrix} 0 & 0 & \frac{-D}{L_1} & 0 \\ 0 & 0 & \frac{-D}{L_1} & 0 \\ \frac{D}{C_1} & \frac{D}{C_1} & 0 & 0 \\ 0 & 0 & 0 & 0 \end{bmatrix}\right) \cdot \begin{bmatrix} I_{123} \\ I_{223} \\ V_{123} \\ V_{223} \end{bmatrix}$$
$$= \begin{bmatrix} 0 & 0 & \frac{-1}{L_1} & 0 \\ 0 & 0 & \frac{-1}{L_1} & 0 \\ \frac{1}{C_1} & \frac{D}{C_1} & 0 & 0 \\ 0 & 0 & 0 & 0 \end{bmatrix} \begin{bmatrix} b_1 I_{122} + b_2 I_{111} + b_3 I_{100} \\ b_1 I_{222} + b_2 I_{211} + b_3 I_{200} \\ b_1 V_{122} + b_2 V_{111} + b_3 V_{100} \\ b_1 V_{222} + b_2 V_{211} + b_3 V_{200} \end{bmatrix} \tag{4.79c}$$

According to Eqs. (4.75), (4.77) and (4.79), the analytic form solutions of the state variables can be obtained by using some symbolic analysis software tools.

4.5.3 Simulations

The circuit parameters of the open-loop Cuk converter is list in Table 4.5.

According to (4.48), when the duty ratio is known, the coefficients b_m can be obtained. Then according to Eqs. (4.75), (4.77) and (4.79), when the circuit parameters are introduced, the approximate solution of the state variable $\mathbf{x}$ can be obtained as follows, where i_1 and i_2 correspond the inductor currents of L_1 and L_2, v_1 and v_2 are the voltages across C_1 and C (output filter capacitor), respectively.

Table 4.5 Circuit parameters of the open-loop Cuk converter

Parameters	Values
Input voltage E	25 V
Switching frequency f	50 kHz
Inductance L_1, L_2	1.9, 0.96 mH
Capacitance C_1, C	850, 47 μF
Load resistance R	30 Ω
Duty ratio D	0.55

Table 4.6 DC values of CCM-Cuk converter from the ESPM and simulations

	DC components of state variables
ESPM	I_1 = 1.306 A, I_2 = 1.0108 A, V_1 = 55.55 V, V_2 = 30.55 V
Numerical simulation	I_1 = 1.250 A, I_2 = 1.030 A, V_1 = 55.56 V, V_2 = 30.56 V

$$
\begin{aligned}
i_1 &= 1.306 + 0.578\cos\tau + 0.0092\sin\tau + 0.0014\cos 2\tau - 0.0044\sin 2\tau \\
&\quad + 0.0052\cos 3\tau + 0.0027\sin 3\tau \qquad \text{(A)} \\
i_2 &= 1.0108 + 0.1144\cos\tau + 0.0181\sin\tau + 0.0028\cos 2\tau - 0.0086\sin 2\tau \\
&\quad + 0.0103\cos 3\tau + 0.0053\sin 3\tau \qquad \text{(A)} \\
v_1 &= 55.55 - 0.0054\cos\tau - 0.0006\sin\tau - 0.0003\cos 2\tau + 0.0004\sin 2\tau \\
&\quad - 0.0005\cos 3\tau - 0.0002\sin 3\tau \qquad \text{(V)} \\
v_2 &= 30.55 - 0.0012\cos\tau + 0.0078\sin\tau + 0.0003\cos 2\tau + 0.0001\sin 2\tau \\
&\quad - 0.0001\cos 3\tau + 0.0002\sin 3\tau \qquad \text{(V)}
\end{aligned}
\tag{4.80}
$$

The DC components obtained from the ESPM and the numerical simulation are listed in Table 4.6, from which it can be seen analytical results are very close to those from simulations.

For the AC components (ripples) of the state variables, the ESP analysis results (dashed line) and the simulation results (solid line) are compared in Fig. 4.10. Figure 4.10a, b show the inductor current ripples, and Fig. 4.10c, d the capacitor voltage ripples. It can be seen that the results from the ESPM are in good coincidence with those from simulations. Thus, for the Cuk converter, although the orders of the circuits are increased to four, the complexity of the analysis is not increased due to the matrix analysis calculation, indicating that even for higher orders converters, the ESPM can analyze the steady-state solutions with sufficient accuracy.

The analysis method for the steady-state periodic solution of the converter operating in discontinuous-conduction-mode (DCM) is similar and will not be repeated in this chapter. We can also refer to Chap. 6, which provides a detailed description of the steady-state periodic solution of a closed-loop converter operating in DCM.

4.6 Transient Analysis of the Open-Loop PWM Converter by ESPM

4.6.1 The Solution Procedure

Similar to the analysis of the steady-state solution of the switching converter by the equivalent small-parameter method, the high-order strong nonlinear system to be solved is described by the state differential equation with the following matrix form as:

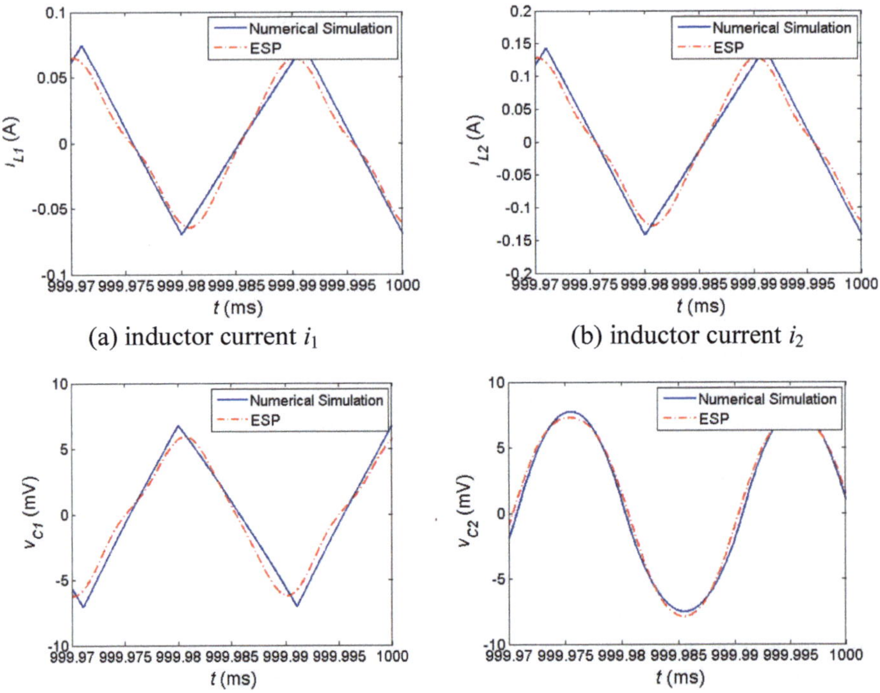

(a) inductor current i_1

(b) inductor current i_2

(c) inductor current v_1

(d) inductor current v_2

Fig. 4.10 Comparison of ripples of state variables of Cuk in CCM

$$G_0(p)\mathbf{x} + G_1(p)\mathbf{f} = \mathbf{u} \tag{4.81}$$

Here the l*1-ordered vector $\mathbf{x} = [x_1, x_2, \cdots x_l]^{Tr}$ (the superscript "Tr" indicates matrix transposition). The l*l-ordered coefficient matrices $\mathbf{G}_0(p)$ and $\mathbf{G}_1(p)$ are all related to the differential operator p and the circuit parameters, where the highest number of p is 1. The term $\mathbf{u}$ a input vector with l*1 order, and the l*l-ordered nonlinear vector function $\mathbf{f}(\mathbf{x})$ is related to the state variable vector $\mathbf{x}$.

According to the method introduced in Chap. 3 and Sect. 4.2, we still assume that the state vector $\mathbf{x}$ could be expanded into the following series:

$$\mathbf{x} = \mathbf{x}_0 + \varepsilon\mathbf{x}_1 + \varepsilon^2\mathbf{x}_2 + \cdots \tag{4.82}$$

Substituting $\mathbf{x}$ into the nonlinear function $\mathbf{f}(\mathbf{x})$ gives:

$$\mathbf{f}(\mathbf{x}) = \mathbf{f}_0 + \varepsilon^1\mathbf{f}_i + \varepsilon^2\mathbf{f}_2 + \cdots \tag{4.83}$$

in which:

$$\begin{cases} \mathbf{f}_0 = \delta_0 \mathbf{x}_0 \\ \mathbf{f}_1 = \delta_0 \mathbf{x}_1 + \delta_1 \mathbf{x}_0 \\ \mathbf{f}_2 = \delta_0 \mathbf{x}_2 + \delta_1 \mathbf{x}_1 + \delta_2 \mathbf{x}_0 \\ \cdots \end{cases} \tag{4.84}$$

In Eqs. (4.82) and (4.83), i is an approximate order; ε is a small parameter marker.

Generally the terms $\mathbf{x}_0$ and $\mathbf{x}_i$ in Eq. (4.82) for state variables $\mathbf{x}$ can be expressed in the Fourier series as follows

$$\begin{aligned} \mathbf{x}_0 &= \sum_{n \in E_0} \mathbf{x}_{n0} = \sum_{n \in E_0} \left(\mathbf{a}_{n0}(\tau) e^{jn\tau} + \bar{\mathbf{a}}_{n0}(\tau) e^{-jn\tau} \right) \\ \mathbf{x}_i &= \sum_{k \in E_i} \mathbf{x}_{ki} = \sum_{k \in E_i} \left(\mathbf{a}_{ki}(\tau) e^{jk\tau} + \bar{\mathbf{a}}_{ki}(\tau) e^{-jk\tau} \right) \end{aligned} \tag{4.85}$$

where the normalized time $\tau = \omega t = 2\pi f_s t$, the terms $\bar{\mathbf{a}}_{n0}(\tau)$ and $\bar{\mathbf{a}}_{ki}(\tau)$ are the conjugate plurals of $\mathbf{a}_{n0}(\tau)$ and $\mathbf{a}_{ki}(\tau)$ respectively. It should be noted that, unlike the steady-state periodic solution, the harmonic coefficients $\mathbf{a}_{n0}(\tau)$ and $\mathbf{a}_{ki}(\tau)$ are variables related to the normalized time τ. In the case of a steady-state periodic solution, the coefficients $\mathbf{a}_{n0}$ and $\mathbf{a}_{ki}$ are unknown constants to be solved, they are independent of time. Similar to the solution of the steady-state periodic solution, in Eq. (4.85), the spectral set E_0 of the principal component is determined by the physical knowledge associated with the system under study. For example, when the switching power converter system is stable, the main component of its output voltage and inductor current is DC, so $E_0 = \{0\}$ can be selected; and for the weak nonlinear system we can choose $E_0 = \{1\}$, which means that only the fundamental wave are contained in the main component. The frequency set E_i of the correction term $\mathbf{x}_i$ is determined step by step during the iteration.

Substituting $\mathbf{x}_\mathrm{i}$ into $\mathbf{f}_\mathrm{i}$ and decomposing $\mathbf{f}_\mathrm{i}$ $(i = 0,\ 1,\ 2\ldots)$ into the sum of the main term and the remainder, the following formula can be obtained.

$$\mathbf{f}_i = \mathbf{f}(\mathbf{x}_0, \mathbf{x}_1, \cdots, \mathbf{x}_i) = \mathbf{f}_{im} + \varepsilon \mathbf{R}_{i+1} \tag{4.86}$$

In which the main term $\mathbf{f}_{im}$ has the same spectral set as the term $\mathbf{x}_\mathrm{i}$, that is, all items in $\mathbf{f}_\mathrm{i}$ with the same harmonic components as $\mathbf{x}_\mathrm{i}$ belong to $\mathbf{f}_{im}$, while other harmonic components belong to the remainder $\mathbf{R}_{i+1}$. Therefore, $\mathbf{R}_{i+1}$ can be regarded as a small amount with higher order than the term $\mathbf{f}_\mathrm{im}$, and in front of which the small amount indicator ε is introduced again. The terms $\mathbf{f}_{im}$ and $\mathbf{R}_{i+1}$ can be expressed into Fourier series as follows:

$$\mathbf{f}_{0m} = \sum_{k \in E_0} \left[\mathbf{h}_{n0}(\tau) e^{jn\tau} + \bar{\mathbf{h}}_{n0}(\tau) e^{-jn\tau} \right] \tag{4.87a}$$

$$\mathbf{f}_{im} = \sum_{k \in E_i} \left[\mathbf{h}_{ki}(\tau)e^{jk\tau} + \bar{\mathbf{h}}_{ki}(\tau)e^{-jk\tau}\right] \tag{4.87b}$$

$$\mathbf{R}_{i+1} = \sum_{k \in E_{i+1}} \left[\mathbf{h}'_{k(i+1)}(\tau)^{jk\tau} + \bar{\mathbf{h}}'_{k(i+1)}(\tau)e^{-jk\tau}\right] \tag{4.87c}$$

It still should be noted that, the amplitude coefficients of each harmonic in the above equations are time-dependent variables, which are quite different with those in the steady-state periodic solution.

Substituting (4.86) into (4.83) gives

$$\mathbf{f} = (\mathbf{f}_{0m} + \varepsilon\mathbf{f}_{1m} + \varepsilon^2\mathbf{f}_{2m} + \cdots) + (\varepsilon\mathbf{R}_1 + \varepsilon^2\mathbf{R}_2 + \cdots) \tag{4.88}$$

Substituting (4.82) and (4.88) into (4.81) and making the coefficients of the same mark ε^i on both sides of the equation equal, you can get the following iterative equations as

$$\begin{cases} \mathbf{G}_0(p)\mathbf{x}_0 + \mathbf{G}_1(p)\mathbf{f}_{0m} = \mathbf{u} \\ \mathbf{G}_0(p)\mathbf{x}_1 + \mathbf{G}_1(p)\mathbf{f}_{1m} = -\mathbf{G}_1(p)\mathbf{R}_1 \\ \mathbf{G}_0(p)\mathbf{x}_2 + \mathbf{G}_1(p)\mathbf{f}_{2m} = -\mathbf{G}_1(p)\mathbf{R}_2 \end{cases} \tag{4.89}$$
$$\cdots\cdots$$

The above iterative Eq. (4.89) is exactly what is needed to solve the transient solution. The main term $\mathbf{x}_0$ and the correction term $\mathbf{x}_i$ can be obtained by solving each equation in (4.89) using the harmonic balance method step by step. It can be seen that Eq. (4.89) is identical in form to the iterative Eq. (3.43) for finding the steady-state periodic solution of the nonlinear system. However, it should be noted that Eq. (3.43) is used to solve the steady-state periodic solution as an algebraic equation, and Eq. (4.89) is used to find the transient solution as a differential equation.

$$p(\mathbf{a}_{ki}e^{jk\omega t}) = \frac{d(\mathbf{a}_{ki}e^{jk\omega t})}{dt} = jk\omega\,(\mathbf{a}_{ki}e^{jk\omega t}) \tag{4.90}$$

Therefore, when finding the coefficients $\mathbf{a}_{ki}$ of the harmonic components with the steady-state solution, the differential operator $p = jk\omega$ in Eq. (3.43), where k = 0, 1 …, represents the order of harmonics.

While in the transient solution, since the coefficients $\mathbf{a}_{ki}(\tau)$ of the harmonic components are time-dependent variables, there exists the following equation about the differential operation.

$$p[\mathbf{a}_{ki}(\tau)e^{jk\omega t}] = \frac{d[\mathbf{a}_{ki}(\tau)e^{jk\omega t}]}{dt} = \frac{d[\mathbf{a}_{ki}(\tau)]}{dt} \cdot e^{jk\omega t} + jk\omega \cdot \mathbf{a}_{ki}(\tau)\, e^{jk\omega t}$$
$$= (p + jk\omega)\mathbf{a}_{ki}(\tau)\, e^{jk\omega t} \tag{4.91}$$

Therefore, when finding the coefficients $\mathbf{a}_{ki}(\tau)$ of the harmonic components with the transient solution, we use $(p + jk\omega)$ instead of the differential operator p in Eq. (4.89).

For open-loop switching converters, the differential equations in Eq. (4.89) are usually linear and can also be expressed as follows.

$$\frac{d\mathbf{x}}{dt} = \mathbf{A}\mathbf{x} + \mathbf{B}\mathbf{g}(t) \tag{4.92}$$

The solution of the equation above is

$$\mathbf{x}(t) = e^{\mathbf{A}t}\mathbf{x}(0) + e^{\mathbf{A}t}\int_0^t e^{-\mathbf{A}\tau}\mathbf{B}\mathbf{g}(\tau)d\tau \tag{4.93}$$

in which the term $\mathbf{x}(0)$ represents the initial values of the state variable vector.

4.6.2 Initial Value Determination

In order to find the transient solution of the nonlinear system, the initial values need to be determined. In the case of accurate solution, the initial values should satisfy the following conditions as

$$\mathbf{x}(0) = \mathbf{x}_0(0) + \varepsilon\mathbf{x}_1(0) + \varepsilon^2\mathbf{x}_2(0) + \cdots = \mathbf{x}_0(0) + \mathbf{x}_1(0) + \mathbf{x}_2(0) + \cdots \tag{4.94}$$

In which the small amount indicator $\varepsilon = 1$.

In the case of simple estimation, $\mathbf{x}_0(0) = \mathbf{x}(0)$ can be used to obtain the transient solution $\mathbf{x}_0(\tau)$ of the first equation in Eq. (4.89), and then the forced solutions $\mathbf{x}_1(\tau)$ and $\mathbf{x}_2(\tau)$ … of the second and the third equations in Eq. (4.89) …, etc. However, the transient solution obtained in this way usually does not satisfy the initial conditions (4.94).

Usually, similar to the literature [3], we determine the initial values as follows:

(1) Let $\mathbf{x}_0(0) = \mathbf{x}(0)$, and find $\mathbf{x}_0(\tau)$ by the first equation of Eq. (4.89);
(2) Solve the second equation of Eq. (4.89) and get the special solution $\mathbf{x}_1(\tau)$, then let $\mathbf{x}_0(0) = \mathbf{x}(0) - \mathbf{x}_1(0)$;
(3) Similarly, after solving Eq. (4.89) to obtain the special solutions $\mathbf{x}_2(\tau)$ and $\mathbf{x}_3(\tau)$, …, etc., let $\mathbf{x}_0(0) = \mathbf{x}(0) - \mathbf{x}_1(0) - \mathbf{x}_2(0) - \mathbf{x}_3(0)\cdots$. Then use the modified $\mathbf{x}_0(0)$ as the initial value to find $\mathbf{x}_0(\tau)$, and according to the correction equation, the special solutions $\mathbf{x}_1(\tau)$ and $\mathbf{x}_2(\tau)$, …, etc. are obtained one by one, and finally the transient solution of Eq. (4.89) is obtained.

It must be noted that when using the equivalent small-parameter method, as long as the appropriate main component is selected, the initial $\mathbf{x}_0(0)$ has little effect on the special solution $\mathbf{x}_i(\tau)$. Thus, if $\mathbf{x}(0) = 0$ is selected, then $\mathbf{x}_i(0) = 0\,(i = 1, 2, \ldots)$ can be made to simplify the analysis process of the transient solution.

4.6.3 Transient Analysis of Open-Loop PWM Boost Converter

We still take the Boost converter shown in Fig. 4.1 as an example to illustrate the analysis of the transient solution of a switching converter using the ESPM. The circuit parameters are chosen as: L = 6 mH, C = 45 μF, R = 30 Ω, E = 37.5 V, f_s = 1 kHz, and the duty ratio is set to be D = 0.25.

According to the analysis in Sect. 4.3, the CCM-operated Boost converter can be described by a nonlinear vector differential equation as shown in (4.14), which is rewritten as follows

$$G_0(p)\mathbf{x} + G_1(p)\mathbf{f} = \mathbf{u} \tag{4.95}$$

where the coefficient matrices and the input vector are shown as follows.

$$G_0(p) = \begin{bmatrix} p & \frac{1}{L} \\ \frac{-1}{C} & p + \frac{1}{RC} \end{bmatrix}, \quad G_1(p) = \begin{bmatrix} 0 & \frac{-1}{L} \\ \frac{1}{C} & 0 \end{bmatrix}, \quad \mathbf{u} = \begin{bmatrix} E/L \\ 0 \end{bmatrix} \tag{4.96a}$$

the nonlinear vector is

$$\mathbf{f} = \delta\mathbf{x} \tag{4.96b}$$

And the switching function $\delta(t)$ is defined as shown in Eq. (4.12).

Considering that the initial condition is zero, there is $\mathrm{x_i}(0) = 0$, so the zero-order approximate solution can be assumed to be

$$\mathbf{x}_0 = a_{00}(t) = [i_{00}, v_{00}]^{Tr}$$

Substituting $\mathbf{x}_0$ into the first equation in Eq. (4.89) gives

$$[\mathbf{G}_0(p) + \mathbf{G}_1(p)b_0]\mathbf{a}_{00}(\tau) = \mathbf{u} \tag{4.97}$$

Equation (4.97) can be written in matrix form as

$$\left(\begin{bmatrix} p & \frac{1}{L} \\ -\frac{1}{C} & p + \frac{1}{RC} \end{bmatrix} + \begin{bmatrix} 0 & -\frac{b_0}{L} \\ \frac{b_0}{C} & 0 \end{bmatrix}\right)\begin{bmatrix} i_{00} \\ v_{00} \end{bmatrix} = \begin{bmatrix} E/L \\ 0 \end{bmatrix} \tag{4.98}$$

in which $b_0 = D$. Equation (4.98) can be expressed as shown in Eq. (4.99), since it is a linear differential equation.

$$\frac{d\mathbf{x}_0}{dt} = \mathbf{A}_0\mathbf{x}_0 + \mathbf{u} \tag{4.99}$$

In which, the coefficient matrix and the input vector are shown as follows.

$$\mathbf{A}_0 = \begin{bmatrix} 0 & -\frac{1-D}{L} \\ \frac{1-D}{C} & \frac{-1}{RC} \end{bmatrix}, \quad \mathbf{u} = \begin{bmatrix} E/L \\ 0 \end{bmatrix}$$

And then the following zero-order solutions, as shown in Eq. (4.100), can be obtained according to Eq. (4.93).

$$\begin{aligned} i_{00} &= 2.222 - \mathrm{e}^{-0.059\tau}[2.222\cos 0.222\tau + 3.89\sin 0.222\tau] \\ v_{00} &= 50 - \mathrm{e}^{-0.059\tau}[50\cos 0.222\tau - 13.276\sin 0.222\tau] \end{aligned} \tag{4.100}$$

in which $\tau = 2\pi f_s t$. The results from Eq. (4.100) (red dashed line) and the simulations (green solid line) are compared in Fig. 4.11, from which it can be seen that the zero-order approximation roughly grasps the characteristics of the transient process.

Similar to Eq. (4.33), the first-order correction term can be chosen as

$$\mathbf{x}_1 = \mathbf{a}_{11}(t)e^{j\tau} + \bar{\mathbf{a}}_{11}(t)e^{-j\tau}$$

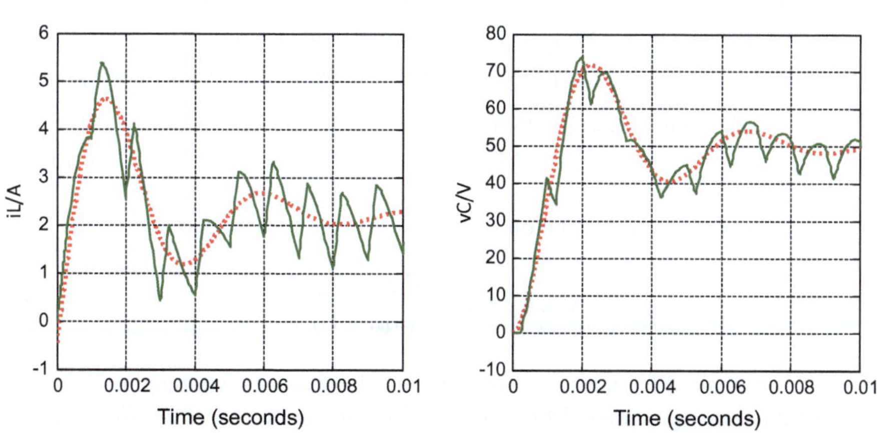

(a) inductor current i_L (b) Capacitor voltage v_C

Fig. 4.11 Zero-order approximation of the transient solution of CCM-Boost converter

Here $\mathbf{a}_{11}(t) = [i_{11}, v_{11}]^{Tr}$. By replacing ($p$ + jω) with the (jω) in Eq. (4.35), the first-order correction equation corresponding to the second equation in Eq. (4.89) can be obtained as

$$[\mathbf{G}_0(p+\mathrm{j}\omega) + \mathbf{G}_1(p+\mathrm{j}\omega)\mathrm{b}_0]\mathbf{a}_{11} = -\mathbf{G}_1(p+\mathrm{j}\omega)\cdot b_1\mathbf{a}_{00} \tag{4.101}$$

which can be rewritten in matrix from as

$$\left(\begin{bmatrix} p+j\omega & \frac{1}{L} \\ -\frac{1}{C} & p+j\omega+\frac{1}{RC} \end{bmatrix} + \begin{bmatrix} 0 & -\frac{D}{L} \\ \frac{D}{C} & 0 \end{bmatrix}\right)\begin{bmatrix} i_{11} \\ v_{11} \end{bmatrix} = \begin{bmatrix} 0 & \frac{1}{L} \\ \frac{-1}{C} & 0 \end{bmatrix}\begin{bmatrix} b_1 i_{00} \\ b_1 v_{00} \end{bmatrix} \tag{4.102}$$

Equation (4.102) is also a linear differential equation, and can be expressed as a standard form as follows:

$$\frac{d\mathbf{a}_{11}(t)}{dt} = \mathbf{A}_1\mathbf{a}_{11}(t) + \mathbf{B}_1\mathbf{x}_0 \tag{4.103}$$

In which the coefficient matrices are shown as

$$\mathbf{A}_1 = \begin{bmatrix} -j\omega & -\frac{1-D}{L} \\ \frac{1-D}{C} & -j\omega-\frac{1}{RC} \end{bmatrix}, \quad \mathbf{B}_1 = \begin{bmatrix} 0 & \frac{b_1}{L} \\ \frac{-b_1}{C} & 0 \end{bmatrix}$$

The initial value $\mathbf{x}_1(0) = 0$, that is, $i_{11}(0) = 0$ and $v_{11}(0) = 0$. Then the special solution $\mathbf{x}_1(\tau)$ can be obtained according to the calculation of Eq. (4.93), as shown in Eq. (4.104).

$$\begin{aligned} i_{11} &= \mathrm{e}^{-0.059\tau}[0.2209\cos(0.222\tau) + 0.2547\sin(0.222\tau) + 0.7298\cos(1.222\tau) \\ &\quad - 0.1333\sin(1.222\tau)] - 0.5007\cos\tau + 0.4031\sin\tau \\ v_{11} &= \mathrm{e}^{-0.059\tau}[-2.188\cos(0.222\tau) + 3.22\sin(0.222\tau) + 1.22\cos(1.222\tau) \\ &\quad + 3.365\sin(1.222\tau)] + 0.968\cos\tau - 3.945\sin\tau \end{aligned} \tag{4.104}$$

Thus the first-order approximation of Eq. (4.95) can be obtained as

$$\mathbf{x} \approx \mathbf{x}_0 + \mathbf{x}_1 = \begin{bmatrix} i_{00} + i_{11} \\ v_{00} + v_{11} \end{bmatrix} \tag{4.105}$$

The results acquired from Eq. (4.105) (red dash line) are shown together with those from simulations (green solid line) in Fig. 4.12. It can be seen from Fig. 4.12 that the first-order approximation is close to the transient process, but the pulsating component of the state variable is still not well estimated.

A similar method can be used to further obtain the second-order correction of the transient solution. Thus the second-order approximate transient solution of Eq. (4.95) is obtained as

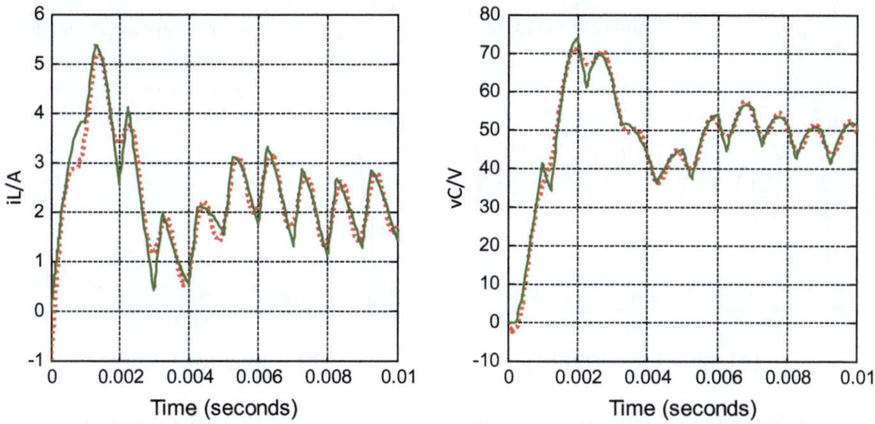

(a) Inductor current (b) capacitor voltage

Fig. 4.12 First-order approximation of the transient solution of CCM-Boost converter

$$\mathbf{x} = \begin{bmatrix} i_L & v_C \end{bmatrix}^{Tr} \approx \mathbf{x}_0 + \mathbf{x}_1 + \mathbf{x}_2 \tag{4.106}$$

The specific approximate solutions are shown in (4.107a) and (4.107b) respectively.

$$\begin{aligned} i_L \approx\; & 2.1734 + \mathrm{e}^{-0.059\tau}[-2.1335\cos(0.222\tau) + 4.1547\sin(0.222\tau) + 0.021\tau\cos(0.222\tau) \\ & + 0.0018\tau\ \sin(0.222\tau) + 0.1810\cos(0.778\tau) - 0.0238\ \sin(0.778\tau) \\ & + 0.2733\ \cos(1.222\tau) - 0.1333\ \sin(1.222\tau)] - 0.5007\ \cos\tau + 0.4031\ \sin\tau \\ & - 0.2051\ \cos 2\tau - 0.002\ \sin 2\tau - 0.0476\cos 3\tau - 0.0408\ \sin 3\tau \end{aligned} \tag{4.107a}$$

$$\begin{aligned} v_C \approx\; & 49.368 + \mathrm{e}^{-0.059\tau}[-51.636\cos(0.222\tau) - 12.352\sin(0.222\tau) \\ & + 0.0438\tau\cos(0.222\tau) + 0.2431\tau\ \sin(0.222\tau) + 0.0768\cos(0.778\tau) \\ & + 0.2672\ \sin(0.778\tau) + 1.22\ \cos(1.222\tau) + 3.3648\ \sin(1.222\tau)] + 0.0968\ \cos\tau \\ & - 3.945\ \sin\tau + 1.2397\ \cos 2\tau - 0.026\ \sin 2\tau + 0.1878\cos 3\tau + 0.3523\ \sin 3\tau \end{aligned} \tag{4.107b}$$

The results acquired from Eq. (4.107a) (red dash line) are shown together with those from simulations (blue solid line) in Fig. 4.13, from which we can see that the second-order approximation in very good coincidence with the transient process, especially it can estimate the ripple at steady state very well.

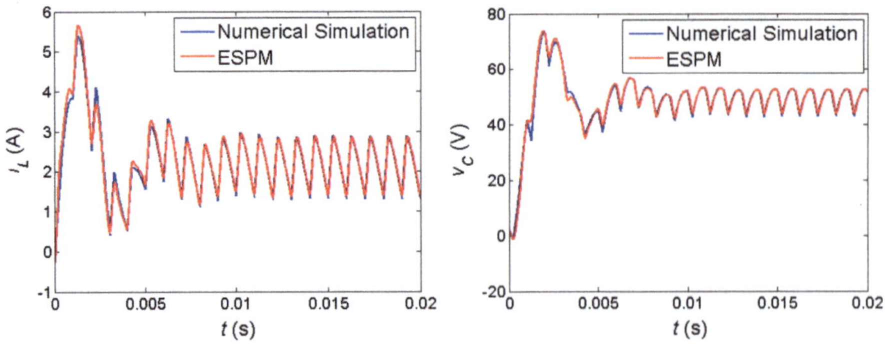

(a) inductor current i_L (b) Capacitor voltage v_C

Fig. 4.13 Second-order approximation of the transient solution of CCM-Boost converter

4.6.4 Simplified Calculation

Since the equations for the first-order and second-order harmonic components are complex, the calculation process involves solving the second-order non-homogeneous linear differential equations, so the solution process is too complicated. Generally, the transient solution of the system can be regarded as the sum of two parts, one is the difference obtained by subtracting the DC component from the principal component of the transient solution, and the other is the steady-state periodic solution of the system. The steady-state solutions of the Boost converter, which can be easily solved by the ESPM, as introduced in Sect. 4.3, are shown again in (4.108).

$$\begin{aligned} i_{steady} &= 2.11 - 0.487\cos\tau + 0.392\sin\tau - 0.205\cos2\tau \\ &\quad - 0.002\sin2\tau - 0.0476\cos3\tau - 0.0408\sin3\tau \\ v_{steady} &= 48.08 + 0.942\cos\tau - 3.945\sin\tau + 1.24\cos2\tau \\ &\quad - 0.026\sin2\tau + 0.1878\cos3\tau + 0.3523\sin3\tau \end{aligned} \tag{4.108}$$

Then, the main component of the transient solution of the system as shown in Eq. (4.100), is obtained by solving Eq. (4.89). The steady-state component (DC component) in Eq. (4.100) is removed, and the remaining result is added to the steady-state solution of Eq. (4.109) as the transient solution of the converter, as shown in (4.109).

$$\begin{aligned} i &\approx e^{-0.059\tau}[-2.222\cos 0.222\tau + 3.89\sin 0.222\tau] + i_{steady} \\ v &\approx e^{-0.059\tau}[-50\cos 0.222\tau - 13.276\sin 0.222\tau] + v_{steady} \end{aligned} \tag{4.109}$$

The results acquired from Eq. (4.109) (red dash line) are shown together with those from simulations (green solid line) in Fig. 4.14, from which we can see that

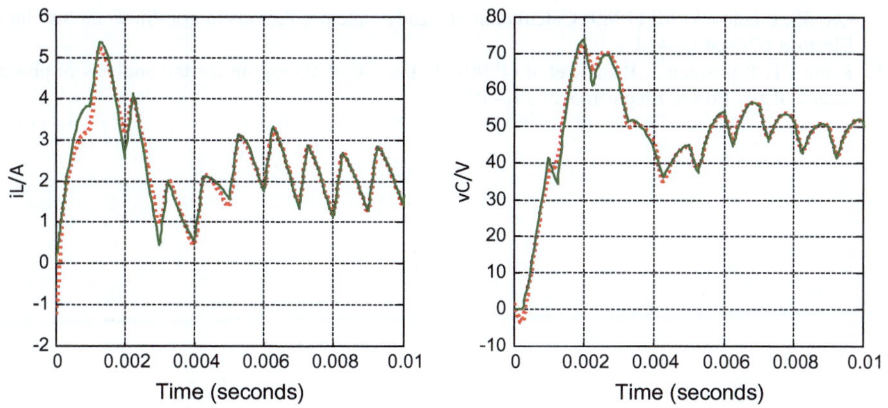

(a) inductor current i_L (b) Capacitor voltage v_C

Fig. 4.14 Approximated transient solution of CCM-Boost converter from simplified method

they agree well with each other. Therefore, the simplified algorithm can solve transient solutions more accurately and can greatly reduce the amount of calculation.

4.7 Summary

In this chapter, the equivalent-small-parameter method is used to analyze the PWM converters operating in CCM (continuous current mode) and DCM (discontinuous current mode) for obtaining the analytical expressions of steady-state periodic solutions and transient solutions. The equivalent-small-parameter method overcomes the shortcoming of dealing with large amount of calculation, which exists in the general average method. It is only necessary to solve the linear algebra or differential equations with lower order. It shows that even if the switching frequency is low and the ripple is large, the ESPM can still be applicable with high precision.

References

1. Hayashi H (1964) Nonlinear oscillations in physical systems. Princeton University Press, Princeton, NJ, Ch. 1
2. Middlebrook RD, Cus S (1976) A general unified approach to modelling switching-converters power stages. In: IEEE power electronics specialist conference record, Cleveland, pp 18–34
3. Sanders SR, Noworolski J, Liu X et al (1991) Generalized averaging method for power conversion circuits. IEEE Trans Power Electron 6(2):251–259

4. Qiu SS, Filanovski MI (1989) Calculation of steady-state oscillations in non-linear circuit. Int J Electron 67(3):403–414
5. Krein PT, Bentsman J, Bass R et al (1990) On the use of averaging for the analysis of power stages. IEEE Power Electron 5(2):182–190

Chapter 5
Analysis of Voltage-Mode Controlled CCM-PWM DC/DC Converters Based on ESPM

5.1 Introduction

State space averaging method [1] is a commonly used method for analyzing switching power converters. But there are obvious limitations [2–7] with the average model as it is independent on the switching frequency. In particular, it is not easy to analyze ripples, or explain the DC offset phenomenon when the switching power converter operates with quite low switching frequency [3]. And the proposed improvement of the average method or the general average method in [2–7] is too cumbersome and difficult to be understood. Equivalent small parameter method (ESPM) [8–11] combines the advantages of the perturbation method and the harmonic balance method, without the need for artificial introduction of small parameters, is a symbolic analysis method with higher precision and relatively simple analysis procedure, and is suitable for analysis of strong nonlinear high-order system. In [12–18], The ESPM is used for analysis of open-loop power converters with continuous current mode (CCM), discontinuous-current-mode (DCM), multi-topology mode operation (including multiple switch converters), as well as in the steady state analysis of Class E amplifiers. And the analytical expression of the converters output ripple can be obtained.

The basic principles of the ESPM together with its applications in the steady-state and transient solutions of open-loop PWM switching power converters are systematically illustrated in Chaps. 3 and 4 respectively. And this chapter begins to apply the equivalent small parameter method (ESP) to the analysis of closed-loop systems. In Chaps. 5 and 6, this symbol analysis method is extended to the closed-loop Voltage-Mode-Controlled (VMC) PWM switching power converter system with CCM and DCM operation. The analytic solutions of the state variable ripples and the duty cycle of the closed-loop system can be obtained. Based on these analytical solutions, the effect of ripple on the duty cycle and the relationship between the DC offset in the converter and the operating frequency can be directly explained. At the same time, the examples show that increasing the integral

Y. Chen and B. Zhang, *Equivalent-Small-Parameter Analysis of DC/DC Switched-Mode Converter*, CPSS Power Electronics Series,
https://doi.org/10.1007/978-981-13-2574-8_5

capacitance in the feedback control loop can properly compensate for the DC drift existing in the system. In view of the fact that the Current-Mode-Control (CMC) has more advantages than the duty cycle programmed control (i.e., voltage-mode-control), we will establish the equivalent small-parameter symbol analysis method of the closed-loop PWM converter system with current-mode-control in Chap. 7. In Chap. 8, we further extend the ESPM to the analysis of frequency-modulated (FM) quasi-resonant converters to show the applicability of the method.

In Sects. 5.2 and 5.3, we first discuss in detail the establishment of the mathematical model of the closed-loop PWM converter system, and then propose the basic algorithm of the closed-loop system based on the equivalent small parameter method. Section 5.4 gives an example analysis and comparison of ESPM and numerical simulations. Section 5.5 proposes a double iterative method to improve the basic algorithm and improve its convergence and accuracy. The experimental results for further verification of the analysis for open-loop and closed-loop converters based on the ESPM are introduced in Sect. 5.6, and the summary of this chapter is given in Sect. 5.7 finally.

5.2 Modeling the Closed-Loop VMC-PWM Converter with CCM Operation

5.2.1 Mathematical Description of the Closed-Loop System

According to Chap. 2, the main circuit of the PWM DC/DC converter operating in CCM mode can be uniformly described as a time-varying state differential equation as

$$\mathbf{g}_1(p)\mathbf{x}_1 + \mathbf{g}_2(p)\mathbf{f}_1 = \mathbf{u}_1 \tag{5.1}$$

The feedback control circuit can also be expressed as a state differential equation:

$$\mathbf{g}_3(p)\mathbf{x}_2 + \mathbf{F}\mathbf{x}_1 = \mathbf{u}_2 \tag{5.2}$$

Here differential operator $p = d/dt$, and the coefficient matrices $\mathbf{g}_1(p)$ and $\mathbf{g}_2(p)$ are associated with a particular circuit parameters and the operator p. The terms $\mathbf{x}_1$ and $\mathbf{x}_2$ are the state variable vectors of the main circuit and the control circuit respectively. $\boldsymbol{F}$ is the feedback coefficient vector, ant the terms $\mathbf{u}_1$ and $\mathbf{u}_2$ represent respectively the input power vectors of the main circuit and control circuit. The strong nonlinear vector function $\mathbf{f}_1 = \delta(t)(\mathbf{x}_1 + \mathbf{e}_1)$, in which $\mathbf{e}_1$ is a constant vector, and $\mathbf{e}_1 = 0$ when $\mathbf{u}_1 \neq 0$. According to (5.1) and (5.2), one can get the following matrix equation as

$$\begin{bmatrix} \mathbf{g}_1(p) & 0 \\ \mathbf{F} & \mathbf{g}_3(p) \end{bmatrix} \begin{bmatrix} \mathbf{x}_1 \\ \mathbf{x}_2 \end{bmatrix} + \begin{bmatrix} \mathbf{g}_2(p) & 0 \\ 0 & 0 \end{bmatrix} \delta(t) \left(\begin{bmatrix} \mathbf{e}_1 \\ 0 \end{bmatrix} + \begin{bmatrix} \mathbf{x}_1 \\ \mathbf{x}_2 \end{bmatrix} \right) = \begin{bmatrix} \mathbf{u}_1 \\ \mathbf{u}_2 \end{bmatrix} \tag{5.3}$$

Equation (5.3) can also be rewritten as

$$\mathbf{G}_0(p)\mathbf{x} + \mathbf{G}_1(\mathbf{p})\mathbf{f} = \mathbf{u} \tag{5.4}$$

In Eq. (5.4) the nonlinear vector $\mathbf{f} = \delta(t)(\mathbf{x} + \mathbf{e})$. The meaning of each item in Eq. (5.4) can be obtained by referring to Eq. (5.3), where $\mathbf{x}$ represents the state variables of the whole closed-loop system (including the power stage main circuit and the feedback control circuit). It can be seen that the mathematical model of the closed-loop system is exactly the same in form as that of the open-loop system. The switch function $\delta(t) = 1(0)$ (hereinafter, $\delta(t)$ is sometimes abbreviated as δ) indicates the on (off) state of the main switch in the circuit, respectively. For CCM-operated converter, the switching function is defined as

$$\delta(t) = \begin{cases} 1 & 0 < t \leq d(t)T \\ 0 & d(t)T < t \leq T \end{cases} \tag{5.5}$$

Here T denotes the switching cycle, and d(t) (abbreviated as d) is duty cycle. In the open loop, d(t) is a constant value of D, while in the case of closed-loop, d(t) is determined by the state variable and the feedback control law.

This method is suitable for higher order converters, and Eqs. (5.1) and (5.2) can be generalized to more complex circuits. For the sake of simplicity, this chapter discusses the case where the main circuit is second order, that is, the main circuit has only one inductor L and one capacitor C, and the inductor current along with the capacitor voltage are taken as the main circuit state variable vector, i.e. $\mathbf{x}_1 = [i_L v_c]^{Tr}$, where the superscript "Tr" indicates that the matrix is transposed.

5.2.2 *Expression of the Duty Cycle* d

5.2.2.1 Expression of *d* with Proportional Feedback Control

In this case, the state variables of the closed-loop system are the same as those in the open loop, i.e. $\mathbf{x} = [i_L v_c]^{Tr}$. When the parasitic parameters of the main circuit filter capacitor are not taken into account, the output voltage is $v_O = v_C$. The feedback circuit is shown in Fig. 5.1a, where V_{ramp} is the sawtooth wave voltage, the control voltage V_k of the comparator in Fig. 5.1b is linear with the state variable vector, and the relationship between the duty cycle d(t) and the state vector $\mathbf{x}$ can be deduced according to Fig. 5.1c. According to Fig. 5.1, we have

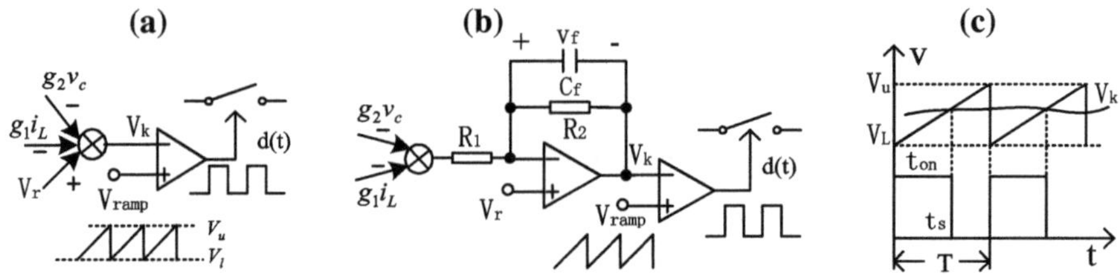

Fig. 5.1 **a** Proportional feedback control, **b** proportional-integral feedback control, **c** diagram of duty cycle

$$V_k = V_r - g_1 i_L - g_2 v_c \tag{5.6}$$

where the terms g_1 and g_2 are the sampling coefficients of the output voltage and the inductor current respectively, and V_r is the reference voltage. Suppose the comparator has a state flip at $t = t_s$, that is, t_s is defined as

$$t_s = t|_{V_k = V_{ramp}} = nT + t_{on} \tag{5.7}$$

In which t_{on} denotes the time that the switch is turned on, here the integer $n = 0$, 1,2,…, means the number of switching cycles. For simplicity, this chapter takes one steady-state cycle for analysis, i.e., $n = 0$. Thus, the duty cycle d can be expressed as

$$d = \frac{t_{on}}{T} = \frac{t_s}{T} = \frac{V_k - V_L}{V_u - V_L} = \frac{V_r - g_1 i_L - g_2 v_c - V_L}{V_u - V_L} = K_0 + \mathbf{K}_1 \mathbf{x}(t_s) \tag{5.8}$$

Here the coefficient $K_0 = \frac{V_r - V_L}{V_u - V_L}$, and the vector $\mathbf{K}_1 = \left[\frac{-g_1}{V_u - V_L} \quad \frac{-g_2}{V_u - V_L}\right]$, in which V_u and V_L are the maximum and minimum values of the sawtooth wave, respectively.

5.2.2.2 Expression of *d* with Proportional-Integral Feedback Control

For the sake of simplicity, the PI adjustment circuit shown in Fig. 5.1b is used as an example. In fact, the analysis method is the same for a more complex feedback circuit. Let v_f denote the voltage over the feedback capacitor, then the relationship between v_f and the reference voltage V_r, along with the output voltage v_C, can be obtained according to the PI feedback network, as shown in Eq. (5.9), it is a polynomial of the differential operator p. It can be seen from Fig. 5.1b that the control voltage signal $V_k = V_r - v_f$, and the duty cycle d can still be deduced from Fig. 5.1c, as shown in Eq. (5.10)

$$\frac{g_1}{R_1C_f}i_L + \frac{g_2}{R_1C_f}v_c + \left(p + \frac{1}{R_2C_f}\right)v_f = -\frac{1}{R_1C_f}V_r \tag{5.9}$$

$$d = \frac{t_{on}}{T} = \frac{V_k - V_L}{V_u - V_L} = \frac{V_r - V_L - v_f}{V_u - V_L} = K_0 + \mathbf{K}_1\mathbf{x}(t_s) \tag{5.10}$$

Here the state variables of the closed-loop system with PI control is chosen as $\mathbf{x} = [\, i_L \quad v_C \quad v_f \,]^{\mathrm{Tr}}$, the coefficient $K_0 = \frac{V_r - V_L}{V_u - V_L}$, and the coefficient row vector is. $\mathbf{K}_1 = \begin{bmatrix} 0 & 0 & \frac{-1}{V_u - V_L} \end{bmatrix}$.

As mentioned previously, the switching converter system has a low pass or bandpass filtering characteristic, so that the steady-state periodic solution $\mathbf{x}$ can be developed into a series as:

$$\mathbf{x} = \mathbf{x}_0 + \varepsilon\mathbf{x}_1 + \varepsilon^2\mathbf{x}_2 + \cdots + \varepsilon^i\mathbf{x}_i + \cdots \tag{5.11}$$

where $\mathbf{x}_0$ contains the main oscillation component of $\mathbf{x}$. The mark ε^i is used to indicate that $|x_i| \ll x_0$, that is $\mathbf{x}_i$ is an i-ordered small amount of $\mathbf{x}_0$. When a specific value is needed for calculation, let $\varepsilon = 1$. Similarly, expand the duty cycle d into the sum of the main item and the correction quantity as follows.

$$d = d_0 + \varepsilon d_1 + \varepsilon^2 d_2 + \cdots \tag{5.12}$$

Substituting (5.12) into (5.8) or (5.10) can give

$$d_0 + \varepsilon d_1 + \varepsilon^2 d_2 + \cdots = K_0 + K_1\mathbf{x}_0 + \varepsilon K_1\mathbf{x}_1(t_s) + \varepsilon^2 K_1\mathbf{x}_2(t_s) + \cdots \tag{5.13}$$

Then according to Fig. 5.1c, the switching instant t_s of the switch S_T can be obtained as

$$t_s = dT = (d_0 + \varepsilon d_1 + \varepsilon^2 d_2 + \cdots)T \tag{5.14}$$

We can define the main term and the correction terms of the time t_s as

$$\begin{aligned} t_{s0} &= d_0T \\ t_{s1} &= (d_0 + \varepsilon d_1)T \\ t_{s2} &= (d_0 + \varepsilon d_1 + \varepsilon^2 d_2)T \\ &\cdots \end{aligned} \tag{5.15}$$

It is easy to know from (5.15) that t_{s0} represents the moment when the comparator is flipped with only the principal component $\mathbf{x}_0$ being contained into the steady-state periodic solution $\mathbf{x}$. And t_{s1} corresponds to the moment that the comparator is flipped when $\mathbf{x}$ contains the main component $\mathbf{x}_0$ and the first-order correction term $\mathbf{x}_1$. Similarly, t_{s2} corresponds to the flip moment when $\mathbf{x}$ contains the terms $\mathbf{x}_0$, $\mathbf{x}_1$ and $\mathbf{x}_2$, and so on.

Let $\tau = \omega t$ and $\tau_{si} = \omega t_{si}$ $(i = 0, 1, 2, \ldots)$, here the angular frequency $\omega = 2\pi/T$, then according to (5.15), we can get that

$$\begin{aligned} \tau_{s0} &= 2\pi d_0 \\ \tau_{s1} &= 2\pi(d_0 + \varepsilon d_1) \\ \tau_{s2} &= 2\pi(d_0 + \varepsilon d_1 + \varepsilon^2 d_2) \\ &\cdots \end{aligned} \tag{5.16}$$

For switching power converters, the duty cycle is determined primarily by t_{s0}. Considering that the general ripple has little effect on the duty cycle, which can be explained from the following example, to simplify the calculation, we use the first-order approximation of the Taylor series to find the term d_{i}. Meanwhile, according to iterative solution, when seeking to find d_i, the terms $d_0, d_1, \ldots d_{i-1}$ are already known. Therefore the following equations can be obtained.

$$\mathbf{x}_1(t_s) = \mathbf{x}_1(t_{s0}) + \left.\frac{\partial \mathbf{x}_1}{\partial t}\right|_{t=t_{s0}} \varepsilon d_1 T + \left.\frac{\partial \mathbf{x}_1}{\partial t}\right|_{t=t_{s0}} \varepsilon^2 d_2 T + \cdots \tag{5.17a}$$

$$\mathbf{x}_2(t_s) = \mathbf{x}_2(t_{s1}) + \left.\frac{\partial \mathbf{x}_2}{\partial t}\right|_{t=t_{s1}} \varepsilon^2 d_2 T + \left.\frac{\partial \mathbf{x}_2}{\partial t}\right|_{t=t_{s1}} \varepsilon^3 d_3 T + \cdots \tag{5.17b}$$

$$\mathbf{x}_i(t_s) = \mathbf{x}_i(t_{s(i-1)}) + \left.\frac{\partial \mathbf{x}_i}{\partial t}\right|_{t=t_{s(i-1)}} \varepsilon^i d_i T + \left.\frac{\partial \mathbf{x}_i}{\partial t}\right|_{t=t_{s(i-1)}} \varepsilon^{i+1} d_{i+1} T + \cdots \tag{5.17c}$$

Substituting (5.17a) into (5.13), and equalizing the coefficients with the same power of ε on both sides of the equation gives:

$$d_0 = K_0 + \mathbf{K}_1 \cdot \mathbf{x}_0 \tag{5.18a}$$

$$d_1 = \mathbf{K}_1 \cdot \mathbf{x}_1(t_{s0}) \tag{5.18b}$$

$$d_2 = \mathbf{K}_1 \cdot \left.\frac{\partial \mathbf{x}_1}{\partial t}\right|_{t=t_{s0}} d_1 T + \mathbf{K}_1 \cdot \mathbf{x}_2(t_{s1}) \tag{5.18c}$$

$$d_3 = \mathbf{K}_1 \left(\left.\frac{\partial \mathbf{x}_1}{\partial t}\right|_{t=t_{s0}} + \left.\frac{\partial \mathbf{x}_2}{\partial t}\right|_{t=t_{s1}} \right) \cdot d_2 T + \mathbf{K}_1 \cdot \mathbf{x}_3(t_{s2}) \tag{5.18d}$$

It can be seen from the above analysis that, for the closed-loop PWM switching converter system, no matter how complex the feedback control circuit is, take the power stage main circuit and feedback control circuit as a whole during analysis, then the state equation for a closed-loop system can be written in a uniform form similar to that of an open-loop system. And the duty cycle can be expressed as a linear combination of state variables (such as (5.8) or (5.10)).

5.2.3 Series Expansion of Switching Function δ(t) for Closed-Loop Systems

Similar to the open-loop system, the switch function $\delta(t)$ defined by (5.5) can still be expanded into a Fourier series as

$$\delta = b_0 + \sum_{m=1}^{\infty} [b_m \exp(jm\tau) + \bar{b}_m \exp(-jm\tau)] \tag{5.19}$$

where the coefficient $\bar{b}_m$ represents the conjugate pair of the term b_m, which can be found by the following equation as

$$\begin{aligned} b_0 &= \frac{1}{T}\int_0^T \delta(t)dt = d(t), \quad b_m = \frac{1}{2}(\alpha_m - j\beta_m), \quad m = 1, 2\ldots \\ \alpha_m &= \frac{2}{T}\int_0^T \delta(t)\cdot\cos(m\omega t)dt, \quad \beta_m = \frac{2}{T}\int_0^T \delta(t)\cdot\sin(m\omega t)dt \end{aligned} \tag{5.20}$$

It should be noted that for closed-loop systems, the duty cycle d is not a known constant, but an unknown number to be determined, which can be approximated by the following series

$$b_0 = d = d_0 + \varepsilon d_1 + \varepsilon^2 d_2 + \cdots \tag{5.21}$$

According to the conduction interval of the switch S_T during one switching cycle, the effective integration interval of Eq. (5.20) can be divided. Moreover, since $d_0 < 1$, there is $d_i \ll 1$. And because of the characteristics of the equivalent small-parameter method, the larger the value of m is, the smaller the corresponding d_i would be. Thus it can be assumed that

$$\begin{aligned} &\sin(\varepsilon m\pi d_1) \approx \varepsilon m\pi d_1 \ (m = 1) \\ &\sin(\varepsilon^2 m\pi d_2) \approx \varepsilon^2 m\pi d_2 \ (m \le 3) \\ &\ldots \end{aligned} \tag{5.22}$$

Therefore the coefficient α_m in Eq. (5.20) can be expressed as

$$\begin{aligned} \alpha_m &= \frac{1}{\pi}\int_0^{2\pi d_0} \cos m\tau d\tau + \frac{1}{\pi}\int_{2\pi d_0}^{2\pi(d_0+\varepsilon d_1)} \cos m\tau d\tau + \frac{1}{\pi}\int_{2\pi(d_0+\varepsilon d_1)}^{2\pi(d_0+\varepsilon d_1+\varepsilon^2 d_2)} \cos m\tau d\tau + \cdots \\ &= \alpha_{m0} + \varepsilon\alpha_{m1} + \varepsilon^2\alpha_{m2} + \cdots \end{aligned} \tag{5.23}$$

And from the equation above one can get

$$\begin{aligned}\alpha_{m0} &= \frac{\sin 2m\pi d_0}{m\pi} \\ \alpha_{mi} &\approx 2d_i \cos 2m\pi(d_0 + \varepsilon d_1 + \cdots + \varepsilon^{i-1} d_{i-1}) \quad (i = 1, 2, \ldots)\end{aligned} \tag{5.24}$$

When deriving (5.24), the following approximation is utilized, i.e.,

$$\frac{1}{2}\varepsilon^i d_i \ll \left(d_0 + \varepsilon d_1 + \varepsilon^2 d_2 + \cdots \varepsilon^{i-1} d_{i-1}\right) \tag{5.25}$$

Similarly, the coefficient β_m in Eq. (5.20) can be expressed as

$$\beta_m = \beta_{m0} + \varepsilon\beta_{m1} + \varepsilon^2\beta_{m2} + \cdots \tag{5.26}$$

where

$$\begin{aligned}\beta_{m0} &= \frac{1 - \cos 2m\pi d_0}{m\pi} \\ \beta_{mi} &\approx 2d_i \sin 2m\pi(d_0 + \varepsilon d_1 + \cdots + \varepsilon^{i-1} d_{i-1}), \quad (i = 1, 2, \ldots)\end{aligned} \tag{5.27}$$

Thus, the coefficient b_m in Eq. (5.20) can also be expressed as

$$\begin{aligned}b_m &= \frac{1}{2}(\alpha_{m0} - j\beta_{m0}) + \varepsilon \cdot \frac{1}{2}(\alpha_{m1} - j\beta_{m1}) + \varepsilon^2 \frac{1}{2}(\alpha_{m2} - j\beta_{m2}) + \cdots \\ &= b_{m0} + \varepsilon^1 b_{m1} + \varepsilon^2 b_{m2} + \cdots\end{aligned} \tag{5.28}$$

where $b_{mi} = \frac{1}{2}(\alpha_{mi} - j\beta_{mi}) \quad (m = 1, 2, \ldots, \; i = 0, 1, 2, \ldots)$

Substituting (5.21), (5.28) into (5.19) gives

$$\begin{aligned}\delta = &\left[d_0 + \sum_{m=1}^{\infty} (b_{m0} e^{jm\tau} + c.c)\right] + \varepsilon \left[d_1 + \sum_{m=1}^{\infty} (b_{m1} e^{jm\tau} + c.c)\right] \\ &+ \varepsilon^2 \left[d_2 + \sum_{m=1}^{\infty} (b_{m2} e^{jm\tau} + c.c)\right] + \cdots\end{aligned} \tag{5.29}$$

Each of the items in square brackets in (5.29) is expanded into the sum of the main item and the small quantity, that is:

$$\begin{aligned}d_i + \sum_{m=1}^{\infty} (b_{mi} e^{jm\tau} + c.c) &= (d_i + b_{1i} e^{j\tau} + c.c) \\ &+ \varepsilon(b_{2i} e^{j2\tau} + b_{3i} e^{j3\tau} + c.c) + \varepsilon^2 (b_{4i} e^{j4\tau} + b_{5i} e^{j5\tau} + c.c) + \cdots\end{aligned} \tag{5.30}$$

The reasons for selection of the main items and small quantities in Eq. (5.30) can be referred to Chap. 4. Substituting (5.30) into (5.29) gives

$$\delta = \delta_0 + \varepsilon\delta_1 + \varepsilon^2\delta_2 + \cdots \tag{5.31}$$

In which the term δ_i is defined as

$$\begin{cases} \delta_0 = d_0 + b_{10}e^{j\tau} + c.c \\ \delta_1 = d_1 + b_{11}e^{j\tau} + b_{20}e^{j2\tau} + b_{30}e^{j3\tau} + c.c \\ \delta_2 = d_2 + b_{12}e^{j\tau} + b_{21}e^{j2\tau} + b_{31}e^{j3\tau} + b_{40}e^{j4\tau} + b_{50}e^{j5\tau} + c.c \\ \cdots \end{cases} \tag{5.32a}$$

From (5.20) to (5.30), and the coefficients in (5.32a) are summarized as follows:

$$\begin{aligned} & b_{mi} = \frac{1}{2}(\alpha_{mi} - j\beta_{mi}) \; (m = 1, 2, \ldots, i = 0, 1, 2, \ldots) \\ & \alpha_{m0} = \frac{\sin 2m\pi d_0}{m\pi}, \quad \beta_{m0} = \frac{1 - \cos 2m\pi d_0}{m\pi} \\ & \alpha_{mi} \approx 2d_i \cos 2m\pi(d_0 + \varepsilon d_1 + \cdots + \varepsilon^{i-1}d_{i-1}) \; (i = 1, 2, \ldots) \\ & \beta_{mi} \approx 2d_i \sin 2m\pi(d_0 + \varepsilon d_1 + \cdots + \varepsilon^{i-1}d_{i-1}) \; (i = 1, 2, \ldots) \end{aligned} \tag{5.32b}$$

So far, the discrete time-varying differential equations represented by (5.4) can be transformed into a state differential equation with continuous time by the transformation of (5.31). Therefore, the equivalent small parameter can be adopted to obtain the analytic solution of the closed-loop system.

5.3 Solution of the Time-Varying Closed-Loop System with CCM Operation

The closed-loop time-varying differential equation (5.4) is solved by using equivalent small parametric method now. Since (5.4) has the same form as the open-loop state equation (5.1), the basic principle of solving is the same as the open-loop system. The difference is that the duty cycle $d(t)$ is not a constant, but a variable associated with the feedback circuit and the state variable (see (5.8) or (5.10)). The switching function defined by (5.5) is expanded in series shown in Eq. (5.31). For the sake of simplicity, we only consider the case of $\mathbf{e} = 0$ in Eq. (5.4).

The steady-state periodic solution $\mathbf{x}$ is expressed as the sum of the main wave $\mathbf{x}_0$ and the corrections $\mathbf{x}_i$, as shown in (5.11). Similarly the nonlinear vector function $\mathbf{f}$ is still expanded into the sum of the main term $\mathbf{f}_0$ and the correction terms $\mathbf{f}_i$, and they are determined by the following expressions as

$$
\begin{aligned}
&\mathbf{f}_0 = \delta_0 \mathbf{x}_0 \\
&\mathbf{f}_1 = \delta_0 \mathbf{x}_1 + \delta_1 \mathbf{x}_0 \\
&\mathbf{f}_2 = \delta_0 \mathbf{x}_2 + \delta_1 \mathbf{x}_1 + \delta_2 \mathbf{x}_0 \\
&\cdots
\end{aligned}
\tag{5.33a}
$$

Using the same method as the analysis of open-loop converter system to solve the closed-loop converter system Eq. (5.4), the same iterative equations as in Eq. (4.26) can be obtained, which is rewritten as follows:

$$
\begin{cases}
\mathbf{G}_0(p)\mathbf{x}_0 + \mathbf{G}_1(p)\mathbf{f}_{0m} = \mathbf{u} \\
\mathbf{G}_0(p)\mathbf{x}_1 + \mathbf{G}_1(p)\mathbf{f}_{1m} = -\mathbf{G}_1(p)\mathbf{R}_1 \\
\mathbf{G}_0(p)\mathbf{x}_2 + \mathbf{G}_1(p)\mathbf{f}_{2m} = -\mathbf{G}_1(p)\mathbf{R}_2 \\
\qquad\cdots
\end{cases}
\tag{5.33b}
$$

5.3.1 Solution of Main Component

If the system can operate stably under the selected parameters, the main component of the state variable is the DC component, so it can be set:

$$\mathbf{x}_0 = \mathbf{a}_{00} \tag{5.34}$$

Substituting (5.32a) and (5.34) into (5.33a) gives

$$\mathbf{f}_0 = \delta_0 \mathbf{x}_0 = d_0 \mathbf{a}_{00} + b_{10}\mathbf{a}_{00}e^{j\tau} + c.c \tag{5.35}$$

The main term and the remaining item are as follows.

$$
\begin{cases}
\mathbf{f}_{0m} = d_0 \mathbf{a}_{00} \\
\mathbf{R}_1 = b_{10}\mathbf{a}_{00}e^{j\tau} + c.c
\end{cases}
\tag{5.36}
$$

Substituting $\mathbf{x}_0$ and $\mathbf{f}_{0m}$ into (5.33b) gives

$$[\mathbf{G}_0(0) + \mathbf{G}_1(0)d_0]\mathbf{a}_{00} = \mathbf{u} \tag{5.37}$$

And according to (5.18a), the main term of the duty cycle is determined by

$$d_0 = K_0 + K_1\mathbf{x}_0 = K_0 + K_1\mathbf{a}_{00} \tag{5.38}$$

The solutions of $\mathbf{a}_{00}$ and d_0 can be obtained by combining Eqs. (5.38) and (5.37). Generally, for closed-loop systems, the solutions of $\mathbf{a}_{00}$ and d_0 are non-linear and depend on the specific main circuit and feedback control law. The specific solution process can be analyzed with reference to the following examples.

5.3.2 Solution of the First-Order Correction

Since R_1 in (5.36) contains the fundamental wave only, the first-order correction $\mathbf{x}_1$ can be assumed that

$$\mathbf{x}_1 = \mathbf{a}_{11}e^{j\tau} + \bar{\mathbf{a}}_{11}e^{-j\tau} \tag{5.39}$$

Substituting $\mathbf{x}_0$, $\mathbf{x}_1$, δ_0 and δ_1 into the equation $\mathbf{f}_1 = \delta_0\mathbf{x}_1 + \delta_1\mathbf{x}_0$, we can get

$$\begin{aligned}\mathbf{f}_{1m} &= (d_0\mathbf{a}_{11} + b_{11}\mathbf{a}_{00}) \cdot e^{j\tau} + c.c \\ \mathbf{R}_2 &= (b_{10}\bar{\mathbf{a}}_{11} + \bar{b}_{10}\mathbf{a}_{11} + d_1\mathbf{a}_{00}) + (b_{10}\mathbf{a}_{11} + b_{20}\mathbf{a}_{00}) \cdot e^{j2\tau} + b_{30}\mathbf{a}_{00} \cdot e^{j3\tau} + c.c\end{aligned} \tag{5.40}$$

Substituting $\mathbf{f}_{1m}$ and $\mathbf{R}_1$ into the first order correction in Eq. (5.33b) gives

$$[\mathbf{G}_0(j\omega) + \mathbf{G}_1(j\omega) \cdot d_0] \cdot \mathbf{a}_{11} = -\mathbf{G}_1(j\omega) \cdot (b_{11} + b_{10}) \cdot \mathbf{a}_{00} \tag{5.41a}$$

The coefficients b_{10} and b_{11} are determined according to (5.32b), i.e.,

$$\begin{aligned}b_{10} &= [\sin 2\pi d_0 - j(1 - \cos 2\pi d_0)]/2\pi \\ b_{11} &= d_1 \cdot (\cos 2\pi d_0 - j\sin 2\pi d_0) = d_1 \cdot e^{-j\tau_{s0}} \ (\tau_{s0} = 2\pi d_0)\end{aligned} \tag{5.41b}$$

And according to (5.18b), the term d_1 is expressed as

$$d_1 = \mathbf{K}_1 \cdot (\mathbf{a}_{11}e^{j\tau_{s0}} + \bar{\mathbf{a}}_{11}e^{-j\tau_{s0}}) \tag{5.41c}$$

The first-order corrections d_1 and $\mathbf{a}_{11}$ can be obtained by combining the Eqs. (5.41a), (5.41b) and (5.41c), i.e.,

$$\begin{aligned}d_1 &= B_1/A_1 \\ \mathbf{a}_{11} &= \mathbf{G}^{-1}(j\omega) \cdot [-\mathbf{G}_1(j\omega) \cdot (d_1e^{-j\tau_{s0}} + b_{10}) \cdot \mathbf{a}_{00}]\end{aligned} \tag{5.42}$$

where the superscript "−1" indicates the inversion, and the symbols are defined as

$$\mathbf{G}(j\omega) = \mathbf{G}_0(j\omega) + \mathbf{G}_1(j\omega)d_0, \quad \mathbf{C}_1 = \mathbf{G}^{-1}(j\omega)\mathbf{G}_1(j\omega)$$

$$A_1 = 1 + \mathbf{K}_1(\mathbf{C}_1 + \bar{\mathbf{C}}_1)\mathbf{a}_{00}, \quad \mathbf{B}_1 = -\mathbf{K}_1(\mathbf{C}_1 b_{10}e^{j\tau_{s0}} + \bar{\mathbf{C}}_1\bar{b}_{10}e^{-j\tau_{s0}})\mathbf{a}_{00}$$

As can be seen from (5.42), in the approximate calculation, the first order correction can be obtained by solving the linear equations.

5.3.3 Solution of the Second-Order Correction

According to the remaining term $\mathbf{R}_2$ in (5.40), one can assume the second-order correction term $\mathbf{x}_2$ can be expressed as

$$\mathbf{x}_2 = \mathbf{a}_{02} + (\mathbf{a}_{22}e^{j2\tau} + \mathbf{a}_{32}e^{j3\tau} + c.c) \tag{5.43}$$

Introducing $\mathbf{x_0}, \mathbf{x_1}, \mathbf{x_2}$ and $\delta_0, \delta_1, \delta_2$ into $\mathbf{f}_2 = \delta_0\mathbf{x}_2 + \delta_1\mathbf{x}_1 + \delta_2\mathbf{x}_0$, one can get $\mathbf{f}_{2m}$ as

$$\begin{aligned}\mathbf{f}_{2m} = {} & (d_0\mathbf{a}_{02} + b_{11}\bar{\mathbf{a}}_{11} + \bar{b}_{11}\mathbf{a}_{11} + d_2\mathbf{a}_{00}) + (d_0\mathbf{a}_{22} + b_{11}\mathbf{a}_{11} + b_{30}\bar{\mathbf{a}}_{11} + b_{21}\mathbf{a}_{00})e^{j2\tau} \\ & + (d_0\mathbf{a}_{32} + b_{10}\mathbf{a}_{22} + b_{20}\mathbf{a}_{11} + b_{31}\mathbf{a}_{00})e^{j3\tau} + c.c\end{aligned} \tag{5.44}$$

In order to simplify the calculation, the higher-order small quantity $\bar{b}_{10}a_{32}$ of the second harmonic is omitted in the second bracket of the above Eq. (5.44). Then introducing (5.44) and (5.40) into (5.33b) gives the three equations of the second order correction term as follows.

$$[\mathbf{G}_0(j2\omega) + \mathbf{G}_1(j2\omega)d_0]\mathbf{a}_{22} = -\mathbf{G}_1(j2\omega)[(b_{21} + b_{20})\mathbf{a}_{00} + (b_{11} + b_{10})\mathbf{a}_{11} + b_{30}\bar{\mathbf{a}}_{11}] \tag{5.45a}$$

$$[\mathbf{G}_0(j3\omega) + \mathbf{G}_1(j3\omega)d_0]\mathbf{a}_{32} = -\mathbf{G}_1(j3\omega) \cdot [(b_{31} + b_{30})\mathbf{a}_{00} + b_{10}\mathbf{a}_{22} + b_{20}\mathbf{a}_{11}] \tag{5.45b}$$

$$[\mathbf{G}_0(0) + \mathbf{G}_1(0)d_0]\mathbf{a}_{02} + \mathbf{G}_1(0)\mathbf{a}_{00}d_2 = -\mathbf{G}_1(0)[(b_{11} + b_{10})\bar{\mathbf{a}}_{11} + (\bar{b}_{11} + \bar{b}_{10})\mathbf{a}_{11} + d_1\mathbf{a}_{00}] \tag{5.45c}$$

The solutions of the terms $\mathbf{a}_{22}$ and $\mathbf{a}_{32}$ can be easily obtained as Eqs. (5.45a) and (5.45b) are linear. While Eq. (5.45c) is nonlinear, which should be solved by combining Eq. (5.18c) for d_2, that is

$$\begin{aligned}\mathbf{a}_{02} &= -[\mathbf{G}_0(0) + \mathbf{G}_1(0)d_0 + \mathbf{G}_1(0)\mathbf{a}_{00} \cdot \mathbf{K}_1]^{-1} \cdot \mathbf{G}_1(0) \cdot [B_1 + \mathbf{a}_{00}\mathbf{B}_2] \\ d_2 &= B_2 + \mathbf{K}_1\mathbf{a}_{02}\end{aligned} \tag{5.46}$$

where the coefficients B_1 and B_2 are determined by

$$\begin{aligned}B_1 &= (b_{11} + b_{10})\bar{\mathbf{a}}_{11} + (\bar{b}_{11} + \bar{b}_{10})\mathbf{a}_{11} + d_1\mathbf{a}_{00} \\ B_2 &= -\mathbf{K}_1[4\pi d_1 \cdot \mathrm{Im}(a_{11}e^{j\tau_{s0}}) - 2\mathrm{Re}(\mathbf{a}_{22}e^{j\tau_{s1}}) - 2\mathrm{Re}(\mathbf{a}_{32}e^{j\tau_{s1}})]\end{aligned}$$

Here the symbols $\mathrm{Im}(\cdot), \mathrm{Re}(\cdot)$ denote the imaginary and real parts of the complex. Generally the ripple has little effect on the duty cycle, especially the second harmonic and third harmonic components, whose amplitudes are usually very small. Therefore the term d_2 is mainly determined by the DC component $\mathbf{a}_{02}$ in the second-order correction term $\mathbf{x}_2$. Equation (5.46) can be simplified as

$$\begin{aligned} \mathbf{a}_{02} &= -[\mathbf{G}_0(0) + \mathbf{G}_1(0)d_0 + \mathbf{G}_1(0)\mathbf{a}_{00} \cdot \mathbf{K}_1]^{-1} \cdot \mathbf{G}_1(0) \cdot \mathbf{B}_1 \\ d_2 &= \mathbf{K}_1 \mathbf{a}_{02} \end{aligned} \tag{5.47}$$

Then according to equations from (5.34) to (5.47), the final expression of the steady-state periodic solution is shown as

$$\mathbf{x} = \mathbf{a}_{00} + \mathbf{a}_{02} + (\mathbf{a}_{11}e^{j\tau} + \mathbf{a}_{22}e^{j2\tau} + \mathbf{a}_{32}e^{j3\tau} + c.c) \tag{5.48a}$$

Which can also expressed in form of trigonometric function as:

$$\mathbf{x} = a_0 + a_1 \cos\tau + b_1 \sin\tau + a_2 \cos 2\tau + b_2 \sin 2\tau + a_3 \cos 3\tau + b_3 \sin 3\tau + \cdots \tag{5.48b}$$

Among which, the steady-state DC solution is

$$\mathbf{x}_{dc} = \mathbf{a}_0 = \mathbf{a}_{00} + \mathbf{a}_{02} \tag{5.48c}$$

5.4 Examples

5.4.1 Boost Regulator with Proportional Control

The circuit diagram is shown in Fig. 5.2, the main circuit parameters are chosen as E = 5 V, L = 50 μH, C = 4.4 μF and R = 28 Ω; and the control circuit parameters are set as V_r = 0.13 V, g_1 = 0.174, g_2 = −0.0435, V_l= 0 and V_u = 1 V. The state differential equation and the duty cycle equation are described as (5.4) and (5.8),

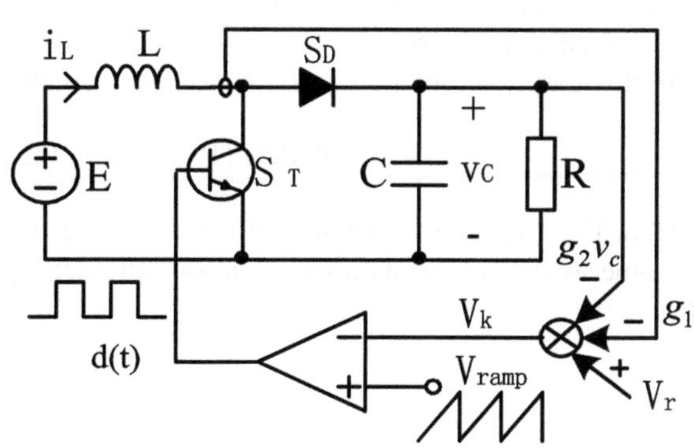

Fig. 5.2 Boost regulator with proportional modulation

respectively, where the state variable vector is $\mathbf{x} = [i_L \quad v_C]^{\mathrm{Tr}}$, the input vector $\mathbf{u} = [E/L \quad 0]^{Tr}$, the nonlinear vector function $\mathbf{f} = \delta\mathbf{x}$; and the feedback proportional coefficient $K_0 = V_r$, and the coefficient matrix $\mathbf{K}_1 = [-g_1, -g_2]$. The matrices $\mathbf{G}_0(p)$ and $\mathbf{G}_1(p)$ are as

$$G_0(p) = \begin{bmatrix} p & \frac{1}{L} \\ \frac{-1}{C} & p + \frac{1}{RC} \end{bmatrix}, \quad G_1(p) = \begin{bmatrix} 0 & \frac{-1}{L} \\ \frac{1}{C} & 0 \end{bmatrix} \tag{5.49}$$

5.4.1.1 Find the Main Term of the Steady State Solution

Similarly to the method introduced for analysis of open-loop converters, the main term of the steady state solution is DC component, that is:

$$\mathbf{x}_0 = \mathbf{a}_{00} = [I_{00}, V_{00}]^{Tr} \tag{5.50}$$

According to (5.37) and (5.38), we can get:

$$V_{00} = \frac{E}{1 - d_0}, \quad I_{00} = \frac{V_{00}}{R(1 - d_0)}, \quad d_0 = V_r - \frac{g_1 E}{(1 - d_0)^2 R} - \frac{g_2 E}{(1 - d_0)} \tag{5.51}$$

From Eq. (5.51) we can see that for the closed-loop Boost converter with the inductor current being contained in the feedback amounts, the solution of d_0 needs to solve a cubic equation, which can be solved by the numerical method or the commonly used formula of the cubic equation. There may be multiple real-value solutions that meet the condition of d_0. In this case, they need to be determined by the stability analysis method of the equilibrium point (see Chap. 9).

However, if only the output voltage feedback is available, for the Boost circuit with the linear PWM feedback control, the solution of d_0 only needs to solve the quadratic equation. Thus under this condition it is easy to determine whether the solution meets the requirements of the converter, and it is also indicates that potential instability factors would exist in the closed-loop converter system with current feedback control.

In this case, we obtain a real solution $d_0 = 0.4091$, and the main terms are

$$V_{00} = 8.4617\,(V), \quad I_{00} = 0.5114\,(A) \tag{5.52}$$

Here V_{00}, I_{00} are independent of the switching frequency of the circuit, which is the same as the steady-state DC solution obtained by the state space method.

5.4.1.2 Find the First-Order Correction

According to the analysis in the previous Sect. 5.3, we can see that only the fundamental component is included in the first-order remainder term $\mathbf{R}_1$, so the first-order correction $\mathbf{x}_1$ can be assumed that

$$\mathbf{x}_1 = \mathbf{a}_{11}e^{j\tau} + \bar{\mathbf{a}}_{11}e^{-j\tau}, \quad \mathbf{a}_{11} = [I_{11}, V_{11}]^{Tr} \tag{5.53}$$

According to (5.42), one can get

$$\begin{aligned} V_{11} &= \frac{(b_{11}+b_{10}) \cdot [j\omega L \cdot I_{00} - (1-d_0) \cdot V_{00}]}{\omega^2 LC - (1-d_0)^2 - j\omega L/R} \approx \frac{(b_{11}+b_{10}) \cdot [j\omega L \cdot I_{00} - (1-d_0) \cdot V_{00}]}{\omega^2 LC - (1-d_0)^2} \\ I_{11} &= \frac{(b_{11}+b_{10}) \cdot V_{00} - (1-d_0) \cdot V_{11}}{j\omega L} \end{aligned} \tag{5.54}$$

$$d_1 = \frac{(1-d_0) \cdot (K_0 - d_0)\sin\tau_{s0} - (1-\cos\tau_{s0})(g_1\omega C \cdot V_{00} - g_2\omega L \cdot I_{00})}{\omega^2 LC - (1-d_0)^2 - 2(1-d_0)(K_0-d_0)} \tag{5.55}$$

With the three switching frequencies, the first-order correction can be obtained by the algorithm in this chapter, as shown in Table 5.1.

As can be seen from the above table, the impact of ripples on the duty cycle is relatively small and inversely proportional to the switching frequency. In fact, Eq. (5.55) can be approximated as

$$d_1 \approx \frac{-(1-\cos\tau_{s0})(g_1 C \cdot V_{00} - g_2 L \cdot I_{00})}{\omega LC} \tag{5.56}$$

Obviously d_1 is inversely proportional to the switching frequency.

5.4.1.3 Find the Second-Order Correction

The amplitudes of the harmonics contained in the second-order correction can be obtained according to the previous mentioned Eq. (5.45a), i.e.,

Table 5.1 First-order ripple component of state variables and correction of duty ratio

Frequency	I_{11}	V_{11}	d_1
fs = 50 kHz	−0.1372 − 0.0692j	0.0589 + 0.1093j	−0.0656
fs = 100 kHz	−0.0742 − 0.0292j	0.0437 + 0.0370j	−0.0325
fs = 1 MHz	−0.0078 − 0.0024j	0.0053 + 0.0018j	−0.0032

$$\begin{array}{ll} V_{22} \approx \frac{j2\omega L\cdot\Delta_{I22}-(1-d_0)\cdot\Delta_{V22}}{4\omega^2 LC-(1-d_0)^2} & I_{22} = \frac{\Delta_{V22}-(1-d_0)\cdot V_{22}}{j2\omega L} \\ V_{32} \approx \frac{j3\omega L\cdot\Delta_{I32}-(1-d_0)\cdot\Delta_{V32}}{9\omega^2 LC-(1-d_0)^2} & I_{32} = \frac{\Delta_{V32}-(1-d_0)\cdot V_{32}}{j3\omega L} \\ V_{02} \approx \frac{(g_1 I_{00}+d_0')\cdot\Delta_{V02}-g_1 V_{00}\cdot\Delta_{I02}}{(g_1 I_{00}+d_0')(g_2 V_{00}+d_0')+M_{02}} & I_{02} \approx \frac{\Delta_{V02}-(g_2 V_{00}+1-d_0)V_{02}}{g_1 V_{00}} \end{array} \tag{5.57}$$

where

$$\begin{aligned} \Delta_{I22} &= (b_{21}+b_{20})I_{00}+(b_{11}+b_{10})I_{11}+b_{30}\bar{I}_{11} \\ \Delta_{V22} &= (b_{21}+b_{20})V_{00}+(b_{11}+b_{10})V_{11}+b_{30}\bar{V}_{11} \\ \Delta_{I32} &= (b_{31}+b_{30})I_{00}+b_{20}I_{11}+b_{10}I_{22} \\ \Delta_{V32} &= (b_{31}+b_{30})V_{00}+b_{20}V_{11}+b_{10}V_{22} \\ \Delta_{I02} &= (b_{11}+b_{10})\bar{I}_{11}+(\bar{b}_{11}+\bar{b}_{10})I_{11}+d_1 I_{00}) \\ \Delta_{V02} &= (b_{11}+b_{10})\bar{V}_{11}+(\bar{b}_{11}+\bar{b}_{10})V_{11}+d_1 V_{00}) \\ d_0' &= 1-d_0,\ M_{02} = g_1 V_{00}(1-g_2 R I_{00})/R \end{aligned}$$

The final expression of the steady-state periodic solution can be obtained from Eqs. (5.51) to (5.57), as shown in Eq. (5.48a), in which the steady-state DC solution is $\mathbf{x}_{dc} = \mathbf{a}_{00} + \mathbf{a}_{02}$, i.e., $I_0 = I_{00} + I_{02}$, $V_0 = V_{00} + V_{02}$.

The results of the steady-state DC solution obtained from the ESPM are compared with those from numerical simulations, as shown in Table 5.2.

For the AC component, the sum of the first three harmonic components of the state variable during one switching cycle can be obtained by using the ESPM, and the results from ESPM (dashed line) are compared with those from numerical simulations (solid line), as shown in Fig. 5.3. The ripples of the inductor current and the capacitor voltage with the converter system operating at the frequency fs = 50 kHz are shown in Fig. 5.3a, b, and those for switching frequency fs = 100 kHz are shown in the Fig. 5.3c, d.

It can be seen from the data in Table 5.2 and the waveforms in Fig. 5.3 that the results obtained by the two methods are close, especially at higher frequencies. The difference is smaller, means that the ESPM introduced in this paper is effective. At the same time, it can be seen from Table 5.2 that when the switching frequency is low, the DC component has a certain change, which is called DC offset.

Table 5.2 Steady-state solutions of state variables in example 1

Frequency	ESPM	Numerical simulation
fs = 50 kHz	I_0 = 0.3325A V_0 = 6.9796 V	I_0 = 0.3750A V_0 = 7.0472 V
fs = 100 kHz	I_0 = 0.4253A V_0 = 7.7490 V	I_0 = 0.4430A V_0 = 7.6504 V
fs = 1 MHz	I_0 = 0.5032A V_0 = 8.3934 V	I_0 = 0.5267A V_0 = 8.3355 V

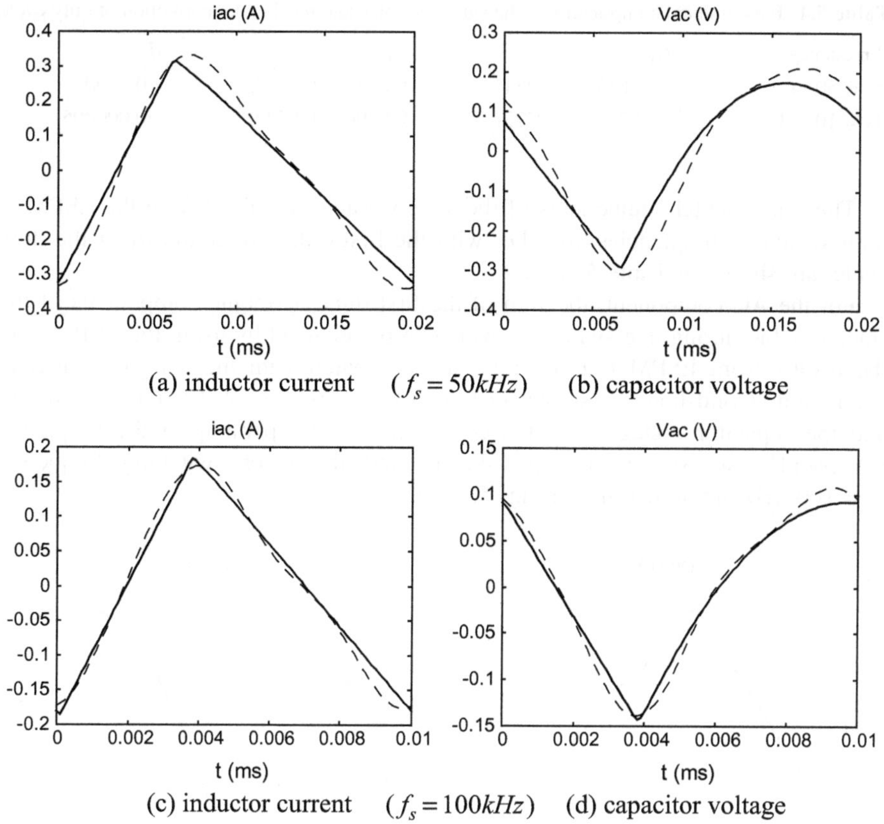

(a) inductor current $(f_s = 50kHz)$ (b) capacitor voltage

(c) inductor current $(f_s = 100kHz)$ (d) capacitor voltage

Fig. 5.3 Steady-state ripples of state variables for Boost with proportional control

5.4.2 Boost Regulator with Proportional-Integral Control

In this example, the main circuit parameters are the same as those in the Sect. 5.4.1, and the proportional-integral control law is adopted with the feedback circuit being shown in Fig. 5.1b. The circuit parameters of the feedback loop are chosen as follows, i.e., R_1 = 0.5 kΩ, R_2 = 500 kΩ, C_f = 5 μF, and the reference voltage is set as V_r = 0.4091 V, the parameters of sawtooth voltage are the same as those in Fig. 5.1a, i.e., $V_l = 0$ and $V_u = 1$ V. As shown in Table 5.3, the steady-state DC solution obtained from the ESPM is compared with the results from simulations.

Table 5.3 Steady-state solutions of state variables

Frequency	ESPM	Simulation
fs = 50 kHz	I_0 = 0.5117A V_0 = 8.4618 V	I_0 = 0.5011A V_0 = 8.2435 V
fs = 100 kHz	I_0 = 0.5115A V_0 = 8.4617 V	I_0 = 0.5166A V_0 = 8.3137 V

Table 5.4 Fundamental components of the state variables and the 1-order correction of duty cycle

Frequency	I_{11}	V_{11}	d_1
fs = 50 kHz	−0.1617 − 0.0431*j*	0.0873 + 0.1032*j*	−0.00002
fs = 100 kHz	−0.0796 − 0.0223*j*	0.0490 + 0.0336*j*	−0.000005

The fundamental components of the state variables obtained from the ESPM at both switching frequencies, together with the first-order correction d_1 of the duty cycle, are shown in Table 5.4.

For the AC component, the sum of the first three harmonic components of the state variable during one switching cycle can be obtained by using the ESPM, and the results from ESPM (dashed line) are compared with those from numerical simulations (solid line), as shown in Fig. 5.4. The ripples of the inductor current and the capacitor voltage with the converter system operating at the frequency fs = 50 kHz are shown in Fig. 5.4a, b, and those for switching frequency fs = 100 kHz are shown in the Fig. 5.4c, d.

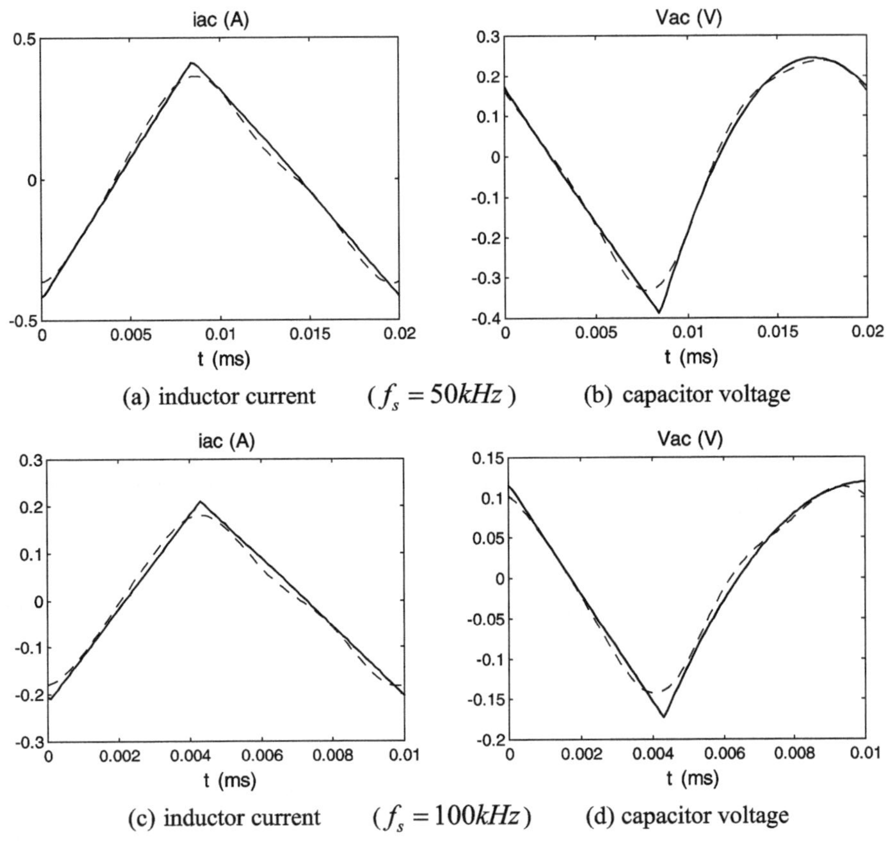

Fig. 5.4 Steady-state ripples of state variables for Boost with PI control

It can be seen from the data in Tables 5.3 and 5.4, and the ripple waveforms in Figs. 5.3 and 5.4, that when there is a compensation capacitor in the feedback control circuit, the results from the ESPM agree quite well with those from numerical simulation, which means that the ESPM has a higher accuracy. The compensation capacitor makes the ripple have a very small effect on the duty cycle because of its filtering characteristics (see Table 5.4). Thus, the errors introduced by some of the assumptions made during the algorithm derivation are quite small.

5.5 Improvement of the Algorithm

5.5.1 Improved Algorithm for Duty Cycle Correction

According to the basic principle of the equivalent small parameter method, when the term d_i is solved, the items that are explicitly related to d_i are moved to the solution equation of higher order correction, so the right side of Eqs. (5.18a) ~ (5.18d) no longer contains the explicit relation with d_i, it is the linear equation of d_i.

If all the d_i-related terms on the right side of (5.13) are taken into account when solving d_i, a more accurate expression for d_i can be obtained, as shown in Eqs. (5.58a) ~ (5.58d).

$$d_0 = K_0 + K_1 \cdot \mathbf{x}_0 \tag{5.58a}$$

$$d_1 = K_1 \cdot \mathbf{x}_1(t_{s0}) + K_1 \cdot \left.\frac{\partial \mathbf{x}_1}{\partial t}\right|_{t=t_{s0}} d_1 T \tag{5.58b}$$

$$d_2 = K_1 \cdot \mathbf{x}_2(t_{s1}) + K_1 \left(\left.\frac{\partial \mathbf{x}_1}{\partial t}\right|_{t=t_{s0}} + \left.\frac{\partial \mathbf{x}_2}{\partial t}\right|_{t=t_{s1}} \right) \cdot d_2 T \tag{5.58c}$$

$$d_3 = K_1 \left(\left.\tfrac{\partial \mathbf{x}_1}{\partial t}\right|_{t=t_{s0}} + \left.\tfrac{\partial \mathbf{x}_2}{\partial t}\right|_{t=t_{s1}} + \left.\tfrac{\partial \mathbf{x}_3}{\partial t}\right|_{t=t_{s2}} \right) \cdot d_3 T + K_1 \cdot \mathbf{x}_3(t_{s2}) \tag{5.58d}$$

$$\cdots$$

Taking the above correction algorithm, the solution of d_1 would be non-linear in (5.42), that is:

$$A d_1^2 + B d_1 + D = 0 \tag{5.59}$$

where the coefficients are determined by the following equations as

$$
\begin{aligned}
A &= K_1 \cdot j2\pi(A_1 - \bar{A}_1)a_{00} \\
B &= 1 + K_1 \cdot j2\pi(A_1 b_{10} \cdot e^{j\tau_{s0}} - \overline{A_1 b_{10}} \cdot e^{-j\tau_{s0}})a_{00} + K_1 \cdot (A_1 + \bar{A}_1)a_{00} \\
D &= K_1 \cdot (A_1 b_{10} \cdot e^{j\tau_{s0}} + \overline{A_1 b_{10}} \cdot e^{-j\tau_{s0}})a_{00} \\
A_1 &= [G_0(j\omega) + d_0 G_1(j\omega)]^{-1} \cdot G_1(j\omega)
\end{aligned}
$$

Obviously, according to (5.59), there would be more than one solution for d_1, thus the correct value of d_1 needs to be selected according to the actual situation of the system.

Similarly, the solutions d_2 and $\mathbf{a}_{02}$ in (5.45a) can be modified to be

$$
\begin{aligned}
a_{02} &= -[G_0(0) + d_0 G_1(0) + G_1(0) \cdot K_1 \cdot a_{00}/C_1]^{-1} \cdot G_1(0)(B_1 + a_{00} \cdot B_2/C_1) \\
d_2 &= (K_1 a_{02} + B_2)/C_1
\end{aligned} \tag{5.60}
$$

where the coefficients are determined by

$$
\begin{aligned}
B_1 &= (b_{11} + b_{10})\bar{a}_{11} + (\bar{b}_{11} + \bar{b}_{10})a_{11} + d_1 a_{00} \\
B_2 &= K_1 \cdot [2R(a_{22}e^{j\tau_{s1}}) + 2R(a_{32}e^{j\tau_{s1}})] \\
C_1 &= 1 + K_1 \cdot [4\pi \cdot I(a_{11}e^{j\tau_{s1}}) + 8\pi \cdot I(a_{22}e^{j\tau_{s1}}) + 12\pi \cdot I(a_{32}e^{j\tau_{s1}})]
\end{aligned}
$$

5.5.2 Correction Algorithm for Series Expansion of the Switching Function δ(t)

In the previous analysis of the state equation of the closed-loop system using the ESPM, it can be found that solving the first-order correction $\mathbf{a}_{11}$ needs to know the coefficients b_{11} and b_{10} in the expansion series of the switching function, and the first-order correction d_1 of the duty cycle. And in the solution of the second-order correction term, it is necessary to know the values of b_{20}, b_{30}, b_{21}, b_{31}, and d_1. As d_1 has been obtained during the solution of the first-order correction, thus it can be utilized in the solutions of b_{20} and b_{30}. Similarly, when solving b_{k0} and $b_{(k+1)0}$ (k = 2,4,6…), we can use d_0, d_1…, $d_{(k-1)}$, which have been obtained in the iterative process. Therefore, in Eq. (5.32b), the expansions of $\alpha_1$and β_1 are unchanged, while the others are modified to be

$$\begin{aligned}\alpha_m &= \frac{1}{\pi}\int_0^{2\pi(d_0+d_1+\cdots d_{m-1})}\cos m\tau d\tau + \frac{1}{\pi}\int_{2\pi(d_0+d_1+\cdots d_{m-1})}^{2\pi(d_0+d_1+\cdots d_{m-1}+\varepsilon^m d_m)}\cos m\tau d\tau \\ &+ \frac{1}{\pi}\int_{2\pi(d_0+d_1+\cdots d_{m-1}+\varepsilon^m d_m)}^{2\pi(d_0+d_1+\cdots d_{m-1}+\varepsilon^m d_m+\varepsilon^{m+1}d_{m+1})}\cos m\tau d\tau + \cdots \\ &= \alpha_{m0}+\varepsilon\alpha_{m1}+\varepsilon^2\alpha_{m2}+\cdots\varepsilon^m\alpha_{mm}+\cdots\end{aligned} \tag{5.61}$$

In which

$$\begin{aligned}\alpha_{m0} &= \frac{\sin 2m\pi(d_0+d_1+d_2+\cdots+d_{m-1})}{m\pi} \\ \alpha_{m1} &= \alpha_{m2} = \cdots \alpha_{m(m-1)} = 0 \\ \alpha_{mi} &\approx 2d_i\cos 2m\pi(d_0+\varepsilon d_1+\cdots+\varepsilon^{i-1}d_{i-1})\end{aligned}$$

Here $m = k$ or $k + 1$, and $k = 2,4,6\ldots$, $i = k, k + 1, k + 2, \ldots$.The same method can be done with the expansion of β_m, i.e.,

$$\beta_m = \beta_{m0}+\varepsilon\beta_{m1}+\varepsilon^2\beta_{m2}+\cdots \tag{5.62}$$

where

$$\begin{aligned}\beta_{m0} &= \frac{1-\cos 2m\pi(d_0+d_1+d_2+\cdots+d_{m-1})}{m\pi} \\ \beta_{m1} &= \beta_{m2} = \cdots \beta_{m(m-1)} = 0 \\ \beta_{mi} &\approx 2d_i\sin 2m\pi(d_0+\varepsilon d_1+\cdots+\varepsilon^{i-1}d_{i-1})\end{aligned}$$

In this way, the series expansion form of the switching function in the closed-loop converter system is unchanged, and the process of solving is exactly the same as the original algorithm, except that the coefficients of some items would become zero, for example $b_{21} = 0$, $b_{31} = 0$.

5.5.3 Double Iterative Symbol Algorithm

In order to distinguish the algorithm to be proposed below, we refer to the algorithm of the closed-loop system proposed in Sect. 5.4 as the single-iteration symbol algorithm (or basic algorithm).

When the ESPM is used to solve the state equation of the closed-loop system, the main oscillation equation should be:

$$\mathbf{G}_0(0)\mathbf{a}_{00} + \mathbf{G}_1(0)(d_0 + \varepsilon d_1 + \varepsilon^2 d_2 + \cdots)\mathbf{a}_{00} = \mathbf{u} \tag{5.63}$$

However, it can be found from the expansion of δ in (5.32a) that in the closed-loop solution, a portion of the DC component in (5.63) is shifted to higher order equations for solution because of the duty cycle and the steady-state periodic solution are both expanded into the sum of the main term and the small correction terms. Similarly this is also the case in the solution of other components. Thus, it is possible to achieve a higher accuracy by using single-iteration symbol algorithm to iterate several times.

When the equivalent small-parameter method is adopted, since the influence of the higher harmonics on the duty ratio is small, it can be considered that when solving the steady-state periodic solution of the closed-loop system, the duty ratio has been determined after three iterations, that is, $d = d_0 + d_1 + d_2$, is considered to be constant. In order to improve the convergence speed, the system can be solved once again by the equivalent small parameter solution method of the open-loop system. This method of solving the closed-loop system of the switching power converter first, after determining the duty ratio, considering the system as an open-loop system, and then solving it again with the ESPM is called double iterative symbol method.

In fact, we find that if d_0 is replaced by the term $(d_0 + d_1 + d_2)$ in the expansion of the switching function δ, the first three iterative equations of the closed-loop system are exactly the same as those of the open-loop system. Therefore, it can be set that:

$$D = d_0 + \varepsilon d_1 + \varepsilon^2 d_2, \quad \tau_{s2} = 2\pi(d_0 + \varepsilon d_1 + \varepsilon^2 d_2) \tag{5.64}$$

Here the meaning of ε is the same as before. According to equations from (5.23) to (5.29) on the method of dividing the integral interval, there are:

$$\alpha_1 = \frac{\sin 2\pi(D + \varepsilon^3 d_3 + \cdots)}{\pi} \approx \frac{\sin 2\pi D}{\pi} + \varepsilon^3 \cdot 2d_3 \cos \tau_{s2} + \cdots \tag{5.65}$$

$$\beta_1 = \frac{1 - \cos 2\pi(D + \varepsilon^3 d_3 + \cdots)}{\pi} \approx \frac{1 - \cos 2\pi D}{\pi} + \varepsilon^3 \cdot 2d_3 \sin \tau_{s2} + \cdots \tag{5.66}$$

Thus the coefficient b_1 can be calculated by the following equations as

$$b_1 \approx \frac{1}{2}(\alpha_1 - j\beta_1) = b_{10} + \varepsilon^3 d_3 e^{-j\tau_{s2}} + \cdots \tag{5.67}$$

Similarly, the series expansion of b_2 and b_3 are given as follows.

$$b_2 = b_{20} + \varepsilon^3 d_3 e^{-j\tau_{s2}} + \cdots, \quad b_3 = b_{30} + \varepsilon^3 d_3 e^{-j\tau_{s2}} + \cdots \tag{5.68}$$

Thus, in the series expansion of switching function δ in (5.32a), the value of the following coefficients is zero, i.e., $b_{11} = b_{12} = b_{21} = b_{22} = b_{31} = b_{32} = 0$, and the coefficient b_{m0} is determined by

$$b_{m0} = \frac{1}{2}(\alpha_{m0} - j\beta_{m0}) \tag{5.69}$$

where

$$\alpha_{m0} = \frac{\sin 2m\pi D}{m\pi}, \quad \beta_{m0} = \frac{1 - \cos 2m\pi D}{m\pi}, \quad (m = 1, 2, 3)$$

It can be seen that the first three terms of the series expansion of δ are exactly the same as those of the open loop, which shows that the joint of the closed-loop solution method and the open-loop solution method are reasonable to determine the steady-state periodic solution of the closed-loop system.

See Chap. 4 for the detailed process of the steady-state periodic solution of the open-loop switching power converter system of the switching power converter by the equivalent small-parameter method. Here we only give the equations for the main component and the first-order and second-order corrections, as shown in Eq. (5.70), which is also available by setting $d_0 = D$, $d_1 = d_2 = 0$, $b_{11} = b_{12} = b_{21} = b_{31} = 0$ in Eqs. (5.37), (5.41a), and (5.45a)~(5.45c).

$$\begin{aligned}
&G_0(0)a_{00} + G_1(0)Da_{00} = u \\
&[G_0(j\omega) + G_1(j\omega) \cdot D] \cdot a_{11} = -G_1(j\omega) \cdot b_{10}a_{00} \\
&[G_0(j2\omega) + G_1(j2\omega)D]a_{22} = -G_1(j2\omega) \cdot [b_{20}a_{00} + b_{10}a_{11} + b_{30}\bar{a}_{11}] \\
&[G_0(j3\omega) + G_1(j3\omega)D]a_{32} = -G_1(j3\omega) \cdot [b_{30}a_{00} + b_{10}a_{22} + b_{20}a_{11}] \\
&[G_0(0) + G_1(0)D]a_{02} = -G_1(0)(b_{10}\bar{a}_{11} + b_{10}a_{11})
\end{aligned} \tag{5.70}$$

Note that when using the double iterative algorithm, the original closed-loop system needs to be decoupled, that is, the power stage main circuit is separated from the feedback compensation network, and only the main circuit state variables are solved. Since the duty cycle is considered constant at this time, the feedback network has no effect on the steady-state periodic solution of the power stage.

5.5.4 Analysis Example

We analyze the Boost converter in Sect. 5.4.1 by the improved symbolic algorithm and then compare it with other methods. When the converter operates with different frequencies, the results from the four methods, i.e., the basic equivalent small parameter method (B-ESPM), the improved equivalent small parameter method (I-ESPM), the numerical simulation, and the Pspice simulation method, are listed in Table 5.5.

Table 5.5 Simulated results of state variables in different methods

fs	Methods	a0	a1	b1	a2	b2	a3	b3
50 kHz $d_0 = 0.4091$ $d_1 = -0.066$ $d_2 = -0.047$	B-ESPM	0.3325	−0.2744	0.1387	−0.0587	−0.0379	−0.0013	0.0039
		6.9796	0.1179	−0.2186	0.0310	0.0407	−0.0151	−0.0052
	I-ESPM	0.3566	−0.1938	0.13640	−0.0649	−0.0196	−0.0040	−0.0095
		7.0651	0.0332	−0.1791	0.0372	0.0124	−0.0084	0.0094
	Numerical	0.3766	−0.2124	0.1349	−0.0631	−0.0289	0.0018	−0.0071
		7.2230	0.0412	−0.1994	0.0351	0.0237	−0.0179	0.0070
	Pspice	0.3750	−0.2295	0.1280	−0.0643	−0.0299	−0.0073	−0.0029
		7.0472	0.0571	−0.1966	0.0294	0.0269	−0.0145	0.0011
100 kHz $d_0 = 0.4091$ $d_1 = -0.032$ $d_2 = -0.022$	B-ESPM	0.4253	−0.1483	0.0584	−0.0211	−0.0210	−0.0032	0.0069
		7.7490	0.0875	−0.0741	0.0098	0.0191	−0.0014	−0.0064
	I-ESPM	0.4287	−0.1286	0.0606	−0.0244	−0.0188	−0.0010	0.0036
		7.7465	0.0651	−0.0698	0.0127	0.0153	−0.0029	−0.0030
	Numerical	0.4298	−0.1295	0.0613	−0.0247	−0.0207	−0.0009	0.0040
		7.7414	0.0641	−0.0732	0.0127	0.0166	−0.0049	−0.0037
	Pspice5	0.4430	−0.1442	0.0481	−0.0220	−0.0222	−0.0071	0.0053
		7.6504	0.0791	−0.0669	0.0083	0.0169	0.0005	−0.0066
1000 kHz $d_0 = 0.4091$ $d_1 = -0.003$ $d_2 = -0.002$	B-ESPM	0.5032	−0.0157	0.0048	−0.0013	−0.0020	−0.0008	0.0009
		8.3984	0.0107	−0.0036	0.0008	0.0014	0.0005	−0.0007
	I-ESPM	0.5023	−0.0155	0.0048	−0.0014	−0.0020	−0.0007	0.0009
		8.3856	0.0104	−0.0036	0.0009	0.0014	0.0005	−0.0006
	Numerical	0.5049	−0.0156	0.0049	−0.0014	−0.0021	−0.0008	0.0010
		8.3750	0.0106	−0.0039	0.0009	0.0014	0.0005	−0.0008
	Pspice5	0.5256	−0.0171	0.0031	−0.0012	−0.0019	−0.0008	0.0008
		8.3348	0.0117	−0.0035	0.0009	0.0011	0.0008	−0.0008

Note The upper row of each set of data in Table 5.5 represents the inductor current (A), and the lower row represents the output capacitor voltage (V)

The closed-loop Boost converter system in Sect. 5.4.1 is used as an analysis example. The basic algorithm of ESPM, the improved algorithm of ESPM, the numerical and the Pspice simulation methods are adopted respectively to analyze the Boost converter operated with different switching frequencies. The resulted DC component and the first three harmonic components of the state variables are listed in Table 5.5, in which a_0 represents the DC component, a_i and b_i represent amplitudes of the cosine and sine components of the *i*th (i = 1,2,3) harmonic, respectively (see Eq. (5.48b)).

When the converter operates at the switching frequency of fs = 50 and 100 kHz respectively, the ripples of the state variables during one cycle obtained from the ESP analysis method (dashed line) are compared with those from numerical simulation method (solid line) in Fig. 5.5, in which the results of the ESPM are obtained by calculation of the sum of the first three harmonic components of the state variables during one cycle. Among Fig. 5.5, (a–d) are the results of comparing the improved algorithm with numerical simulation; and (e–h) are the results of the comparison between the original single-iteration algorithm (basic algorithm) and the numerical simulation.

It can be seen from Table 5.5 and the waveforms in Fig. 5.5, that using the improved symbol algorithm to analyze the closed-loop switching power system can get more accurate results.

When the switching frequency is low and the ripple is large, the values of d_1 and d_2 in the direct ripple feedback control mode are larger. Therefore, when the single-iteration algorithm is used, the obtained DC component would have a larger error. However, this situation will be improved, and more accurate results can be obtained by using the improved symbol algorithm.

It can be seen from the above examples and Table 5.5, that the basic symbol algorithm can get more accurate results when the switching frequency is higher, or when the average feedback control law is used (i.e., the appropriate error compensation capacitor is added in the control loop to smooth the effects of ripples on the duty cycle). Because under these situations, the values of the first-order correction d_1 and the second-order d_2 are quite small, they have less influence on the duty cycle d.

5.6 Experiments and Verification

5.6.1 Diagram of the Experimental Circuit

A Boost regulator and its feedback-controlled circuit is shown in Fig. 5.6, where r_1 and R_s are the parasitic resistance and the current sensing resistor of the inductor respectively. In order to verify that the symbolic Equivalent-Small-Parameters (ESP) analysis method is still applicable in the case of high output ripple, we choose a lower switching frequency and a smaller output capacitance.

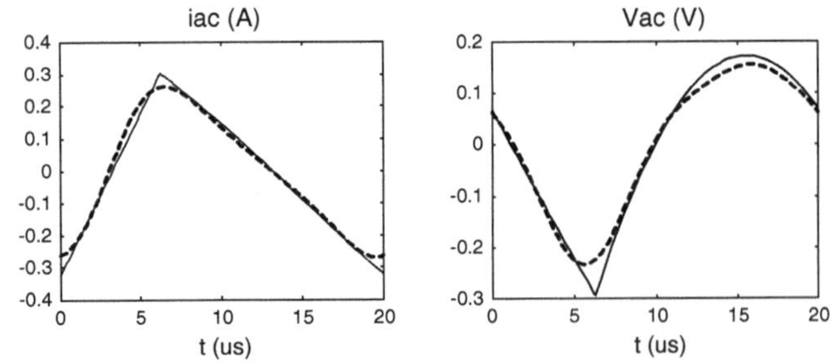

(a) inductor current ($f_s = 50kHz$) (b) capacitor voltage

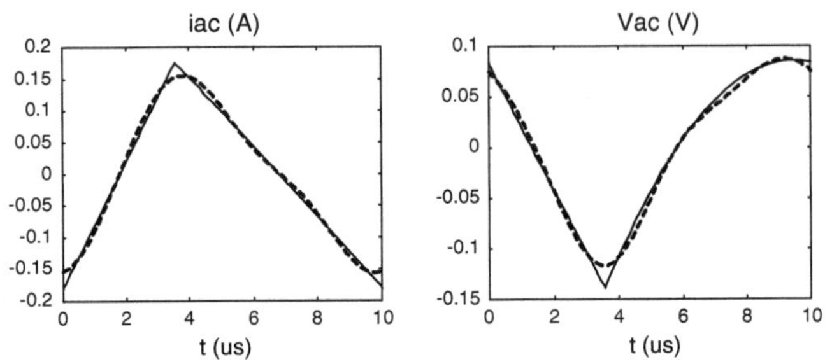

(c) inductor current ($f_s = 100kHz$) (d) capacitor voltage

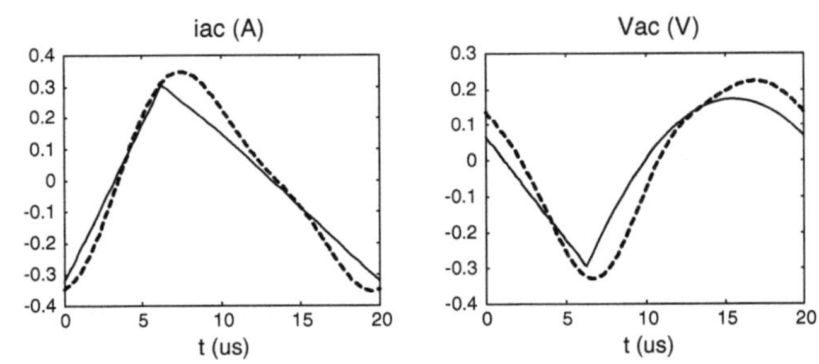

(e) inductor current ($f_s = 50kHz$) (f) capacitor voltage

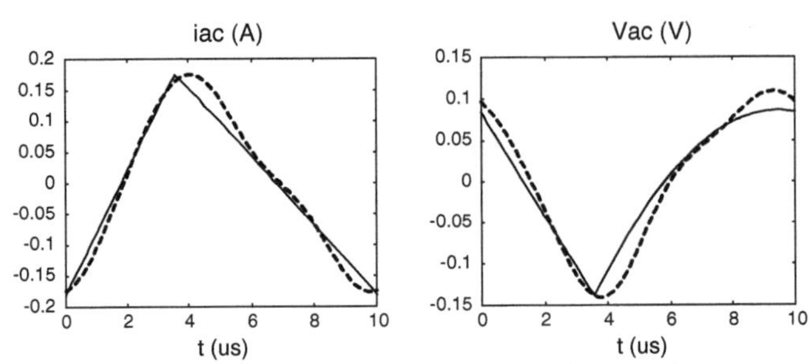

(g) inductor current ($f_s = 100kHz$) (h) capacitor voltage

Fig. 5.5 Comparison of simulated state ripple waveforms between numerical simulations and the ESPM (**a–d**: with the improved algorithm; **e–h**: with the original algorithm)

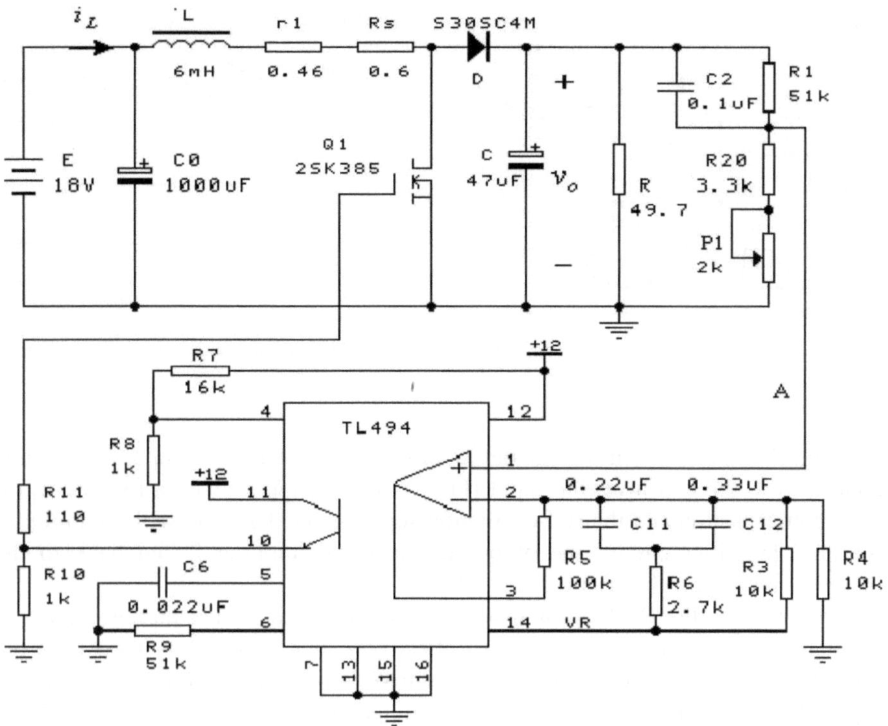

Fig. 5.6 Experimental circuit of Boost regulator

The TL494 PWM control chip is considered for experiments, its detailed internal block diagram, pin configuration and functions, and the operation principle can be found in related power integrated control circuit datasheet. The voltage of Pin-4 is used to limit the maximum width of the pulse, which determines the dead time control of the main circuit. The internal voltage reference $V_R = +5V$ is provided by Pin-14. The resistors R_3 and R_4 set the reference of about 2.5 V on the negative input port (Pin-2) of the error amplifier (EA), and the output voltage of the main circuit is divided by resistors R_1 and R_2, setting the sampling voltage on the positive input port of the error amplifier (Pin-1). The error amplifier compares and amplifies the difference between the sampled and the reference voltage, and the resulted output error voltage is then compared with the sawtooth wave voltage signal of a constant frequency to generate a pulse with a certain width (the width of the pulse is directly determined by the error voltage). The pulse is shaped and amplified by TL-494 and then output via Pin-8, Pin-9 or Pin 10 and Pin-11 with a certain drive capability to turn on the power switch. Pins 5 and 6 are connected to the capacitor C_6 and the resistor R_9 respectively, which determine the oscillating frequency of sawtooth wave, i.e., $f_s = 1.2/R_9C_6\,(\mathrm{kHz})$. The oscillation amplitude of the sawtooth wave is given by the relevant datasheet or measured by Pin-5. The maximum and minimum values measured in the experiment are $v_u = 3.12V$ and $v_l = 0$

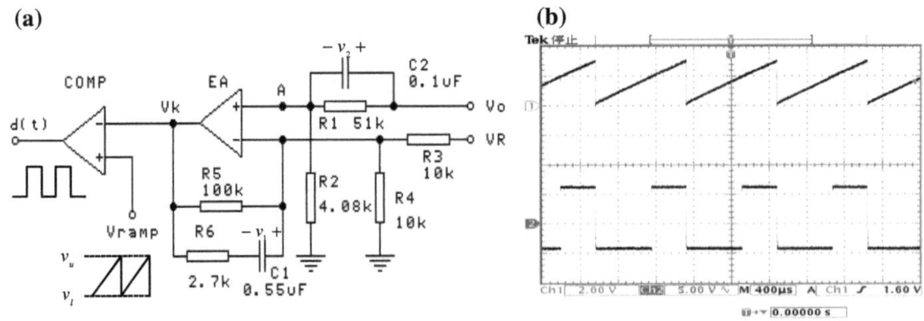

Fig. 5.7 **a** Diagram of feedback controlled circuit, **b** measured sawtooth ramp and switching signal waveforms

(see Fig. 5.7b below). Pin-3 is the output of the TL-494 internal error amplifier. It is connected to the non-inverting input (Pin-2) of the amplifier with a RC circuit for error compensation and the self-oscillation suppression. To increase the stability of the error-amplifier circuit, the output of the error amplifier is fed back to the inverting input port through a resistor R_5.

According to the PWM principle of the TL-494 chip, when the circuit works normally, i.e., the circuit is controlled effectively, that is, the duty ratio is in the unsaturated state. The schematic of the control circuit in Fig. 5.6 can be shown in Fig. 5.7a, where EA indicates the operational amplifier, COMP indicates the comparator, R_2 is the sum of the resistance of resistor R_{20} and potentiometer P_1 in Fig. 5.6, and C_1 is the sum of C_{11} and C_{12} in Fig. 5.6.

5.6.2 Comparison of Experiment, ESPM and Simulation for Open-Loop System

The input filter capacitor C_0 in Fig. 5.6 has such a large value that it can be replaced with an open circuit when the circuit operates in steady state, which means it does not affect the dynamic characteristics of the circuit, so it was ignored in symbolic analysis. Thus the main circuit is considered to be a second-order circuit. Disconnect the point A in Fig. 5.6, and connect a constant voltage V_A independent of the output voltage of the main circuit at point A, then the circuit is in open-loop operation. The experimental results are $V_A = 2.46(V)$, $V_R = 4.97(V)$; the current sensing resistor of the inductor $R_s = 0.595\,\Omega$. According to Fig. 5.7, the output voltage V_k of the error amplifier and the duty cycle D_0 can be calculated as:

$$\begin{aligned} V_k &= V_A(2R_5/R_3 + 1) - V_R \cdot R_5/R_3 = 21V_A - 10V_R = 1.96(V) \\ D_0 &= 1 - V_k/(v_u - v_l) = 0.3718 \end{aligned} \tag{5.71}$$

And the operating frequency of the switch calculated from the circuit parameters is:

$$f_{s0} = 1.0695\,(\text{kHz}) \tag{5.72}$$

Figure 5.7b shows the measured sawtooth ramp and switching signal waveforms when the circuit is in open-loop operation. The figure shows that the switching period $T = 961.6\,\mu\text{s}$, $t_{on} = 368.2\,\mu\text{s}$, thus, the actual switching frequency and duty cycle would be

$$f_s = 1.039933\,(\text{kHz}),\ D = 0.3829 \tag{5.73}$$

From the comparison of Eqs. (5.71)–(5.73), it can be seen that the measured values are close to the theoretically calculated values.

According to the analysis of the Chap. 4, the Boost converter in CCM can be described by the following general vector state variable differential equation:

$$\mathbf{G}_0(p)\mathbf{x} + \mathbf{G}_1(\mathbf{p})\mathbf{f} = \mathbf{u} \tag{5.74}$$

where

$$\mathbf{G}_0(p) = \begin{bmatrix} p + \frac{r1 + R_s}{L} & \frac{1}{L} \\ \frac{-1}{C} & p + \frac{1}{RC} \end{bmatrix},\quad \mathbf{G}_1(p) = \begin{bmatrix} 0 & \frac{-1}{L} \\ \frac{1}{C} & 0 \end{bmatrix},\ \mathbf{x} = \begin{bmatrix} i_L \\ v_o \end{bmatrix},$$
$$\mathbf{u} = \begin{bmatrix} E/L \\ 0 \end{bmatrix},\quad \mathbf{f} = \delta\mathbf{x},\ \delta = \begin{cases} 1 & 0 \le t \le DT \\ 0 & DT \le t \le T \end{cases}$$

The switching frequency and duty cycle are determined by Eq. (5.73). Use the analysis method in Chap. 4, the approximated steady-state periodic analytic solution of state variables are acquired:

$$\begin{aligned} i_L &= 0.8907 - 0.4157\cos\tau + 0.1448\sin\tau - 0.0527\cos 2\tau - 0.0490\sin 2\tau \\ &\quad - 0.0130\cos 3\tau + 0.0200\sin 3\tau \\ v_o &= 27.5296 + 1.2535\cos\tau - 1.5518\sin\tau + 0.0862\cos 2\tau + 0.3408\sin 2\tau \\ &\quad - 0.0398\cos 3\tau - 0.1254\sin 3\tau \end{aligned} \tag{5.75}$$

where $\tau = \omega t$, $\omega = 2\pi f_s$.

Figure 5.8 shows waveforms of capacitor output voltage and switching signal and sensed inductor current signal, Fig. 5.9 shows waveforms of measured state ripples and comparison of state ripples obtained by symbolic analysis (solid line) and Pspice simulation (dotted line), detailed data comparison is shown in Table 5.6.

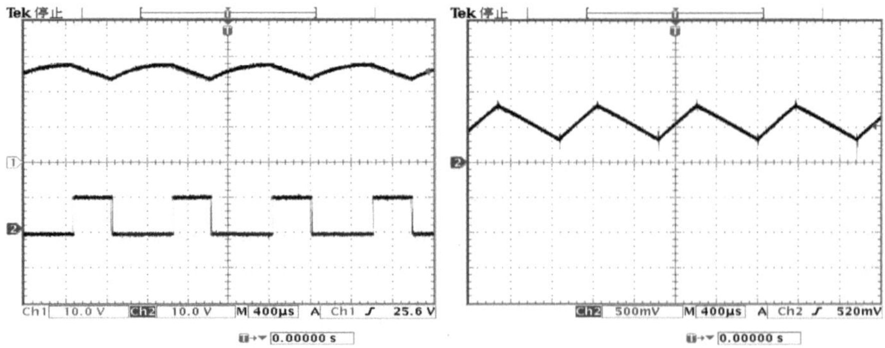

Fig. 5.8 Waveforms of capacitor output voltage and switching signal (left) and sensed inductor current signal (right) for open-loop Boost converter

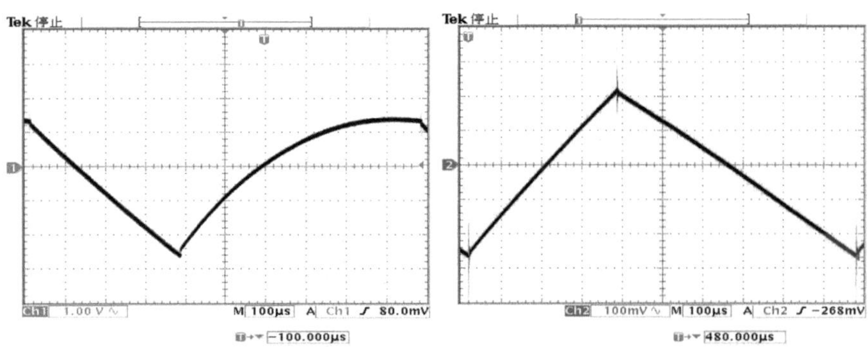

(a) Output Voltage $v_o(t)$ (b) Inductor current $i_L(t)$*Rs

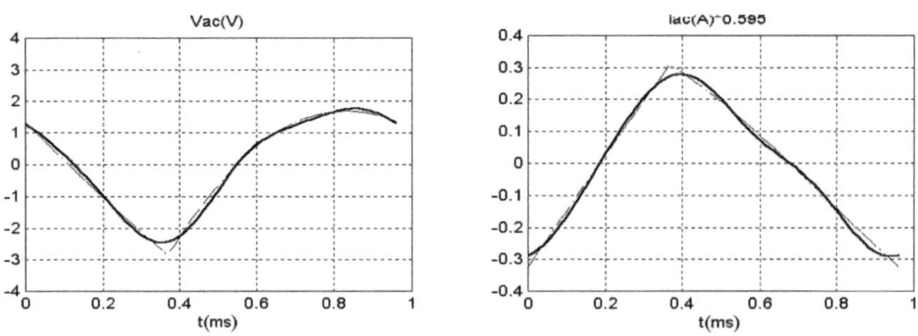

(c) Output Voltage $v_o(t)$ (d) Inductor current $i_L(t)$*Rs

Fig. 5.9 Comparison of state ripples for open-loop Boost converter: **a** and **b** measured results; **c** and **d** ESPM analysis (solid line) and Pspice simulation (dotted line)

Table 5.6 Comparison of symbolic (ESPM) analysis, Pspice simulation and experiment for open-loop Boost converter

E = 18.30 V	DC values		Ripples					
	I_L (A)	V_o (V)	I_LR_s (V)			V_o (V)		
			Max	Min	Amp.	Max	Min	Amp.
ESPM	0.8907	27.5296	0.2390	−0.2493	0.4883	1.7654	−2.4454	4.2108
Pspice5	0.8792	27.2668	0.2616	−0.2783	0.5399	1.6904	−2.8187	4.5090
Exp.	0.9529	26.0	0.228	−0.278	0.506	1.5	−2.64	4.14

5.6.3 Comparison of Experiment, ESPM and Simulation for Closed-Loop System

According to Fig. 5.7 and the analysis method in Chap. 5, the time-variant differential equation of the closed-loop system can still be expressed as (5.74), where

$$\mathbf{G}_0(p) = \begin{bmatrix} p+\frac{r1+R_s}{L} & \frac{1}{L} & 0 & 0 \\ -\frac{1}{C} & p+\frac{1}{RC} & 0 & 0 \\ 0 & \frac{m_1m_2R_5}{C_1} & p+\frac{m_1}{C_1} & -\frac{m_1m_2R_5}{C_1} \\ 0 & -\frac{1}{R_2C_2} & 0 & p+\frac{m_3}{C_2} \end{bmatrix}, \quad \mathbf{G}_1(p) = \begin{bmatrix} 0 & -\frac{1}{L} & 0 & 0 \\ \frac{1}{C} & 0 & 0 & 0 \\ 0 & 0 & 0 & 0 \\ 0 & 0 & 0 & 0 \end{bmatrix},$$

$$\mathbf{x} = [i_L \quad v_o \quad v_1 \quad v_2]', \quad \mathbf{u} = [E/L \quad 0 \quad V_R R_5 m_1/(R_3C_1) \quad 0]', \quad \mathbf{f} = \delta\mathbf{x},$$

where

$$m_1 = 1/(R_5+R_6), \quad m_2 = 1/R_3 + 1/R_4, \quad m_3 = 1/R_1 + 1/R_2$$

and the switching function is:

$$\delta = \begin{cases} 1 & 0 \le t \le d(t)T \\ 0 & d(t)T \le t \le T \end{cases}$$

where $d(t)$ is the duty cycle, which can be expressed as a linear function of state variable:

$$d = K_0 + \mathbf{K}_1\mathbf{x}(t_s) \; (t_s = dT) \tag{5.76}$$

where

$$K_0 = 1 + R_5R_6m_1V_R/(R_3v_u)$$

$$\mathbf{K}_1 = [0 \quad 1 + m_1m_2R_5R_6) \quad -R_5m_1 \quad -(1+m_1m_2R_5R_6)]/v_u$$

The switching frequency of the circuit is the same as that of the open loop. Using the double iterative symbol algorithm in Sect. 5.5.3, the steady-state periodic

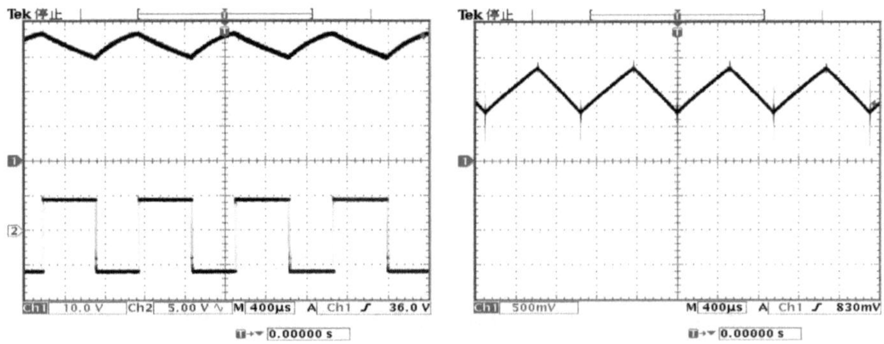

Fig. 5.10 Waveforms of output voltage and switching signal (left) and sensed inductor current signal (right) for closed-loop Boost converter

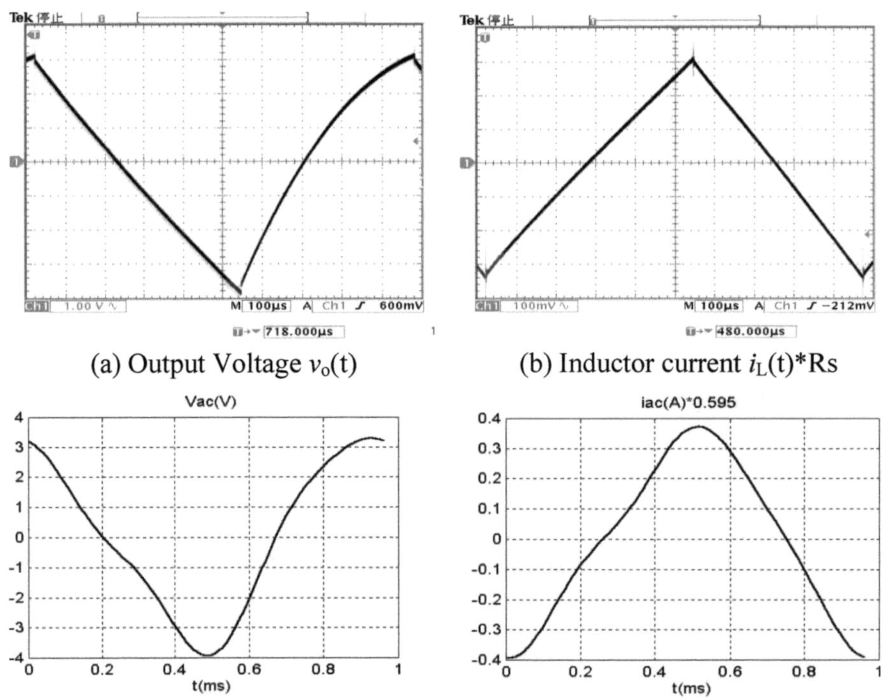

(a) Output Voltage $v_o(t)$ (b) Inductor current $i_L(t)$*Rs

(c) Output Voltage $v_o(t)$ (d) Inductor current $i_L(t)$*Rs

Fig. 5.11 Comparison of state ripples for closed-loop Boost converter: **a** and **b** measured results; **c** and **d** symbolic analysis

solution of the inductor current and the output voltage containing the DC component and the first three harmonics can be calculated as Eq. (5.77), where $\tau = \omega t$, $\omega = 2\pi f_s$.

Table 5.7 Comparison of symbolic (ESPM) analysis and experiment for closed-loop Boost converter

E 18.00 V	DC values		Ripples					
	I_L (A)	V_o (V)	$I_L R_s$ (V)			V_o (V)		
			Max	Min	Amp.	Max	Min	Amp.
ESPM	1.5146	34.7894	0.3733	−0.3921	0.7654	3.2941	−3.9366	7.2247
Exp.	1.7142	33.7	0.318	−0.344	0.662	3.28	−3.96	7.24

$$\begin{aligned} i_L &= 1.5146 - 0.5747\cos\tau - 0.0962\sin\tau - 0.0279\cos 2\tau + 0.0372\sin 2\tau \\ &\quad - 0.0533\cos 3\tau - 0.0212\sin 3\tau \\ v_o &= 34.7894 + 3.2657\cos\tau - 0.6497\sin\tau - 0.3595\cos 2\tau - 0.3204\sin 2\tau \\ &\quad + 0.2960\cos 3\tau + 0.1295\sin 3\tau \end{aligned} \tag{5.77}$$

Figure 5.10 shows waveforms of capacitor output voltage and switching signal and sensed inductor current signal when the circuit is in closed-loop operation, waveforms of measured state ripples and comparison of state ripples obtained by symbolic analysis (solid line) and Pspice simulation (dotted line) are shown in Fig. 5.11, detailed data comparison is shown in Table 5.7. As shown in Fig. 5.10a, the experimentally measured duty cycle is $d = 0.5467$, the duty cycle calculated by symbolic analysis method is $d = 0.5420$, one can see that the measured result and the analytic calculation are in a good agreement.

5.7 Summary

In this chapter, the equivalent small parameter method (ESP) is applied to the steady-state analysis of the closed-loop converter system with CCM (continuous-conduction-mode) operation. A single-iterative and a double iterative symbol algorithm for analyzing the steady-state solution of the closed-loop system are proposed. It is shown that the equivalent small parameter method can be extended to the steady-state analysis of the closed-loop PWM switching converter system and still possesses the advantages of simple algorithm and high accuracy. The results are all analytical expressions, and from which the working mechanism of the circuit can be easily mastered. And moreover, the analytical expressions of the ripples would have obvious applications in the engineering design of circuit and computer symbolic simulation analysis.

In addition, the principle of the method and the analysis results in this chapter show that: (1) The effect of ripple on the duty cycle of the PWM closed-loop switching power converter system is quite small, especially at a higher switching frequency; (2) The DC offset presented in the system can be suppressed by adding the appropriate integral compensation in the feedback circuit; (3) The data comparison in Table 5.5 shows that using a single-iterative algorithm to analyze the closed-loop system with PI feedback control law has a higher accuracy (the actual application circuit is generally the case).

The comparison between experiments and the results of symbolic analysis further verified that the ESP symbol analysis method has high accuracy for the steady state analysis of open-loop and closed-loop systems of power switching converters even if the output contains large ripples.

References

1. Middlebrook RD, Cuk S (1976) A general unified approach to modelling switching-converters power stages. IEEE PESC Rec 18–34
2. Krein PT, Bentsman J, Bass RM et al (1990) On the use of averaging for the analysis of power electronic systems. IEEE Trans on PE 5(2):182–190
3. Lehman B, Bass RM (1996) Switching frequency dependant averaged models for PWM DC/DC converters. IEEE Trans PE 11(1):89–98
4. Lehman B, Bass RM (1996) Extensions of averaging theory for power electronic systems. IEEE Trans PE 11(4):542–553
5. Sanders, SR, Noworolski, JM, Liu XZ et al (1991) Generalized averaging method for power conversion circuits. IEEE Trans PE 6(2):251–258
6. Caliskan VA, Verghese GC, Stankovic AM (1999) Multifrequency averaging of DC/DC converters. IEEE Trans PE 14(1):124–133
7. Mahdavi, J, Emaddi, A, Bellar MD et al (1997) Analysis of power electronic converters using the generalized state-space averaging approach. IEEE Trans CAS-I 44(8):767–770
8. Qiu S-S, Filanovsky IM (1987) Periodic solutions of the Van der Pol equation with moderate values of damping coefficient. IEEE Trans Circuits Syst CAS-34:913–918
9. Qiu SS, Filanovsky IM, Stromsmoe KA (1985) An equivalent small parameter method for strongly nonlinear systems. In: Proceedings 28th Midwest symposium on circuits and systems, Louisille, pp 61–64
10. Qiu SS, Filanovsky IM (1989) Calculation of steady-state oscillation in nonlinear circuits. Int J Electronics 67(3):403–414
11. Qiu SS, Filanovsky IM, Stromsmoe KA (1987) On one combination of the harmonic balance method and perturbation techniques. Control-Theory Adv Technol 3:189–195
12. Qiu SS, Filanovsky IM, Lin BT (1999) A new method of analysis for PWM switching power converters. Int J of Electronics 86(11):1395–1410
13. Qiu S (1997) Principle of symbolic analysis of PWM power converters. Acta Electronica Sinica 25(1):5–10
14. Lin B, Qiu SS (1996) Symbolic analysis of PWM switching power converters. Acta Electronica Sinica 24(9):83–87
15. Qiu S-S, Filanovsky IM (2000) Harmonic analysis of PWM converters. Circuits Syst I: Fundam Theory and Appl IEEE Trans 47:1340–1349; ISSN 1057-7122
16. Wu Y, Hu AP, Budgett D, Malpas SC, Dissanayake T (2011) Efficient power-transfer capability analysis of the TET system using the equivalent small parameter method. IEEE Trans Biomed Circuits Syst 5(3):272–282
17. Chen W, Floberg H, Qiu S (1999) A new analytical method for analysis and design of class E power amplifiers taking into account the switching device on resistance. Int J Circuit Theory Appl 27(4):421–436
18. Chen W (1997) Symbolic analysis for class E amplifiers. J South China Univ Technol 25(8):88–92

Chapter 6
Analysis of Voltage-Mode Controlled DCM-PWM DC/DC Converters Based on ESPM

6.1 Introduction

Discontinuous conduction mode (DCM) typically occurs when the DC/DC converter operates at a light load. For low-power applications, it is necessary for the converter to operate at DCM, even when fully loaded [1]. In addition the power-factor-correction (PFC) circuit is intentionally designed to operate at DCM [2–4] for better performances. Hence, modeling and analysis of the converter operating at DCM are very important.

At present, the method of analyzing PWM converters with DCM operation is mainly the state-space-averaging method [5]. In the case of discontinuous inductor current mode, the inductor current decreases to zero before the end of the switching cycle and remains at zero until the beginning of the next cycle. Hence, the inductor current is always zero at the beginning and the end of each cycle. The traditional state space averaging method [6] uses this as a constraint to get an order-reduced model that no longer contains the inductor current. For the dynamic characteristics of the system, the order-reduced model is not accurate [1]. As the inductor current in PFC is the ultimate control objective, the disappearance of the inductor current in the model is not what we expect.

Thus a new model for analyzing the full-order equivalent duty cycle model of DCM-operated converter is proposed in [5], which is finally experimentally proved to analyze the dynamic characteristics of AC small signal accurately. And in the reference [7], the average switching model applies the averaging technique to the nonlinear part of the converter circuit and directly obtains the DC and AC small signal analysis models of the PWM converter. However, for large signal characteristics analysis, it is necessary to resort to the more complicated circuit Model [8] (i.e., the *n*th harmonic three-port model).

In the late 1980s, power factor correction (PFC) technology in switching power supply has drawn many scholars' attention all over the world [9–18]. The PFC circuit is a strong nonlinear periodic time-varying system. Although several

Y. Chen and B. Zhang, *Equivalent-Small-Parameter Analysis of DC/DC Switched-Mode Converter*, CPSS Power Electronics Series,
https://doi.org/10.1007/978-981-13-2574-8_6

engineering analysis and numerical modeling methods have appeared in recent years, the analytical solution of periodic time-varying system has not been found [14–18]. In Chap. 2, we establish a unified nonlinear large-signal model of PWM converters operating at CCM and DCM based on the switching function. In this chapter, the Equivalent-Small-Parameter method is used to analyze this model for DCM-operated converters, and the analytic solution to the steady-state of the closed-loop PWM switching converter system is obtained. It is believed that this method is easy to be applied to the analysis of the PFC system.

6.2 Time-Varying Equation for Closed-Loop DCM-Operated Converter System

Discontinuous inductor current waveform is shown in Fig. 6.1b. For convenience, in this chapter, d_1 represents the duty cycle when the transistor switch is on (the inductor current rises in this period), d_2 represents the duty cycle when the inductor current drops and d_3 represents the duty cycle when the inductor current remains at zero. The corresponding switching functions are $\delta_1(t)$, $\delta_2(t)$ and $\delta_3(t)$, or abbreviated as δ_1, δ_2 and δ_3. Their meaning is a little different from that of Chap. 4.

According to the modeling method in Chap. 2, the PWM converter, operating at discontinuous inductor current, can be described by time-varying differential Eq. (6.1):

$$\mathbf{G}_0(p)\mathbf{x} + \mathbf{G}_1(p)\cdot\mathbf{f}^{(1)}(\mathbf{x}) + \mathbf{G}_3(p)\cdot\mathbf{f}^{(2)}(\mathbf{x}) = \mathbf{u} \tag{6.1}$$

where the nonlinear vector functions are defined as follows.

$$\mathbf{f}^{(1)} = \delta_1(\mathbf{x} + \mathbf{e}_1) \tag{6.2}$$

$$\mathbf{f}^{(2)} = \delta_3(\mathbf{x} + \mathbf{e}_2) \tag{6.3}$$

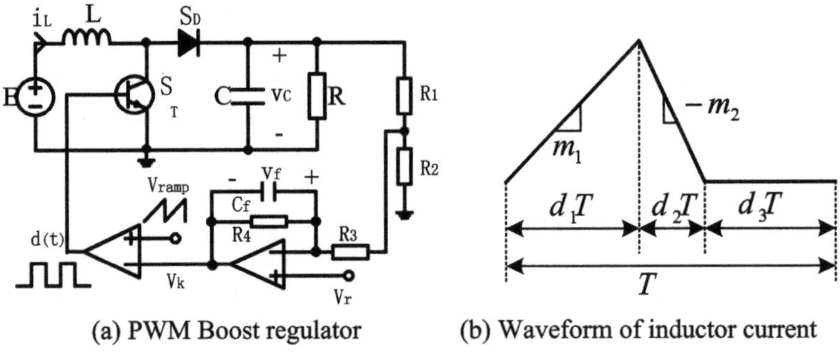

(a) PWM Boost regulator (b) Waveform of inductor current

Fig. 6.1 DCM boost regulator and the inductor current waveform where the meaning of the square matrices $\mathbf{G}_i(p)$ (i = 0, 1, 3) are similar to those in Chap. 5, and $\mathbf{e}_1$, $\mathbf{e}_2$ are constant vectors. Switching functions δ_1, δ_3 are defined as:

$$\delta_1(t) = \begin{cases} 1 & 0 < t \leq d_1(t)T \\ 0 & d_1(t)T < t \leq T \end{cases} \tag{6.4}$$

$$\delta_3(t) = \begin{cases} 0 & 0 < t \leq [1 - d_3(t)]T \\ 1 & [1 - d_3(t)]T < t \leq T \end{cases} \tag{6.5}$$

where δ_1 represents the on-off state of the main switch, d_1 represents the duty cycle when the main switch is on, which is determined by the feedback control algorithm. According to the discussion in Chap. 5, the duty cycle d_1 can be expressed as a linear function of the system's state variables, namely:

$$d_1(t) = K_0 + \mathbf{K}_1\mathbf{x}(d_1T) \tag{6.6}$$

where the meaning of each coefficient is the same with that in Chap. 5.

The switching function $\delta_3 = 1$ indicates that the main switch and diodes are disconnected, and $\delta_3 = 0$ indicates that either the main switch conducts or the diode conduct. The symbol d_3 represents the duty cycle when the inductor current remains zero (or when both switches are turned off). The reason why we choose δ_3 to establish the time-varying differential equations of the switching converter system operating at discontinuous conduction mode is as follows: (1) When $\delta_3 = 0$ is constantly true during one switching cycle, Eq. (6.1) is the time-varying differential equations of the switching converter system under CCM, in the meantime, $\mathbf{G}_0(p)$, $\mathbf{G}_1(p)$, $\mathrm{f}^{(1)}(\mathbf{x})$ remain unchanged; (2) As can be seen from the following analysis, δ_3 can be transformed into series forms as δ_1. According to Sect. 5.2, using the Equivalent-Small-Parameter Method to solve Eq. (6.1), the following iterative equations can be obtained.

$$\begin{cases} \mathbf{G}_0(p)\mathbf{x}_0 + \mathbf{G}_1(p)\mathbf{f}_{0m}^{(1)} + \mathbf{G}_3(p)\mathbf{f}_{0m}^{(2)} = \mathbf{u} & \text{(a)} \\ \mathbf{G}_0(p)\mathbf{x}_1 + \mathbf{G}_1(p)\mathbf{f}_{1m}^{(1)} + \mathbf{G}_3(p)\mathbf{f}_{1m}^{(2)} = -[\mathbf{G}_1(p)\mathbf{R}_1^{(1)} + \mathbf{G}_3(p)\mathbf{R}_1^{(2)}] & \text{(b)} \\ \mathbf{G}_0(p)\mathbf{x}_2 + \mathbf{G}_1(p)\mathbf{f}_{2m}^{(1)} + \mathbf{G}_3(p)\mathbf{f}_{2m}^{(2)} = -[\mathbf{G}_1(p)\mathbf{R}_2^{(1)} + G_3(p)\mathbf{R}_2^{(2)}] & \text{(c)} \\ \quad\cdots \end{cases} \tag{6.7}$$

6.3 Determination of Switching Function and Duty Cycle

The method used to determine δ_1, d_1 is the same as the CCM analysis method in Chap. 5. The following focuses on the discussion of the method of determining d_3, δ_3. We propose two different methods, known as the waveform determination method, and traditional determination method respectively.

6.3.1 The Waveform-Based Determination Method for $\mathbf{d}_3$

Assuming that the inductor current is piecewise linear (note that there is only the assumption of the linear waveform, without existence of small ripple). The slope of the ascending phase is m_1, and the slope of the descent phase is m_2 (shown in Fig. 6.1b). The average value of the inductor current in one cycle during steady the state is:

$$\bar{i}_L = \frac{1}{T}\int_0^{d_1T} (I_0 + m_1 t)dt + \int_0^{d_2T} [(I_0 + m_1 d_1 T - m_2 t]dt \tag{6.8}$$

where I_0 is the initial value of the inductor current at the beginning of a switching cycle T. During the steady state, there exists $m_1 d_1 T = m_2 d_2 T$. From the above formula, the following relationship can be obtained.

$$\bar{i}_L = I_0(d_1 + d_2) + \frac{m_1 T}{2} d_1^2 + \frac{m_1 T}{2} d_1 d_2 \tag{6.9}$$

Setting m = $m_1T/2$, then, the following relationships can be obtained.

$$d_2 = \frac{\bar{i}_L - (md_1^2 + I_0 d_1)}{I_0 + md_1}, \quad d_3 = 1 - d_1 - \frac{\bar{i}_L - (md_1^2 + I_0 d_1)}{I_0 + md_1} \tag{6.10}$$

For the basic second-order PWM converter, such as Boost, Buck, Buck-boost converter, when the inductor current is discontinuous, generally, $I_0 = 0$. Not taking into account the circuit's parasitic parameters, m_1 and m_2 usually are functions of the average input power and the average output voltage (capacitor voltage). Thus, d_3 can be expressed as a function of the power supply voltage, the average state variables $\bar{\mathbf{x}}$ and d_1, i.e.,

$$d_3 = h_1(E, d_1, \bar{\mathbf{x}}) \tag{6.11}$$

According to the series expansion of $\bar{\mathbf{x}}$ and d_1, which are defined as follows.

$$\bar{\mathbf{x}} = \bar{\mathbf{x}}_0 + \varepsilon\bar{\mathbf{x}}_1 + \varepsilon^2\bar{\mathbf{x}}_2 + \cdots \tag{6.12}$$

$$d_1 = d_{10} + \varepsilon d_{11} + \varepsilon^2 d_{12} + \cdots \tag{6.13}$$

Assuming that d_3 can be transformed into the series expansion that consists of the main term and small terms, the following relationship can be obtained.

$$d_3 = d_{30} + \varepsilon d_{31} + \varepsilon^2 d_{32} + \cdots \tag{6.14}$$

The function $h_1(E, d_1, \bar{\mathbf{x}})$ is approximated by the linear terms of its Taylor series expanded at $(d_{10}, \bar{\mathbf{x}}_0)$, then d_{3i} (i = 1, 2, …) can be approximated by the following equations as

$$\begin{aligned}
&d_{30} = h_1(E, d_{10}, \bar{\mathbf{x}}_0) \\
&d_{31} = \left(\left.\frac{\partial h_1}{\partial d_1}\right|_{\substack{d_1 = d_{10} \\ \bar{\mathbf{x}} = \bar{\mathbf{x}}_0}} \right) \cdot d_{11} + \left(\left.\frac{\partial h_1}{\partial \bar{x}}\right|_{\substack{d_1 = d_{10} \\ \bar{\mathbf{x}} = \bar{\mathbf{x}}_0}} \right) \cdot \bar{\mathbf{x}}_1 = H_{11}^1 \cdot d_{11} + H_{12}^1 \cdot \bar{\mathbf{x}}_1 \\
&d_{32} = \left(\left.\frac{\partial h_1}{\partial d_1}\right|_{\substack{d_1 = d_{10} + d_{11} \\ \bar{\mathbf{x}} = \bar{\mathbf{x}}_0 + \bar{\mathbf{x}}_1}} \right) \cdot d_{12} + \left(\left.\frac{\partial h_1}{\partial \bar{x}}\right|_{\substack{d_1 = d_{10} + d_{11} \\ \bar{\mathbf{x}} = \bar{\mathbf{x}}_0 + \bar{\mathbf{x}}_1}} \right) \cdot \bar{\mathbf{x}}_2 = H_{21}^1 \cdot d_{12} + H_{22}^1 \cdot \bar{\mathbf{x}}_2 \\
&\cdots
\end{aligned} \tag{6.15}$$

where H_{11}^1 and H_{21}^1 are real numbers or vectors with $1 \times n$ dimensions (if $\mathbf{x}$ is $1 \times n$-dimensional). In this way, δ_3 can be transformed into series expansion the same as δ_1, then the following relationship can be obtained.

$$\delta_3 = \delta_{30} + \varepsilon \delta_{31} + \varepsilon^2 \delta_{32} + \cdots \tag{6.16}$$

In which each term in (6.16) is chosen as

$$\begin{aligned}
\delta_{30} &= d_{30} + q_{10} e^{j\tau} + c.c \\
\delta_{31} &= d_{31} + q_{11} e^{j\tau} + q_{20} e^{j2\tau} + q_{30} e^{j3\tau} + c.c \\
\delta_{32} &= d_{32} + q_{12} e^{j\tau} + q_{21} e^{j2\tau} + q_{31} e^{j3\tau} + q_{40} e^{j4\tau} + q_{50} e^{j5\tau} + c.c \\
&\cdots
\end{aligned}$$

where the coefficients in Eq. (6.16), according to the improved algorithm proposed in Sect. 5.5, can be deduced as:

$$\begin{aligned}
&q_{mi} = \frac{1}{2}(\gamma_{mi} - j\eta_{mi}) \quad (m = 1, 2, \ldots, i = 0, 1, 2, \ldots) \\
&\gamma_{m0} = \frac{\sin 2m\pi d_{30}}{m\pi}, \quad \eta_{m0} = -\frac{1 - \cos 2m\pi d_{30}}{m\pi} \\
&\gamma_{mi} \approx 2d_{3i} \cos 2m\pi (d_{30} + \varepsilon d_{31} + \cdots + \varepsilon^{i-1} d_{3(i-1)}) \quad (i = 1, 2, \ldots) \\
&\eta_{mi} \approx -2d_{3i} \sin 2m\pi (d_{30} + \varepsilon d_{31} + \cdots + \varepsilon^{i-1} d_{3(i-1)}) \quad (i = 1, 2, \ldots)
\end{aligned} \tag{6.17}$$

Table 6.1 Relationship between d_1 and d_2 in steady-state basic converter

	Buck	Boost	Buck/boost	Cuk
k_c	$k_c = 2Lf_s/R$	$k_c = 2Lf_s/R$	$k_c = 2Lf_s/R$	$k_c = 2L_e f_s/R$
d_2	$\frac{2k_c}{d_1(1+\sqrt{1+4k_c/d_1^2})}$	$\frac{k_c(1+\sqrt{1+4d_1^2/k_c})}{2d_1}$	$\sqrt{k_c}$	$\sqrt{k_c}$

6.3.2 Traditional Determination Method for $\mathbf{d}_3$

According to the basic principle of DC/DC converters, when the converter is at steady state, d_2 can be represented by d_1, as shown in Table 6.1.

Where $L_e = L_1L_2/(L_1 + L_2)$, denoting the equivalent inductance of the Cuk converter.

As shown in Table 6.1, using the DC analysis method of converters, the duty cycle d_2 can be described as the function of d_1. Then, d_3 can also be expressed as the function of d_1 as follows.

$$d_3 = 1 - d_1 - d_2 = h_2(d_1)$$

Similarly, d_3 can be expanded into the series form as formula (6.14), and the term d_{3i} is determined as follows.

$$\begin{aligned} & d_{30} = h_2(d_{10}), d_{31} = \left(\frac{\partial h_2}{\partial d_1}\bigg|_{d_1 = d_{10}} \right) \cdot d_{11} = H_{11}^2 \cdot d_{11}, \\ & d_{32} = \left(\frac{\partial h_2}{\partial d_1}\bigg|_{d_1 = d_{10} + d_{11}} \right) \cdot d_{12} = H_{21}^2 \cdot d_{12}, \\ & \cdots \end{aligned} \tag{6.18}$$

where H_{11}^2 and H_{21}^2 are real numbers. According to (6.18), δ_3 can also be expanded into series form as Eq. (6.16). Thus, both methods have the same solving process, except for applying different methods to get the main term and the small terms of the duty cycle d_3.

6.4 Solution of Time-Varying Equation for Closed-Loop DCM-Operated System

Utilizing the equivalent small-parameter method, the time-varying equation of the closed-loop system under DCM operation can be solved in the same way as that of the closed-loop system in CCM operation.

6.4.1 Solution of the Main Component

The main component of state variable is DC component, the following assumptions are made, i.e.,

$$\mathbf{x}_0 = \mathbf{a}_{00} \tag{6.19}$$

Then the nonlinear vectors are determined by

$$\begin{aligned} \mathbf{f}_0^{(1)} &= \delta_{10}\mathbf{x}_0 \\ \mathbf{f}_0^{(2)} &= \delta_{30}(\mathbf{x}_0 + \mathbf{e}_2) \end{aligned} \tag{6.20}$$

Then the main term and the remainder belonging to $\mathbf{f}_0^{(1)}$ can be obtained as

$$\begin{cases} \mathbf{f}_{0m}^{(1)} = d_{10}\mathbf{a}_{00} \\ \mathbf{R}_1^{(1)} = b_{10}\mathbf{a}_{00}e^{j\tau} + c.c \end{cases} \tag{6.21}$$

And those of $\mathbf{f}_0^{(2)}$ are shown as

$$\begin{cases} \mathbf{f}_{0m}^{(2)} = d_{30}(\mathbf{a}_{00} + \mathbf{e}_2) \\ \mathbf{R}_1^{(2)} = q_{10}(\mathbf{a}_{00} + \mathbf{e}_2)e^{j\tau} + c.c \end{cases} \tag{6.22}$$

Substituting $\mathbf{x}_0$, $\mathbf{f}_{0m}^{(1)}$, and $\mathbf{f}_{0m}^{(2)}$ into (6.7a) with the above formulas, the following matrix equation can be obtained, i.e.,

$$\mathbf{G_0}(0)\mathbf{a}_{00} + \mathbf{G}_1(0)d_{10}\mathbf{a}_{00} + \mathbf{G}_3(0)d_{30}(\mathbf{a}_{00} + \mathbf{e_2}) = \mathbf{u} \tag{6.23}$$

In which, d_{30} determined by Eq. (6.15) or (6.18). And according to (5.38a), the term d_0 satisfies the following equation as

$$d_0 = K_0 + \mathbf{K_1}\mathbf{x}_0 = K_0 + \mathbf{K}_1\mathbf{a}_{00} \tag{6.24}$$

As can be seen from the above discussion that the formulas for solving $\mathbf{a}_{00}$ and d_{10}, d_{30} are all nonlinear, requiring the application of symbol method or numerical method.

6.4.2 Solution of First-Order Correction

According to Eqs. (6.21) and (6.22), the first-order correction can be assumed by the following formula as

$$\mathbf{x}_1 = \mathbf{a}_{11}e^{j\tau} + \bar{\mathbf{a}}_{11}e^{-j\tau} \tag{6.25}$$

And the first-order corrections for nonlinear vectors can be deduced as

$$\begin{cases} \mathbf{f}_1^{(1)} = \delta_{11}\mathbf{x}_0 + \delta_{10}\mathbf{x}_1 \\ \mathbf{f}_1^{(2)} = \delta_{31}(\mathbf{x}_0 + \mathbf{e}_2) + \delta_{30}\mathbf{x}_1 \end{cases} \tag{6.26}$$

Then the main terms and the remainders belonging to $\mathbf{f}_1^{(1)}$ and $\mathbf{f}_1^{(2)}$ can be obtained as

$$\begin{cases} \mathbf{f}_{1m}^{(1)} = (d_{10}\mathbf{a}_{11} + b_{11}\mathbf{a}_{00}) \cdot e^{j\tau} + c.c \\ \mathbf{R}_2^{(1)} = (b_{10}\bar{\mathbf{a}}_{11} + \bar{b}_{10}\mathbf{a}_{11} + d_{11}\mathbf{a}_{00}) + (b_{10}\mathbf{a}_{11} + b_{20}\mathbf{a}_{00}) \cdot e^{j2\tau} + b_{30}\mathbf{a_{00}} \cdot e^{j3\tau} + c.c \end{cases}$$
$$\begin{cases} \mathbf{f}_{1m}^{(2)} = [d_{30}\mathbf{a}_{11} + q_{11}(\mathbf{a}_{00} + \mathbf{e_2})] \cdot e^{j\tau} + c.c \\ \mathbf{R}_2^{(2)} = [q_{10}\bar{\mathbf{a}}_{11} + \bar{q}_{10}\mathbf{a}_{11} + d_{31}(\mathbf{a}_{00} + e_2)] + [q_{10}\mathbf{a}_{11} + q_{20}(\mathbf{a}_{00} + \mathbf{e}_2)] \cdot e^{j2\tau} \\ \qquad + q_{30}(\mathbf{a}_{00} + e_2) \cdot e^{j3\tau} + c.c \end{cases} \tag{6.27}$$

Substituting $\mathbf{f}_{1m}^{(1)}$, $\mathbf{f}_{1m}^{(2)}$ and $\mathbf{R}_1^{(1)}$, $\mathbf{R}_1^{(2)}$ into Eq. (6.7b) would give the following matrix equation as

$$\begin{aligned} &[\mathbf{G}_0(j\omega) + \mathbf{G}_1(j\omega) \cdot d_{10} + \mathbf{G}_3(j\omega) \cdot d_{30}] \cdot \mathbf{a}_{11} \\ &\quad = -\mathbf{G}_1(j\omega) \cdot (b_{11} + b_{10}) \cdot \mathbf{a}_{00} - \mathbf{G}_3(j\omega)[(q_{11} + q_{10})(\mathbf{a}_{00} + \mathbf{e_2})] \end{aligned} \tag{6.28a}$$

According to Eqs. (5.32b) and (6.17), the following equations can be obtained.

$$b_{11} = d_{11} \cdot (\cos 2\pi d_{10} - j \sin 2\pi d_{10}) = d_{11} \cdot e^{-j\tau_{s0}} \quad (\tau_{s0} = 2\pi d_{10}) \tag{6.28b}$$

$$q_{11} = d_{31} \cdot (\cos 2\pi d_{30} + j \sin 2\pi d_{30}) = d_{31} \cdot e^{j\sigma_{s0}} \quad (\sigma_{s0} = 2\pi d_{30}) \tag{6.28c}$$

Since $\bar{\mathbf{x}}_1 = 0$, then according to Eqs. (6.15) or (6.18), one can get the following formula as

$$d_{31} = H_1 d_{11}$$

in which $H_1 = H_{11}^1$ or H_{11}^2.

And the term d_{11} can be obtained according to the Eq. (5.18b), that is

$$d_{11} = \mathbf{K}_1 \cdot (\mathbf{a}_{11}e^{j\tau_{s0}} + \bar{\mathbf{a}}_{11}e^{-j\tau_{s0}}) \tag{6.29a}$$

Then setting the following symbolic expressions as

$$\begin{aligned} \mathbf{G}_1 &= \mathbf{G}_0(j\omega) + \mathbf{G}_1(j\omega) \cdot d_{10} + \mathbf{G}_3(j\omega) \cdot d_{30} \\ \mathbf{A}_{12} &= -\mathbf{G}_1^{-1} \cdot \mathbf{G}_1(j\omega), \mathbf{A}_{13} = -\mathbf{G}_1^{-1} \cdot \mathbf{G}_3(j\omega) \\ \mathbf{B}_{10} &= \mathbf{A}_{12} \cdot b_{10} \cdot e^{j\tau_{s0}}, \mathbf{Q}_{10} = \mathbf{A}_{13} \cdot q_{10} \cdot e^{j\tau_{s0}} \\ F &= H_1 \cdot e^{j(\tau_{s0} + \sigma_{s0})} \\ B_1 &= K_1[(\mathbf{B}_{10} + \bar{\mathbf{B}}_{10})a_{00} + (\mathbf{Q}_{10} + \bar{\mathbf{Q}}_{10})(\mathbf{a}_{00} + \mathbf{e_2})] \\ A_1 &= 1 - \mathbf{K}_1[(\mathbf{A}_{12} + \bar{\mathbf{A}}_{12})\mathbf{a}_{00} + (\mathbf{A}_{13}F + \bar{\mathbf{A}}_{13}\bar{F})(\mathbf{a}_{00} + \mathbf{e_2})] \end{aligned} \tag{6.29b}$$

And considering Eqs. (6.28) and (6.29), the following equations for solving the first-order corrections can be obtained, i.e.,

$$\begin{aligned}
&d_{11} = B_1/A_1, \quad d_{31} = H_1 \cdot d_{11}, \\
&b_{11} = d_{11} \cdot e^{-j\tau_{s0}}, \quad q_{11} = d_{31} \cdot e^{j\sigma_{s0}}, \\
&a_{11} = -\mathbf{G}^{-1}(j\omega)[\mathbf{G}_1(j\omega)(b_{11} + b_{10})a_{00} + \mathbf{G}_3(j\omega)(q_{11} + q_{10})(\mathbf{a}_{00} + \mathbf{e}_2)]
\end{aligned} \tag{6.30}$$

6.4.3 Solution of Second-Order Correction

According to (6.27), the second-order correction can be assumed by the following formula as

$$\mathbf{x}_2 = \mathbf{a}_{02} + (\mathbf{a}_{22}e^{j2\tau} + \mathbf{a}_{32}e^{j3\tau} + c.c) \tag{6.31}$$

And the second-order corrections for nonlinear vectors can be deduced as

$$\begin{cases} \mathbf{f}_2^{(1)} = \delta_{12}\mathbf{x}_0 + \delta_{11}\mathbf{x}_1 + \delta_{10}\mathbf{x}_2 \\ \mathbf{f}_2^{(2)} = \delta_{32}(\mathbf{x}_0 + \mathbf{e}_2) + \delta_{31}\mathbf{x}_1 + \delta_{30}\mathbf{x}_2 \end{cases}$$

Then the main terms belonging to $\mathbf{f}_2^{(1)}$ and $\mathbf{f}_2^{(2)}$ can be obtained as

$$\begin{cases} \mathbf{f}_{2m}^{(1)} = (d_{10}\mathbf{a}_{02} + b_{11}\bar{\mathbf{a}}_{11} + \bar{b}_{11}\mathbf{a}_{11} + d_{12}\mathbf{a}_{00}) + (d_{10}\mathbf{a}_{22} + b_{11}\mathbf{a}_{11} + b_{30}\bar{\mathbf{a}}_{11} + b_{21}\mathbf{a}_{00})e^{j2\tau} \\ \qquad + (d_0\mathbf{a}_{32} + b_{10}\mathbf{a}_{22} + b_{20}\mathbf{a}_{11} + b_{31}\mathbf{a}_{00})e^{j3\tau} + c.c \\ \mathbf{f}_{2m}^{(2)} = [d_{30}\mathbf{a}_{02} + q_{11}\bar{\mathbf{a}}_{11} + \bar{q}_{11}\mathbf{a}_{11} + d_{32}(\mathbf{a}_{00} + \mathbf{e}_2)] + [d_{30}\mathbf{a}_{22} + q_{11}\mathbf{a}_{11} \\ \qquad + q_{30}\bar{\mathbf{a}}_{11} + q_{21}(\mathbf{a}_{00} + \mathbf{e}_2)]e^{j2\tau} + [d_{30}\mathbf{a}_{32} + q_{10}\mathbf{a}_{22} + q_{20}\mathbf{a}_{11} + q_{31}(\mathbf{a}_{00} + \mathbf{e}_2)]e^{j3\tau} + c.c \end{cases}$$

It should be noted that the remainders of $\mathbf{f}_2^{(1)}$ and $\mathbf{f}_2^{(2)}$ are no longer listed here, as they are used to solve the third-order correction $\mathbf{x}_3$. However, since the ESPM is adopted, it usually only needs to be iterated three times, i.e., only the terms $\mathbf{x}_0$, $\mathbf{x}_1$ and $\mathbf{x}_2$ are solved, and the obtained steady-state periodic solutions of the state variables already have sufficient accuracy.

Then substituting $\mathbf{f}_{2m}^{(1)}, \mathbf{f}_{2m}^{(2)}, \mathbf{R}_2^{(1)}, \mathbf{R}_2^{(2)}$ in Eq. (6.7c) with the above formulas, we can get equations for solving the second-order correction, that is,

$$a_{22} = -\mathbf{G}_2^{-1} \cdot (\mathbf{B}_{22} + \mathbf{Q}_{22}) \tag{6.32}$$

where the symbols are defined as

$$\begin{aligned}
\mathbf{G}_2 &= \mathbf{G}_0(j2\omega) + \mathbf{G}_1(j2\omega)\cdot d_{10} + \mathbf{G}_3(j2\omega)\cdot d_{30} \\
\mathbf{B}_{22} &= \mathbf{G}_1(j2\omega)\cdot[(b_{10}+b_{11})\mathbf{a}_{11} + (b_{20}+b_{21})\mathbf{a}_{00} + b_{30}\bar{\mathbf{a}}_{11}] \\
\mathbf{Q}_{22} &= \mathbf{G}_3(j2\omega)\cdot[(q_{10}+q_{11})\mathbf{a}_{11} + (q_{20}+q_{21})(\mathbf{a}_{00}+\mathbf{e}_2) + q_{30}\bar{\mathbf{a}}_{11}]
\end{aligned}$$

And

$$\mathbf{a}_{32} = -\mathbf{G}_3^{-1}\cdot(\mathbf{B}_{32}+\mathbf{Q}_{32}) \tag{6.33}$$

In which the symbols are defined as

$$\begin{aligned}
\mathbf{G}_3 &= \mathbf{G}_0(j3\omega) + \mathbf{G}_1(j3\omega)\cdot d_{10} + \mathbf{G}_3(j3\omega)\cdot d_{30} \\
\mathbf{B}_{32} &= \mathbf{G}_1(j3\omega)\cdot[b_{10}\mathbf{a}_{22} + b_{20}\mathbf{a}_{11} + (b_{30}+b_{31})\mathbf{a}_{00}] \\
\mathbf{Q}_{32} &= \mathbf{G}_3(j3\omega)\cdot[q_{10}\mathbf{a}_{22} + q_{20}\mathbf{a}_{11} + (q_{30}+q_{31})(\mathbf{a}_{00}+\mathbf{e}_2)]
\end{aligned}$$

As we can see, that the formulas for solving $\mathbf{a}_{22}$ and $\mathbf{a}_{32}$ are linear.

And the DC component $\mathbf{a}_{02}$ in the second-order correction can be solved by the following equation as

$$\mathbf{G}_0\cdot\mathbf{a}_{02} + \mathbf{G}_1(0)\cdot\mathbf{a}_{00}\cdot d_{12} + \mathbf{G}_3(0)\cdot(\mathbf{a}_{00}+\mathbf{e}_2)\cdot d_{32} = -(\mathbf{B}_{02}+\mathbf{Q}_{02}) \tag{6.34}$$

In which the symbols are defined as

$$\begin{aligned}
\mathbf{G}_0 &= \mathbf{G}_0(0) + \mathbf{G}_1(0)\cdot d_{10} + \mathbf{G}_3(0)\cdot d_{30} \\
\mathbf{B}_{02} &= \mathbf{G}_1(0)\cdot[(b_{10}+b_{11})\bar{\mathbf{a}}_{11} + (\bar{b}_{10}+\bar{b}_{11})\mathbf{a}_{11} + d_{11}\mathbf{a}_{00}] \\
\mathbf{Q}_{02} &= \mathbf{G}_3(0)\cdot[(q_{10}+q_{11})\bar{\mathbf{a}}_{11} + (\bar{q}_{10}+\bar{q}_{11})\mathbf{a}_{11} + d_{31}(\mathbf{a}_{00}+\mathbf{e}_2)]
\end{aligned}$$

And the term d_{12} in Eq. (6.34) can be determined by (5.38c). We assume that:

$$\begin{aligned}
&\mathbf{P}_{11} = \mathbf{a}_{11}e^{j\tau_{s0}}, \mathbf{P}_{22} = \mathbf{a}_{22}e^{j\tau_{s1}}, \mathbf{P}_{32} = \mathbf{a}_{32}e^{j\tau_{s1}}, \quad (\tau_{s0} = 2\pi d_{10}, \tau_{s1} = 2\pi(d_{10}+d_{11})) \\
&M = -\mathbf{K}_1\cdot 4\pi d_{11}\mathrm{Im}(\mathbf{P}_{11}) + 2\mathbf{K}_1[\mathrm{Re}(\mathbf{P}_{22}) + \mathrm{Re}(\mathbf{P}_{32})]
\end{aligned}$$

Among them, Im(.) and Re(.) respectively represent the imaginary part and the real part of the complex number. Then according to (5.38c), the term d_{12} can be solved by the equation as follows.

$$d_{12} = M + \mathbf{K}_1\cdot\mathbf{a}_{02} \tag{6.35}$$

And according to (6.15), the term d_{32} can be obtained from the following equation as

$$d_{32} = H_2 \cdot d_{12} + \mathbf{N}_2 \cdot \mathbf{a}_{02} \tag{6.36}$$

In which there would be $H_2 = H_{21}^1$ and $N_2 = H_{22}^1$ for the first method, or $H_2 = H_{21}^2$ and $N_2 = 0$ for the second method of determining d_3.

Then considering equations from (6.34) to (6.36), one can get the following equation for solving the term $\mathbf{a}_{02}$, as shown in Eq. (6.37).

$$\mathbf{a}_{02} = \mathbf{B}_2 / \mathbf{A}_2 \tag{6.37}$$

In which the coefficients are determined by

$$\begin{aligned}
\mathbf{A}_2 &= \mathbf{G}_0 + \mathbf{G}_1(0) \cdot \mathbf{a}_{00} \cdot \mathbf{K}_1 + \mathbf{G}_3(0) \cdot (\mathbf{a}_{00} + \mathbf{e}_2) \cdot (H_2 \cdot \mathbf{K}_1 + N_2) \\
\mathbf{B}_2 &= \mathbf{B}_{02} + \mathbf{Q}_{02} - \mathbf{G}_1(0) \cdot a_{00} \cdot M - \mathbf{G}_3(0)(\mathbf{a}_{00} + \mathbf{e}_2) \cdot H_2 \cdot M
\end{aligned}$$

6.5 Analysis Example

Figure 6.1 shows the basic circuit of a Boost regulator along with the typical waveform of inductor current when operating at DCM. Then this circuit will be used as an example to illustrate the process of solving the steady-state periodic solution by ESPM.

The parameters of the main circuit are chosen as E = 16 V, L = 208 uH, C = 222 uF, R = 12.5 Ω; and parameters of the control circuit are chosen as V_R = 2.5 V, R_1 = 18 kΩ, R_2 = 2 kΩ, R_3 = 0.5 kΩ, R_4 = 500 kΩ, C_f = 2 uF; the amplitude and frequency of the sawtooth wave are set respectively as V_p = 7.5 V and $f_s = 2.5$ kHz, $(T = 1/f_s)$. And state differential equation in matrix form of the system is the same as (6.1), in which the coefficient matrices are determined according to the topologies and parameters of the converter system, i.e.,

$$\mathbf{G}_0(p) = \begin{bmatrix} p & \frac{1}{L} & 0 \\ -\frac{1}{C} & p + \frac{1}{RC} & 0 \\ 0 & -\frac{g_2}{R_3 C_f} & p + \frac{1}{R_4 C_f} \end{bmatrix}, \quad \mathbf{G}_1(p) = \mathbf{G}_3(p) = \begin{bmatrix} 0 & -\frac{1}{L} & 0 \\ -\frac{1}{C} & 0 & 0 \\ 0 & 0 & 0 \end{bmatrix}$$

The state variable vector, nonlinear function vectors, the constant vector, and the input vector are determined by

$$\mathbf{x} = \begin{bmatrix} i_L & v_c & v_f \end{bmatrix}^{Tr}, \quad \mathbf{f}^{(1)}(\mathbf{x}) = \delta_1 \mathbf{x}, \quad \mathbf{f}^{(2)}(\mathbf{x}) = \delta_3(\mathbf{x} + \mathbf{e}_2),$$

$$\mathbf{e}_2 = \begin{bmatrix} 0 & -E & 0 \end{bmatrix}^{Tr} \quad \text{and} \quad \mathbf{u} = \begin{bmatrix} \frac{E}{L} & 0 & \frac{-V_r g_1}{R_3 C_f} \end{bmatrix}^{Tr}$$

And the coefficients g, g_1 and g_2 are got from:

$$g = \frac{1}{R_1} + \frac{1}{R_2} + \frac{1}{R_3}, \quad g_1 = 1 - \frac{1}{R_3 \cdot g}, \quad g_2 = \frac{1}{R_1 \cdot g}$$

The duty cycle d_1 is determined by the Eq. (6.6), in which

$$K_0 = V_r/V_p \quad \text{and} \quad \mathbf{K}_1 = \begin{bmatrix} 0 & 0 & -1/V_p \end{bmatrix}$$

When using the first method to determine the term d_3, the following equation can be obtained.

$$d_3 = 1 - d_1 - (\mathbf{K}_{11}\bar{\mathbf{x}} - md_1^2)/(md_1) \tag{6.38}$$

In which the coefficients are got by

$$K_{11} = \begin{bmatrix} 1 & 0 & 0 \end{bmatrix}, \quad m = (E \cdot T)/(2L)$$

And when using the second method, the term d_3 can be obtained according to Table 6.1.

Then when using the first method of determining d_3, the steady-state periodic solutions to the state variables in the main circuit can be deduced as shown in (6.39), in which the DC components can be got as I_{dc} = 3.2793 A, and V_{dc} = 25.0182 V.

$$\begin{aligned} i_L &= 3.2793 - 2.2452\cos\tau + 3.4186\sin\tau - 0.6579\cos 2\tau - 0.2409\sin 2\tau \\ &\quad - 0.0078\cos 3\tau - 0.3045\sin 3\tau \\ v_c &= 25.0182 - 0.3962\cos\tau - 1.0736\sin\tau + 0.2067\cos 2\tau + 0.0931\sin 2\tau \\ &\quad - 0.0318\cos 3\tau + 0.0835\sin 3\tau \end{aligned} \tag{6.39}$$

And the duty cycle d_1 and d_3 can be solved as d_1 = 0.2819 and d_3 = 0.2438 respectively. Comparison of the state variable ripples got from the ESPM (dashed line) and the numerical simulations (solid line) is shown in Fig. 6.2a, b, in which the waveform obtained from the ESPM is the calculation of the first three harmonics of state variables in the main circuit during one steady-state switching cycle.

When the second method of determining d_3 is adopted, the steady-state periodic solutions to the state variables can be deduced as shown in (6.40), in which the DC components can be got as I_{dc} = 3.0266 A, and V_{dc} = 25.0196 V.

$$\begin{aligned} i_L &= 3.0266 - 2.1535\cos\tau + 3.4044\sin\tau - 0.6708\cos 2\tau - 0.2228\sin 2\tau \\ &\quad - 0.0337\cos 3\tau - 0.3177\sin 3\tau \\ v_c &= 25.0196 - 0.4144\cos\tau - 1.0576\sin\tau + 0.2091\cos 2\tau + 0.0799\sin 2\tau \\ &\quad - 0.0242\cos 3\tau + 0.0868\sin 3\tau \end{aligned} \tag{6.40}$$

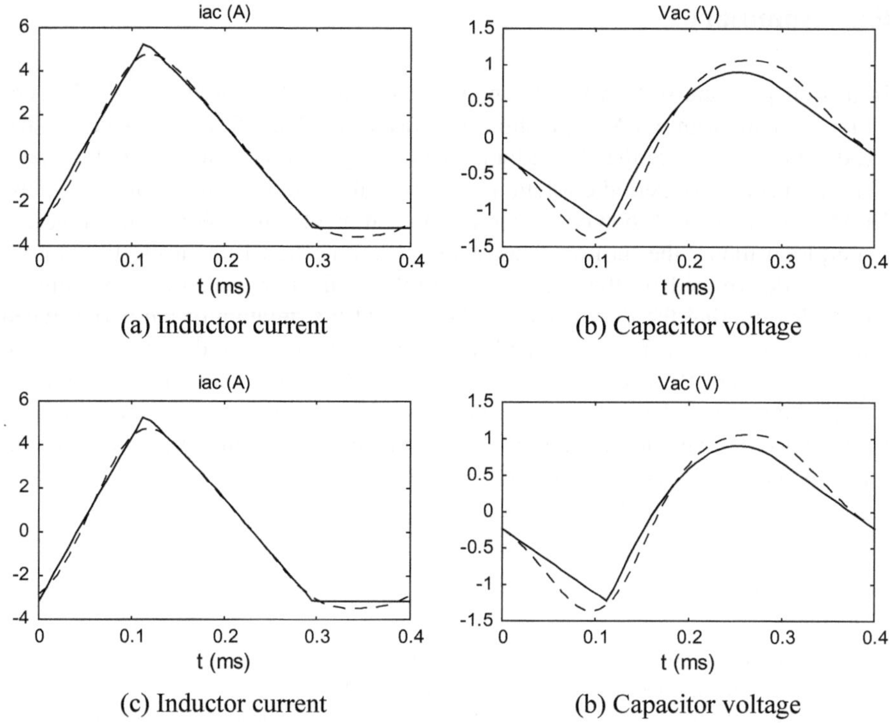

Fig. 6.2 Ripple comparison between the ESPM (dashed line) and numerical simulation (solid line)

Under this situation, the duty cycle d_1 and d_3 can be solved as $d_1 = 0.2776$ and $d_3 = 0.2475$ respectively. And similarly, comparison of the state variable ripples got from the ESPM (dashed line) and the numerical simulations (solid line) is shown in Fig. 6.2c, d, in which the waveform obtained from the ESPM is the calculation of the first three harmonics of state variables in the main circuit during one steady-state switching cycle.

The DC components got from numerical simulation are $I_{dc} = 3.151$ A and $V_{dc} = 24.939$ V, which are consistent with the results of the above two methods, i.e., $I_{dc} = 3.2793$ A, and $V_{dc} = 25.0182$ V for method1, and $I_{dc} = 3.0266$ A, and $V_{dc} = 25.0196$ V for method 2. The ripples and the DC components from both of the ESP methods are close to each other, and they are also in good agreement with the results from numerical simulation. But the first ESP method, deriving from the waveform of the state variable, is of more generality and easy for algorithm improvement, that is, when considering the rising or falling slope of the inductor current, the influence of the harmonic component can be taken into account at the cost of that duty ratio d_2 or d_3 cannot be explicitly expressed, which increases the computational complexity.

6.6 Summary

In this chapter, the ESP symbolic analysis method of the closed-loop PWM converter system with CCM operation proposed in Chap. 5 is extended to the steady-state analysis of the closed-loop converter system with DCM operation. And thus the steady-state periodic solution in analytic form of the state variables with the DCM-operated closed-loop converter system can be obtained. When operating in a closed-loop mode, the turn-on time of the switching transistor in the PWM converter is determined by the feedback control circuit. For converters operating at DCM, determining the duty cycle d_3, the ratio of the duration of the period when inductor current remains at zero and the entire switching period, is the key to the Equivalent-Small-Parameter Analysis method. With the help of waveform of inductor current and the DC analysis at steady-state, this chapter presents two methods for determining the term d_3. The simulation shows that the results of both methods are consistent and have high accuracy.

References

1. Sun J, Mitchell DM, Grenel MF et al (1998) Modeling of PWM converters in discontinuous conduction mode a reexamination. APEC Rec 615–622
2. Sebastian J, Cobos JA et al (1995) The determination of the boundaries between continuous and discontinuous conduction modes in PWM DC-to-DC converters used as power factor preregulators. IEEE Trans PE 10(5):574–582
3. Wang J, Dunford WD, Mauch K (1997) Analysis of a ripple-free input-current boost converter with discontinuous conduction characteristics. IEEE Trans PE 12(4):684–694
4. Lee YS, Wang SJ, Hui SYR (1997) Modeling, analysis, and application of buck converters in discontinuous-input-voltage mode operation. IEEE Trans PE 12(2):350–360
5. Maksimovic D, Cuk S (1991) A unified analysis of PWM converters in discontinuous modes. IEEE Trans PE 6(3):476–490
6. Cuk S, Middlebrook RD (1977) A general unified approach to modelling switching DC-to-DC converters in discontinuous conduction mode. IEEE PESC Rec 36–57
7. van Dijk E, Spruijt HJN, O'Sullivan DM, et al (1995) PWM-switch modeling of DC/DC converters. IEEE Trans PE 10(6):659–664
8. Tymerski R, Vorperian V, Lee FC et al (1988) Nonlinear modelling of the PWM switch. IEEE PESC Rec 968–976
9. Ma K-W, Lee Y-S (1991) A novel uninterruptible DC/DC converter for UPS applications. IEEE Trans Ind Appl 28(4):808–815
10. Kim H, Seo G, Cho B, Choi H (2015) A simple average current control with on-time doubler for multiphase CCM PFC converter. IEEE Trans Power Electron 30(3):1683–1693
11. Madigan M, Erickson R Ismail E (1992) Integrated high quality rectifier-regulators. IEEE PESC Rec 1043–1051
12. Srinivasan R, Oruganti R (1996) Single-phase parallel power processing scheme with power factor control. Int J Electronics 80(2):291–306
13. Khan A, Batarseh I (1995) Zero-voltage switching boost converter for power factor correction. Int J Electronics 78(6):1177–1188
14. Kim Y-S, Sung W-Y, Lee B-K (2014) Comparative performance analysis of high density and efficiency PFC topologies. IEEE Trans Power Electron 29(6):2666–2679

15. Simonetti DSL, Viera JLF, Sousa GCD (1999) Modeling of the high-power-factor discontinuous boost rectifiers. IEEE Trans Ind Electron 46:788–795. ISSN 0278-0046
16. Ridley RB (1989) Averaged small-signal analysis of the boost power factor correction circuit. In: Proceedings of VPEC seminar, pp 108–120
17. Sanders SR et al (1986) Non linear control laws for switching power converters. In Proceedings of 25th IEEE conference on decision and control, pp 46–53
18. Mahabir K et al (1990) Linear averaged and sampled data models for large signal control of power factor AC-DC converters. IEEE PESC Rec 372–381

Chapter 7
Analysis of Current-Mode Controlled PWM DC/DC Converters Based on ESPM

7.1 Introduction

As a single-loop control system, voltage-mode control usually has only one voltage feedback loop [1]. While the current-mode control [2–6], a dual-loop control system, contains both negative current feedback loop and voltage feedback control. The basic DC/DC converter (Buck/Boost/Buck-boost) is a second-order system with two state variables, i.e., the output filter capacitor voltage and the inductor current. Therefore, from the perspective of control theory, the double closed-loop system is the most optimal control strategy.

Compared with the voltage-mode control technology, the current-mode control technology has the advantages of fast dynamic response, automatic overcurrent protection, being suitable for parallel operation of multiple converters [4–6], etc., and has been widely put into use. It is shown in practice that the output ripple of the current-mode controlled converter is usually about 25 mV, which makes it far superior to similar products adopting the voltage-mode control. However, due to the double closed-loop (voltage loop and current loop) feedback and the impossibility of determining duty cycle explicitly, current-mode control increases the difficulty of modeling and analysis [5, 6]. At present, the analysis of this kind of converter is basically based on the average modeling method, assuming that inductor current changes linearly and ripple is small [5–11]. However, the averaged circuit by applying this kind of method, no longer contains the ripple information of the original circuit and fails to predict the stability of the original system [10]. Recently, symbolic analysis has always been a research topic of power electronics. Compared with numerical simulation [12–15], symbolic analysis helps to investigate circuit behavior and produce an analytical model for circuit optimization and control design. In this chapter, the equivalent-small-parameter method introduced in the previous chapters is extended to the steady-state analysis of the closed-loop current-mode controlled converter system. The analytical solutions of the state variables of the closed-loop converter system, expressed with symbols, are obtained

Y. Chen and B. Zhang, *Equivalent-Small-Parameter Analysis of DC/DC Switched-Mode Converter*, CPSS Power Electronics Series,
https://doi.org/10.1007/978-981-13-2574-8_7

containing DC component and ripple. And this kind of application make it is easy to understand the working mechanism of the circuit, and can provide guidance for engineering design, such as selection of filter capacitor and inductance parameters [16, 17].

The following sections of this chapter are arranged as follows. Section 7.2 introduces the basic principles of current-mode control. And Sect. 7.3 proposes the symbolic analysis method, i.e. equivalent-small-parameter analysis method for the closed-loop current-mode controlled converter. Section 7.4 presents The analysis cases are presented in Sect. 7.4 and finally the summary of this analytical methods in this chapter is given in Sect. 7.5.

7.2 The Basic Principle of Constant Frequency Current-Mode Control

The diagram of constant frequency current-mode control scheme is showed as Fig. 7.1, where the module CSP (current signal processor) is used to detect the value of the inductor current or the current of the switching devices, which may be a transient value or an average value. In Fig. 7.1 the symbol v_p denotes a ramp compensation signal; the error amplifier (EA) processes the difference between the output voltage and the reference voltage and obtains a control signal v_k, forming a voltage feedback outer loop to realize automatic voltage adjustment; the current feedback control constitutes the inner loop to achieve automatic current regulation. The voltage loop and current loop constitute the double closed-loop feedback controlled system.

In traditional duty-ratio programmed (see Fig. 1.1) DC/DC converter, an analog comparator compares the control signal to a ramp signal, whose period and rising

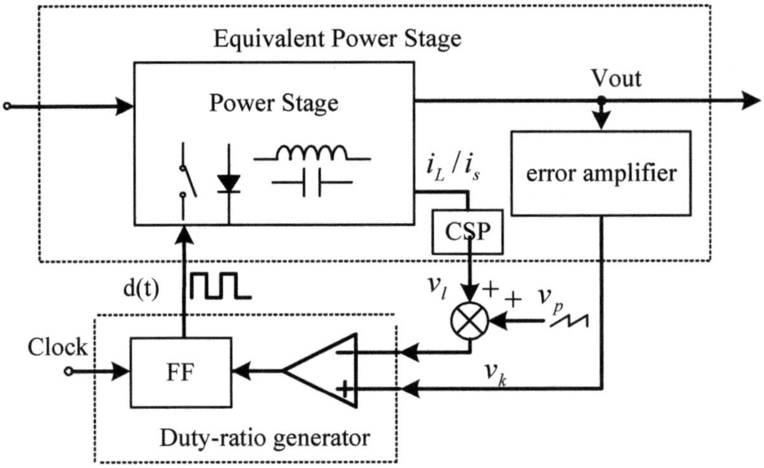

Fig. 7.1 Basic diagram of current-mode controlled system

slope is constant. When the value of the ramp signal rises to the control voltage, the comparator flips and the state of the switch changes. The ramp signal is independent of the state variables in power stage. Therefore, the duty cycle is only determined by the control signal. When the control signal is independent of the output variable of the circuit, the open-loop control of the system can be realized. In a switching power converter with constant frequency current-mode control, the conduction of the switch is controlled by a clock signal with a constant frequency, and the turn-off is determined by the moment at which the inductor current signal reaches the control signal. Thus, the duty cycle is not only related to the control signal, but also to the initial value of the inductor current and the rising slope. Therefore, the current-mode control itself introduces a feedback loop [18].

The basic principle of current mode control [5] will be illustrated via an example of constant frequency peak current feedback control. Figure 7.2 shows typical waveforms of the duty cycle and control signal without (Fig. 7.2a) and with (Fig. 7.2b) slope compensation respectively. In the figure, m_1, m_2, m_c represent the rising and falling slope of the inductor current and the slope of the ramp compensation signal respectively. In the peak-current-mode control, the detected current value $v_l = M * i_L$, where M is the gain of the CSP.

Consider the case without slope compensation shown as Fig. 7.2a. Let T denote the switching period. When $t = \mathrm{n}T$, the clock signal sets the trigger (output high level), turning on the main switch of the circuit in the power stage, and thus the inductor current rises linearly. And at the moment of $t = \mathrm{n}T + \mathrm{d}T$, i.e., $v_l(\mathrm{t}) = v_k(\mathrm{t})$, the clock signal resets the trigger (output low level). Consequently, the main switch of circuit in the power stage is turned off and the inductor current drops. At the moment of $t = (\mathrm{n} + 1)T$, the next clock signal arrives and the switch is turned on again. It is pointed out in [5] that the constant frequency control has potential instability (sub-harmonic oscillation), as shown as the dotted line in Fig. 7.3. Assume that, in steady state, the rising and falling slope of the inductor current are m_1 and m_2 respectively, and the converter operates in the continuous inductor current mode. Then if the inductor current has a disturbance ΔI_0 at the beginning of a switching cycle, there will be a disturbance ΔI_1 at the end of the cycle. From Fig. 7.3, it can be seen that $\Delta I_1 = -\left(\frac{d}{1-d}\right)\Delta I_0$. Thus, after n cycles, the disturbance

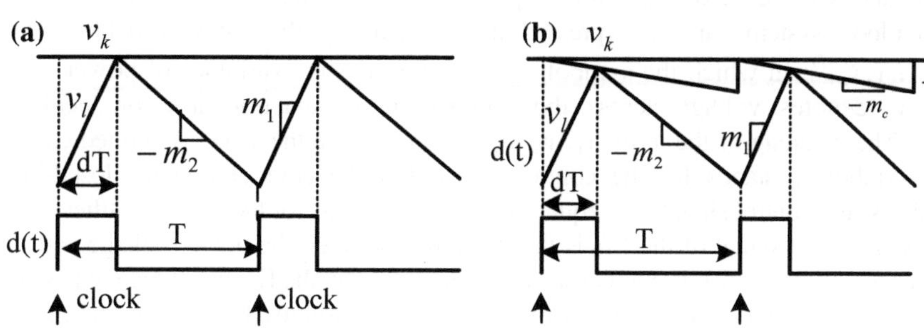

Fig. 7.2 Typical waveform of current-mode controlled system

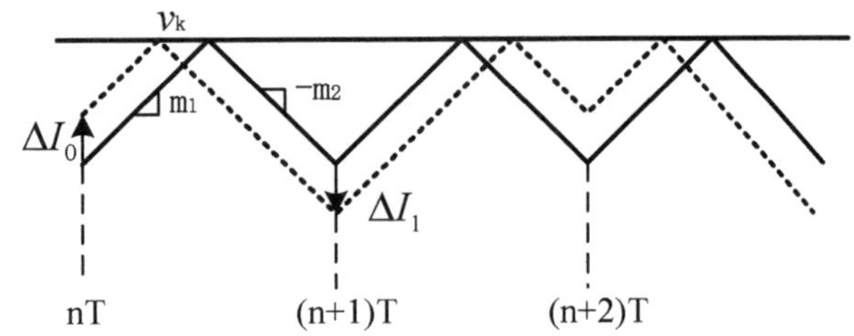

Fig. 7.3 Waveform of sub-harmonic oscillation

at the nth cycle would be $\Delta I_n = -\left(\frac{d}{1-d}\right)^n \Delta I_0$. Obliviously, the error would gradually increases when the duty cycle $d > 0.5$, which means that the system is unstable. When the duty cycle $d = 0.5$, a typical sub-harmonic oscillation occurs, as shown in Fig. 7.3.

The solution to eliminate sub-harmonic oscillation is to artificially integrate a ramp function, called ramp compensation, into the detected current ripple of the switching converter. As shown in Fig. 7.2b, let the term $-m_c$ denotes the falling slope of the function, then the disturbance would be $\Delta I_n = -\left(\frac{m_2 - m_c}{m_1 + m_c}\right)^n \Delta I_0$. It is obviously that ΔI_n decreases gradually for any duty cycle. Therefore, after several cycles, the disturbance in the inductor current will become negligible and the system remains stable. Particularly, if $m_2 = m_c$, the disturbance in the inductor current will disappear after one cycle, which not only ensures the stability of the inner loop but also provides the fastest dynamic response for the current-mode control. Therefore, proper selection of the slope of the ramp compensation function can appropriately improve the dynamic response characteristics of the system [5, 18–20].

7.3 Symbolic Analysis of Closed-Loop Current-Mode Controlled Converter System

Figure 7.1 can be simplified as shown in Fig. 7.4a, and the control law shown in Fig. 7.2b is adopted in Fig. 7.1, which is redrawn as Fig. 7.4b. Combining the power stage and voltage feedback error amplifiers in the system as a whole, then the closed-loop system can be expressed as two parts, with a state variable vector $\mathbf{x} = [i, v, v_a]^{Tr}$, in which the symbols i and v are the state variables of power stage and v_a denotes voltage across the compensation capacitor in error feedback loop. The meaning of the terms v_p, v_l, m_1, m_2 and m_c are the same as defined earlier. The symbol v_k stands for the control signal. For the case of double closed-loop feedback, the control signal v_k is equal to the output voltage of the error amplifier (EA), where the input signal of the EA is the difference between the output voltage and the reference voltage, which is compensated and amplified by the EA by taking some kinds of adjustment method (such as proportional-integral algorithm). In this case, the control signal is related to the output voltage, thus introducing a voltage feedback loop. And

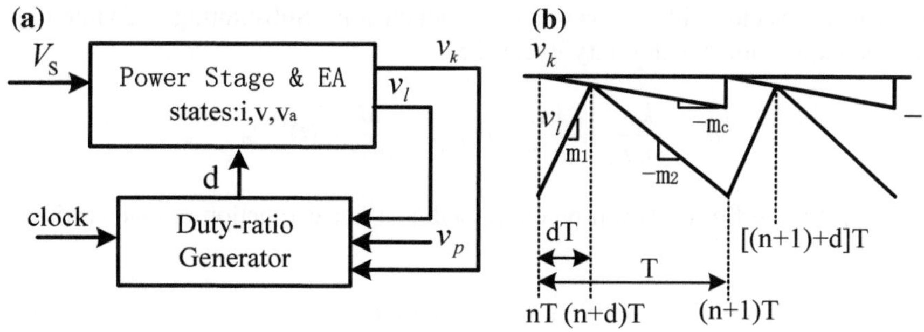

Fig. 7.4 Simplified block diagram of current-controlled system

when v_k is a constant, i.e. $v_k = I_{ref}$, there is only a current feedback loop in the system, which is called single-loop current-mode control, or open-loop current-mode control. In this case, the term I_{ref} represents a reference current signal, and the control signal v_k is independent of the output voltage.

In the simplified symbolic analysis below, three assumptions, which are also the basic assumptions made by the state-space averaging method in fact, would be proposed as follows.

(i) The waveform of inductor current is linear;
(ii) The control voltage changes slowly during one switching cycle, i.e. $v_k(t) = \bar{v}_k$;
(iii) The ripple of output voltage is small compared to the steady-state value.

7.3.1 Expression of the Duty Cycle for Closed-Loop CMC System

7.3.1.1 Accurate Solution of Duty Cycle

According to the waveform in Fig. 7.4b, the following equation can be deduced as

$$v_k(dT) = v_l(dT) + m_c dT \tag{7.1}$$

where $v_l(t) = M \cdot i_L$, M is the gain of CSP. According to the discussion in Chap. 5, the control signal $v_k(t)$ generally can be expressed as a linear function of the whole system state variables, i.e.

$$v_k(t) = k_0 + \mathbf{k}_1 \cdot \mathbf{x} \tag{7.2}$$

where $k_0, \mathbf{k}_1$ are constants associated with the circuit parameters, here the term k_0 is a scalar, and $\mathbf{k}_1$ is a row vector. If $\mathbf{x}$ is an column vector with $n \times 1$ dimensions, then $\mathbf{k}_1$ is row vector with the size of $1 \times n$ dimensions. Substituting (7.2) into (7.1) gives the expression for the duty cycle, i.e.,

$$d = \frac{k_0}{m_c T} + \frac{\mathbf{k}_1}{m_c T}\mathbf{x}(t)\big|_{t=dT} - \frac{M}{m_c T} i_L(t)\big|_{t=dT} \tag{7.3}$$

Obviously, the Eq. (7.3) can be expressed as a linear function of state variables as

$$d = K_0 + K_1 \cdot \mathbf{x}(dT) \tag{7.4}$$

The coefficients $K_0, \mathbf{K}_1$ are determined according to the specific circuit. The duty cycle is expressed in the same way as the voltage-mode-controlled PWM converter. Thus, the analysis method for duty cycle, is also the same. The detailed solution can be found in Chap. 5 or the example below.

According to the analysis before, this modeling method doesn't require any assumption of the original system. Yet, most analysis methods for current-mode controlled converter depend on the geometric relationship of the inductor current waveform to establish the equation with duty cycle. Assumptions of linear waveform and small ripple are needed. Therefore, the proposed symbolic analysis method for closed-loop current-mode controlled converter system is more general.

What's more, when the feedback doesn't contain the ramp compensation, the duty cycle cannot be expressed as an explicit function using this method, which is the limitation of this method. Then the following simplified method can be adopted. Since the ramp compensation is usually applied, the method is applicable for the actual circuit.

7.3.1.2 Simplified Solution of Duty Cycle

When the three assumptions (1), (2) and (3) are satisfied, the duty cycle can be obtained by adopting the following simplified method. According to the waveform in Fig. 7.4b, the following relationship can be deduced.

$$\bar{v}_l = \bar{v}_k - \frac{1}{2} m_1 dT - m_c dT \tag{7.5}$$

where $\bar{v}_l = M \cdot \bar{i}_L$ is the average value of inductor current in a switching period. Equation (7.5) is the mathematical model of the duty cycle for most current-mode controlled converters. With the assumptions of linear waveform and small ripple, the term m_1 can generally be expressed as a function of the input voltage and the average value of voltage across filter capacitor in power stage. Assumed that the input voltage V_s remains constant, m_1 can be expressed as

$$m_1 = g'(\bar{v}_o) \tag{7.6}$$

Substituting (7.2) and (7.6) into (7.5), one can get the expression of the duty cycle as shown in Eq. (7.7), which is a function of mean values of the state variables.

$$d = \frac{k_0 + \mathbf{k}_1\bar{\mathbf{x}}(dT) - M \cdot \bar{i}_L(dT)}{(\frac{1}{2}m_1 + m_c) \cdot T} = g(\bar{\mathbf{x}}) \tag{7.7}$$

When the equivalent-small-parameter method is put into use, due to the correction of the DC component by the high-order terms, $\bar{\mathbf{x}}$ may contains both the zero-order component and the DC component of the higher-order correction term. According to the analysis in Chap. 5, suppose that the steady-state periodic solution $\mathbf{x}$ is an n-dimensional vector, i.e., $\mathbf{x} = [x^{(1)}, x^{(2)}, \ldots x^{(n)}]^{Tr}$, it can be expanded into series as follows.

$$\mathbf{x} = \mathbf{x}_0 + \varepsilon\mathbf{x}_1 + \varepsilon^2\mathbf{x}_2 + \cdots \tag{7.8}$$

And then the average state variable vector can also expanded as

$$\bar{\mathbf{x}} = \bar{\mathbf{x}}_0 + \varepsilon\bar{\mathbf{x}}_1 + \varepsilon^2\bar{\mathbf{x}}_2 + \cdots \tag{7.9}$$

where $\mathbf{x}_i$ is the ith-order small parameter of $\mathbf{x}$, $|\mathbf{x}_i| \ll \mathbf{x}_0$. Similarly, the duty cycle can be expanded into

$$d = d_0 + \varepsilon d_1 + \varepsilon^2 d_2 + \cdots \tag{7.10}$$

Then combining with (7.7) leads to

$$d_0 + \varepsilon d_1 + \varepsilon^2 d_2 + \cdots = g(\bar{\mathbf{x}}_0 + \varepsilon\bar{\mathbf{x}}_1 + \varepsilon^2\bar{\mathbf{x}}_2 + \cdots) \tag{7.11}$$

Because $|\bar{\mathbf{x}}_i| \ll \mathbf{x}_0$, then according to the linear approximation of the Tailor Series expansion, the increment of d caused by $\bar{\mathbf{x}}_i$ can be approximated as

$$\Delta d = d_i = \left(\left.\frac{\partial g}{\partial \bar{x}_i{}^{(1)}}\right|_{\bar{\mathbf{x}}=\bar{\mathbf{x}}_0 + \varepsilon\bar{\mathbf{x}}_1 + \cdots + \varepsilon^{i-1}\bar{\mathbf{x}}_{i-1}} \right) \cdot \bar{x}_i^{(1)} + \cdots + \left(\left.\frac{\partial g}{\partial \bar{x}_i{}^{(n)}}\right|_{\bar{\mathbf{x}}=\bar{\mathbf{x}}_0 + \varepsilon\bar{\mathbf{x}}_1 + \cdots + \varepsilon^{i-1}\bar{\mathbf{x}}_{i-1}} \right) \cdot \bar{x}_i^{(n)} \tag{7.12}$$

Thus as can be seen from (7.12), d_1 is the increment caused by the DC component in the first-order correction, and d_2 is the increment caused by the DC component in the second-order correction, and so on.

7.3.2 Solution of the Closed-Loop Equation of the CMC Converter

Similar to the voltage-mode controlled converter, the state equation of the current-mode controlled system can still be expressed as

$$G_0(p)\mathbf{x} + G_1(p)\mathbf{f} = \mathbf{u} \tag{7.13}$$

where the nonlinear function

$$f = \delta(t)(\mathbf{x} + \mathbf{e}) \tag{7.14}$$

The meaning of each item in (7.13) and (7.14) is the same as that of in the voltage-mode controlled converter, where $\mathbf{x}$ is the state variable of the entire closed-loop system. The switching function $\delta(t) = 1(0)$ represent the on/off state of the main switch. If the switch has two states during one cycle, then the switching function can be defined as

$$\delta(t) = \begin{cases} 1 & 0 < t \leq d(t)T \\ 0 & d(t)T < t \leq T \end{cases} \tag{7.15}$$

where T is the switch period, $d(t)$ is the duty cycle, decided by (7.3) or (7.7) in the current-mode controlled converters.

Regardless of whether the duty cycle is an exact solution or a simplified solution, the duty cycle can be expressed as the sum of the main term and the small corrections in both cases, so the steady-state solution process of the current-mode controlled converter is exactly the same as that of the voltage-mode controlled converter. The only difference is that the solution of the duty cycle is different.

In order to facilitate the analysis, equivalent-small-parameter method is adopted to obtain the first three components (zero-order, first-order, second-order) of the steady-state periodic solution of the closed-loop system, as shown in (7.16). And for simplicity, in the analysis of the closed-loop current-mode controlled system, the constant vector is set to be zero, i.e., $\mathbf{e} = 0$ in (7.14) is taken into account.

$$\begin{cases} G_0(0)\mathbf{a}_{00} + G_1(0)d_0\mathbf{a}_{00} = \mathbf{u} & \text{(a)} \\ [G_0(j\omega) + G_1(j\omega)\cdot d_0]\mathbf{a}_{11} = -G_1(j\omega)[(b_{11} + b_{10})\mathbf{a}_{00}] & \text{(b)} \\ [G_0(j2\omega) + G_1(j2\omega)d_0]\mathbf{a}_{22} = -G_1(j2\omega)[(b_{21} + b_{20})\mathbf{a}_{00} + (b_{11} + b_{10})\mathbf{a}_{11} + b_{30}\bar{\mathbf{a}}_{11}] & \text{(c)} \\ [G_0(j3\omega) + G_1(j3\omega)d_0]\mathbf{a}_{32} = -G_1(j3\omega)[(b_{31} + b_{30})\mathbf{a}_{00} + b_{10}\mathbf{a}_{22} + b_{20}\mathbf{a}_{11}] & \text{(d)} \\ [G_0(0) + G_1(0)d_0]\mathbf{a}_{02} = -G_1(0)[(b_{11} + b_{10})\bar{\mathbf{a}}_{11} + (\bar{b}_{11} + \bar{b}_{10})\mathbf{a}_{11} + (d_1 + d_2)\mathbf{a}_{00}] & \text{(e)} \end{cases} \tag{7.16}$$

In (7.16) the coefficient b_{mi} can be determined by the following equations as

$$
\begin{aligned}
b_{mi} &= \frac{1}{2}(\alpha_{mi} - j\beta_{mi}) \quad (m = 1, 2, \ldots, i = 0, 1, 2, \ldots) \\
\alpha_{m0} &= \frac{\sin 2m\pi d_0}{m\pi}, \quad \beta_{m0} = \frac{1 - \cos 2m\pi d_0}{m\pi} \\
\alpha_{mi} &\approx 2d_i \cos 2m\pi(d_0 + \varepsilon d_1 + \cdots + \varepsilon^{i-1} d_{i-1}) \qquad (i = 1, 2, \ldots) \\
\beta_{mi} &\approx 2d_i \sin 2m\pi(d_0 + \varepsilon d_1 + \cdots + \varepsilon^{i-1} d_{i-1}) \qquad (i = 1, 2, \ldots)
\end{aligned} \tag{7.17}
$$

where $d_0, d_1, d_2 \cdots d_i$ represent the main component, the first- and second-order,…, and the ith-order corrections of the duty cycle respectively. They can be approximately obtained according to the specific duty cycle equation.

In the following content, the application of ESPM in analysis of closed-loop current-mode controlled converter system will be illustrated with examples.

7.4 Examples

7.4.1 *Double-Loop Current-Mode Controlled Boost Converter*

The typical closed-loop current-mode controlled Boost converter is shown in Fig. 7.5, and the parameters of the PCS (power circuit stage) the FN (feedback network) are set respectively as listed in Table 7.1, the switching cycle $T = 1/f_s$.

According to the principle of the Boost converter with CCM operation, one can get the state differential equation describing the system as shown in (7.13), where the state variable vector is chosen as $\mathbf{x} = [i_L, v_o, v_a]^{Tr}$, and the square coefficient matrices are determined by

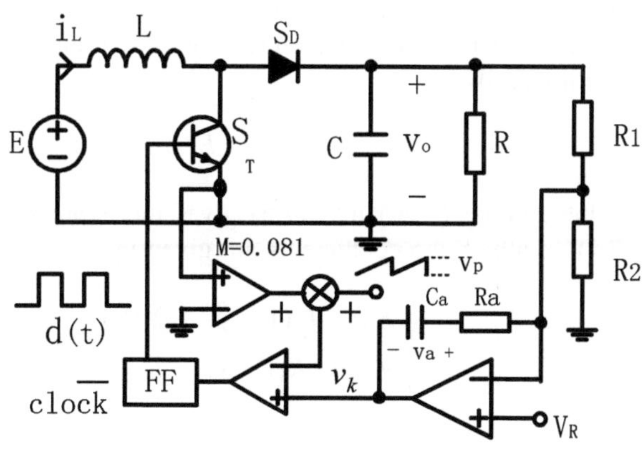

Fig. 7.5 Closed-loop current-mode controlled Boost regulator

Table 7.1 Circuit parameters of the dual-loop CMC-Boost converter

Parameters	Values	Parameters	Values
Input voltage E	28 V	Reference voltage V_R	1.8 V
Switching frequency f_s	25 kHz	Resistance R_1	47.5 kΩ
Inductance L	195 μH	Resistance R_2	2.5 kΩ
Capacitance C	2000 μF	Resistance R_a	72.2 kΩ
Load resistance R	11.2 Ω	Feedback capacitance C_a	0.23 μF
Current coefficient M	0.081	Peak ramp voltage V_p	0.25 V

$$G_1(p) = \begin{bmatrix} p & \frac{1}{L} & 0 \\ -\frac{1}{C} & p+\frac{1}{RC} & 0 \\ 0 & -\frac{1}{R_1 C_a} & p \end{bmatrix} \text{ and } G_2(p) = \begin{bmatrix} 0 & -\frac{1}{L} & 0 \\ \frac{1}{C} & 0 & 0 \\ 0 & 0 & 0 \end{bmatrix} \tag{7.18a}$$

The nonlinear vector function **f** and the input vector **u** are given by

$$f = \delta \mathbf{x} \text{ and } \mathbf{u} = \begin{bmatrix} \frac{E}{L} & 0 & \frac{-1}{C_a}\left(\frac{1}{R_1}+\frac{1}{R_2}\right)V_R \end{bmatrix}^{Tr} \tag{7.18b}$$

The control signal $v_k(t)$ can be derived as

$$v_k(t) = \left(1 + \frac{R_a}{R_1} + \frac{R_a}{R_2}\right)V_R - \frac{R_a}{R_1}v_0 - v_a \tag{7.19}$$

7.4.1.1 Symbolic Analysis with Simplified Solution of Duty Cycle

Due to the smoothing effect of the error amplifier, it can be considered that the control signal changes little during one cycle, that is, the assumption (ii) can be satisfied. In the simplified solution, the duty cycle can be deduced by Eq. (7.5), where the rising slope of the inductor current and the slope of the ramp signal are determined by

$$m_1 = E/L, \quad m_c = v_p/T \tag{7.20}$$

Introducing (7.20) into (7.5) gives the expression of duty cycle as shown in (7.4), in which the coefficients are given by

$$\begin{cases} K_0 = \frac{1}{F_m}\left(1+\frac{R_a}{R_1}+\frac{R_a}{R_2}\right)V_R \\ \mathbf{K}_1 = \begin{bmatrix} K_{11} & K_{12} & K_{13} \end{bmatrix} = \begin{bmatrix} \frac{-M}{F_m} & \frac{-R_a}{R_1 F_m} & \frac{-1}{F_m} \end{bmatrix} \end{cases} \tag{7.21}$$

Here the symbol $F_m = (\frac{1}{2}m_1 + m_c)T$.

According to above discussion, the duty cycle d can be expanded into the series shown in (7.10). Then based on the Eq. (7.12), the following equations for determining d_0 and d_i (i = 1, 2....) can be can be obtained as

$$\begin{cases} d_0 = K_0 + \mathbf{K}_1\bar{\mathbf{x}}_0 \\ d_1 = \mathbf{K}_1\bar{\mathbf{x}}_1 \\ d_2 = \mathbf{K}_1\bar{\mathbf{x}}_2 \\ \cdots \end{cases} \tag{7.22}$$

where $\bar{\mathbf{x}}_i$ is the average value of the ith correction of $\mathbf{x}$ during one switching cycle.

(1) **Solution of the Main Term**

Based on the principle of the ESPM, the main term (DC component) can be assumed as

$$\mathbf{x}_0 = \mathbf{a}_{00} = \begin{bmatrix} I_{00} & V_{00} & V_{a00} \end{bmatrix}^{Tr} \tag{7.23}$$

According to (7.16a) and (7.22), the following equation for solutions of $\mathbf{a}_{00}$ and d_0 can be obtained.

$$G_0(0)\mathbf{a}_{00} + G_1(0)d_0\mathbf{a}_{00} = \mathbf{u} \tag{7.24a}$$

$$d_0 = K_0 + \mathbf{K}_1\mathbf{a}_{00} \tag{7.24b}$$

Then, the steady-state solution can be deduced from (7.24a) and (7.24b), as shown follows.

$$\begin{aligned} &V_{00} = \left(1 + \frac{R_1}{R_2}\right)V_R,\ d_0 = 1 - \frac{E}{V_{00}},\quad I_{00} = \frac{V_{00}}{R(1-d_0)} \\ &V_{a00} = F_m K_0 - \frac{R_a}{R_1}V_{00} - M \cdot I_{00} \end{aligned} \tag{7.25}$$

When the circuit parameters are introduced, the values of $\mathbf{a}_{00}$ and d_0 are given by

$$a_{00} = \begin{bmatrix} 4.1327 & 36 & 0.8817 \end{bmatrix}^{Tr}, d_0 = 0.2222. \tag{7.26}$$

(2) **Solution of First-order Correction**

The first-order correction term can be assumed as

$$\mathbf{x}_1 = \mathbf{a}_{11}e^{j\tau} + c.c,\ \mathbf{a}_{11} = \begin{bmatrix} I_{11} & V_{11} & V_{a11} \end{bmatrix}^{Tr} \tag{7.27}$$

Since the first-order correction only contains the fundamental, its average value during one switching period is zero, and the duty cycle increment caused by it is also zero, i.e., $\bar{\mathbf{x}}_1 = 0$ and $d_1 = 0$. Then according to (7.17), the coefficient $b_{m1} = 0$

(m = 1, 2, 3…) can be determined, and thus the solution of $\mathbf{a}_{11}$ can be derived by the equation as shown in (7.28).

$$[G_0(j\omega) + G_1(j\omega) \cdot d_0] \cdot \mathbf{a}_{11} = -G_1(j\omega) \cdot b_{10} \cdot \mathbf{a}_{00} \tag{7.28}$$

By solving the above matrix equation, the symbolic expressions for each element of the column vector $\mathbf{a}_{11}$ can be obtained as

$$\begin{aligned} V_{11} &= \frac{b_{10} \cdot [j\omega L \cdot I_{00} - (1 - d_0) \cdot V_{00}]}{\omega^2 LC - (1 - d_0)^2 - j\omega L/R} \\ I_{11} &= \frac{b_{10} \cdot V_{00} - (1 - d_0) \cdot V_{11}}{j\omega L} \\ V_{a11} &= \frac{V_{11}}{j\omega R_1 C_a} \end{aligned} \tag{7.29}$$

(3) **Solution of Second-order Correction**

Based on the ESPM, the second-order correction term can be assumed as

$$\mathbf{x}_2 = \mathbf{a}_{02} + \mathbf{a}_{22} e^{j2\tau} + \mathbf{a}_{32} e^{j3\tau} + c.c \tag{7.30}$$

Since the second-order correction contains the DC component, the second- and the third-order harmonics, its average value during one switching period should be the DC values, i.e., $\bar{\mathbf{x}}_2 = \mathbf{a}_{02}$, and then the duty cycle increment caused by it would be $d_2 = \mathbf{K}_1 \mathbf{a}_{02}$. Thus according to (7.16) and (7.22), the equations for the solution of the second-order correction can be obtained as

$$[G_0(j2\omega) + G_1(j2\omega) d_0] \mathbf{a}_{22} = -G_1(j2\omega)[b_{20}\mathbf{a}_{00} + b_{10}\mathbf{a}_{11} + b_{30}\bar{\mathbf{a}}_{11}] \tag{7.31a}$$

$$[G_0(j3\omega) + G_1(j3\omega) d_0] \mathbf{a}_{32} = -G_1(j3\omega)[b_{30}\mathbf{a}_{00} + b_{10}\mathbf{a}_{22} + b_{20}\mathbf{a}_{11}] \tag{7.31b}$$

$$[G_0(0) + G_1(0) d_0 + G_1(0) a_{00} \cdot \mathbf{K}_1] \mathbf{a}_{02} = -G_1(0)[b_{10}\bar{\mathbf{a}}_{11} + \bar{b}_{10}\mathbf{a}_{11}] \tag{7.31c}$$

where the coefficient $b_{m1} = 0$ (m = 1, 2, 3…) can be determined by (7.17), and thus the solutions of $\mathbf{a}_{22}$, $\mathbf{a}_{32}$ and $\mathbf{a}_{02}$ can be derived by the above equations shown in (7.31).

It is worth noting that all the equations in (7.31) are linear. Then by solving the equations in (7.31) one by one, the symbolic expressions for each element of the column vectors $\mathbf{a}_{22}$, $\mathbf{a}_{32}$ and $\mathbf{a}_{02}$ can be obtained as

$$\begin{aligned} &V_{22} \approx \frac{j2\omega L \cdot \Delta_{I22} - (1-d_0) \cdot \Delta_{V22}}{4\omega^2 LC - (1-d_0)^2} && I_{22} = \frac{\Delta_{V22} - (1-d_0) \cdot V_{22}}{j2\omega L} && V_{a22} = \frac{V_{22}}{j2\omega R_1 C_a} \\ &V_{32} \approx \frac{j3\omega L \cdot \Delta_{I32} - (1-d_0) \cdot \Delta_{V32}}{9\omega^2 LC - (1-d_0)^2} && I_{32} = \frac{\Delta_{V32} - (1-d_0) \cdot V_{32}}{j3\omega L} && V_{a32} = \frac{V_{32}}{j3\omega R_1 C_a} \\ &V_{02} \approx 0 && I_{02} \approx \frac{\Delta_{V02} + \Delta_{I02}}{1-d_0} && V_{a32} = \frac{\Delta_{V02} - K_{11} I_{02}}{K_{13}} \end{aligned} \tag{7.32}$$

where

$$\begin{aligned} \Delta_{I22} &= b_{20}I_{00} + b_{10}I_{11} + b_{30}\bar{I}_{11} & \Delta_{V22} &= b_{20}V_{00} + b_{10}V_{11} + b_{30}\bar{V}_{11} \\ \Delta_{I32} &= b_{30}I_{00} + b_{20}I_{11} + b_{10}I_{22} & \Delta_{V32} &= b_{30}V_{00} + b_{20}V_{11} + b_{10}V_{22} \\ \Delta_{I02} &= b_{10}\bar{I}_{11} + \bar{b}_{10}I_{11} & \Delta_{V02} &= b_{10}\bar{V}_{11} + \bar{b}_{10}V_{11} \end{aligned}$$

When the circuit parameters are introduced into (7.25), (7.29) and (7.32), the steady-state periodic solution (DC component + first three harmonics) of the state variable in the main circuit can be finally obtained as

$$\begin{aligned} i_L &= 4.1327 - 0.3093\cos\tau + 0.3684\sin\tau - 0.1814\cos 2\tau + 0.0320\sin 2\tau \\ &\quad - 0.0624\cos 3\tau + 0.0005\sin 3\tau \\ v_o &= 36.0000 + 0.0025\cos\tau - 0.0049\sin\tau + 0.0020\cos 2\tau - 0.0005\sin 2\tau \\ &\quad + 0.0007\cos 3\tau + 0.0005\sin 3\tau \end{aligned} \tag{7.33}$$

where $\tau = \omega t, \omega = 2\pi f_s$.

Comparison between the DC components from (7.33), which is obtained based on the symbolic ESPM with simplified duty cycle solution, and that of numerical simulation and Pspice simulation are shown in Table 7.2.

And the comparison of the state variable ripples resulted from ESPM (dashed line) and Pspice simulation (solid line) would be shown in Fig. 7.6e, f, where the dashed lines from ESPM are the calculation of the sum of the first three harmonic components in the steady-state period solutions shown in (7.32).

7.4.1.2 Symbolic Analysis with Accurate Solution of Duty Cycle

The time-varying differential equations of closed-loop system are still expressed as (7.13) and (7.18). And the duty cycle can be expressed as the linear function of the state vector as shown in (7.4), and rewritten as $d = K_0 + \mathbf{K}_1 \cdot \mathbf{x}(dT)$, where the coefficients can be determined by the following equations according to the Eq. (7.3), i.e.,

$$\begin{aligned} K_0 &= \frac{1}{V_p}\left(1 + \frac{R_a}{R_1} + \frac{R_a}{R_2}\right)V_R \\ \mathbf{K}_1 &= [\,K_{11} \quad K_{12} \quad K_{13}\,] = \left[\frac{-M}{V_p} \quad \frac{-R_a}{R_1 V_p} \quad \frac{-1}{V_p}\right] \end{aligned} \tag{7.34}$$

where V_p is the peak-to-peak value of ramp compensation signal.

Table 7.2 Comparison of DC component of the CCM CMC-Boost resulted from ESPM, numeric and Pspice5 simulation

Variables	Symbolic	Numerical	Pspice
Inductor current	$I_{dc} = 4.1327$	$I_{dc} = 4.1982$	$I_{dc} = 4.3052$
Output voltage	$V_{dc} = 36.0000$	$V_{dc} = 36.2386$	$V_{dc} = 36.2237$
Duty cycle	$d = 0.222229$	$d = 0.222229$	$d = 0.2429$

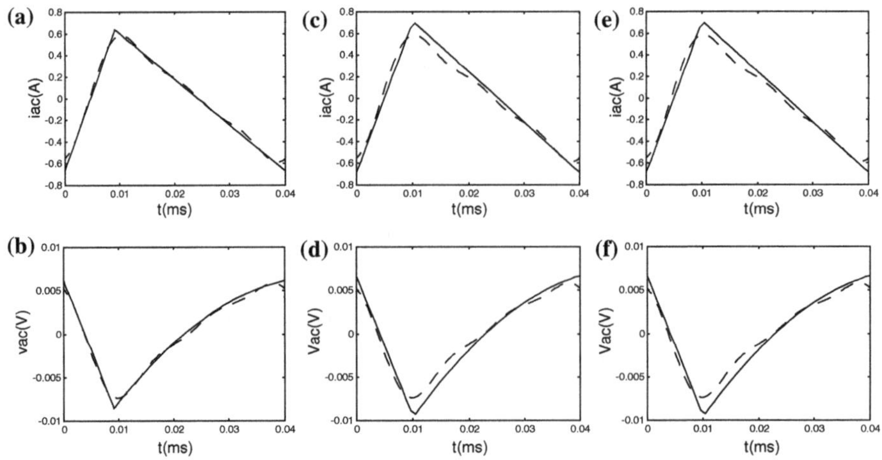

Fig. 7.6 Ripple comparisons of the CCM CMC-Boost: **a** and **b** for the ESPM with accurate duty cycle and the numerical simulation, **c** and **d** for the same ESPM and Pspice, **e** and **f** for the ESPM with simplified duty cycle and Pspice

As can be seen from (7.22), the zero-order component d_0 of duty cycle is determined by the average value of inductor current. But actually the duty cycle is determined by the peak value of inductor current according to the peak current-mode control law. It means that when the accurate solution of duty cycle method is adopted, more components are shifted to the high-order equation to be solved. This will slow down the convergence speed of the ESP algorithm, resulting in the need to iterate more times when using ESPM to obtain a steady-state periodic solution with sufficient accuracy. In order to improve this situation, the improved double iteration algorithm mentioned in Chap. 5 is adopted. Then steady-state periodic solution, including DC component and the first three harmonics, in main circuit, can be finally obtained as

$$\begin{aligned} i_L &= 4.1326 - 0.3093\cos\tau + 0.3684\sin\tau - 0.1814\cos 2\tau + 0.0320\sin 2\tau \\ &\quad - 0.0623\cos 3\tau + 0.0005\sin 3\tau \\ v_o &= 35.9999 + 0.0025\cos\tau - 0.0049\sin\tau + 0.0020\cos 2\tau - 0.0005\sin 2\tau \\ &\quad + 0.0007os3\tau + 0.0005\sin 3\tau \end{aligned} \tag{7.35}$$

Compare (7.33) and (7.35), it can be found that the AC components are almost the same except a small difference in DC component. The main term of duty cycle obtained in this case is $d_0 = 0.2222226$, which is very close to those of accurate solution and numerical simulation, as shown in Table 7.2.

As for the ripples of the main circuit state variables during a steady-state period, the comparisons among different methods are shown in Fig. 7.6, where the solid line is for the numerical simulation or the Pspice simulation, and the dashed line is for the ESPM, which is the calculation of the sum of the first three harmonic components in the steady-state period solutions shown in (7.33) or (7.35). Figure 7.6a, c and e show the ripple waveform of inductor current, and the Fig. 7.6b, d and f show the ripple waveform of the output voltage. Figure 7.6a, b show the comparison between the ESPM algorithm under accurate solution of duty cycle and the numerical simulation, Fig. 7.6c, d show the comparison between the same ESPM algorithm and the Pspice simulation. And the comparison of the ESPM algorithm under simplified solution of duty cycle and the Pspice simulation is shown in Fig. 7.6e, f. It can be seen from all the figures that the results from the ESPM and those from simulations agree quite well with each other, regardless of whether the duty cycle is an exact solution or a simplified solution.

Therefore, even with a simplified solution for the duty cycle, a steady-state periodic solution with sufficient accuracy can be obtained using the ESPM. Since the assumptions (i)–(iii) mentioned in Sect. 7.3 can be satisfied generally in the actual circuit, the simplified method is often applied.

7.4.2 *Single-Loop Current-Mode Controlled Buck Converter*

The typical single-loop current-mode controlled Buck converter is shown in Fig. 7.7, and the parameters of the PCS (power circuit stage) the FN (feedback network) are set respectively as follows, i.e., $E = 24$ V, $L = 200$ μH, $C = 47$ μF and $R = 2\ \Omega$ for the PCS; the reference current $I_{ref} = 6$ A for the FN, and the switching frequency is chosen as $f_s = 10$ kHz, and the switching cycle $T = 1/f_s$.

According to the principle of the Buck converter with CCM operation, one can get the square state differential equation describing the system as shown in (7.36),

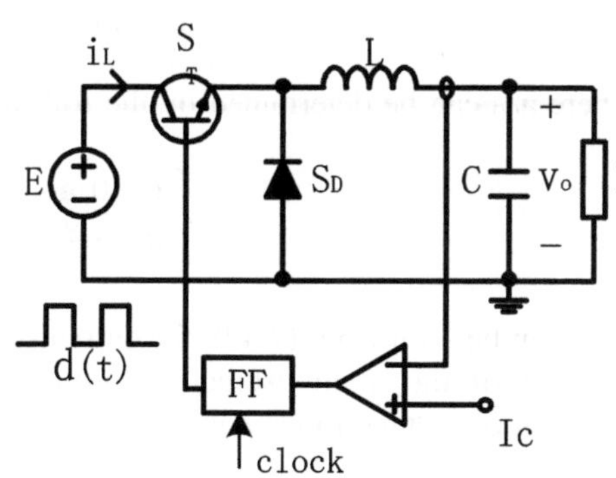

Fig. 7.7 Current-mode controlled Buck regulator

where the state variable vector is chosen as $\mathbf{x} = [i_L, v_o]^{Tr}$, and the square coefficient matrix is determined by (7.37)

$$G_0(p)\mathbf{x} + \mathbf{f} = 0 \quad (7.36)$$

$$G_0(p) = \begin{bmatrix} p & \frac{1}{L} \\ -\frac{1}{C} & p + \frac{1}{RC} \end{bmatrix} \quad (7.37)$$

The nonlinear vector function $\mathbf{f}$ and the input vector $\mathbf{u}$ are given by

$$\mathbf{f} = \delta\mathbf{u} \text{ and } \mathbf{u} = \begin{bmatrix} -\frac{E}{L} & 0 \end{bmatrix}^{Tr} \quad (7.38)$$

The solution of duty cycle is determined using simplified method mentioned in Sect. 7.4.1.1. According to Fig. 7.7, the control signal is a constant, i.e., $v_k = I_{ref.}$ The duty cycle can be obtained by (7.5) as

$$d = \frac{I_{ref} - \bar{i}_l}{m_1 T/2} \quad (7.39)$$

where the rising slope of the inductor current $m_1 = \frac{E - \bar{v}_o}{L}$, then the duty cycle in (7.39) can be expressed as the function of the mean value of state variable, i.e.

$$d = \frac{2Lf_s(I_c - \bar{i}_L)}{E - \bar{v}_0} = g(\bar{i}_L, \bar{v}_o) \quad (7.40)$$

According to equations from (7.10) to (7.12), the duty cycle in (7.40) can also be expanded into the sum of main item and small corrections. Thus, the state Eq. (7.36) can be solve by the same method with ESPM, the main term equation and each correction are shown as follows.

(1) **Solution of the Main term**

Based on the principle of the ESPM, the main term (DC component) can be assumed as

$$\mathbf{x}_0 = \mathbf{a}_{00} = \begin{bmatrix} I_{00} & V_{00} \end{bmatrix}^{Tr} \quad (7.41)$$

where $\mathbf{a}_{00}$ can be determined by the following equations as

$$\begin{cases} G_1(0)\mathbf{a}_{00} + d_0\mathbf{u} = 0 \\ d_0 = \frac{2Lf_s(I_{ref} - I_{00})}{(E - V_{00})} \end{cases} \quad (7.42)$$

It can be seen from (7.42), for current-mode controlled Buck converter, that the solution of the main oscillation component is nonlinear (the main oscillation equation of voltage-mode controlled Buck converter is linear). Equation (7.42) can be solved by the symbolic method or numerical method. It may have multiple solutions, so it is necessary to determine which solution is true according to the situation of the actual converter. In this example, the solution can be obtained by numerical method as

$$d_0 = 0.3542, I_{00} = 4.251, V_{00} = 8.5020.$$

(2) **Solution of the First-order Term**

The first-order correction term can be still assumed as

$$\mathbf{x}_1 = \mathbf{a}_{11}e^{j\tau} + c.c \tag{7.43}$$

Since the first-order correction only contains the fundamental, its average value during one switching period is zero, and the duty cycle increment caused by it is also zero, i.e.,$\bar{\mathbf{x}}_1 = 0$ and $d_1 = 0$. Then according to (7.17), the coefficient $b_{m1} = 0$ (m = 1, 2, 3…) can be determined, and thus the solution of $\mathbf{a}_{11}$ can be derived by the equation as shown in (7.44).

$$G_0(j\omega)\mathbf{a}_{11} = -b_{10} \cdot \mathbf{u} \tag{7.44}$$

Obviously, the solution of $\mathbf{a}_{11}$ is linear. By solving the above matrix equation, the symbolic expressions for each element of the column vector $\mathbf{a}_{11}$ can be obtained as

$$\begin{aligned} V_{11} &= \frac{b_{10} \cdot E}{1 + j\omega L/R - \omega^2 LC} \\ I_{11} &= \left(j\omega C + \frac{1}{R}\right) V_{11} \end{aligned} \tag{7.45}$$

(3) **Solution of the Second-order Term**

Based on the ESPM, the second-order correction term for the Buck converter can be assumed as

$$\mathbf{x}_2 = \mathbf{a}_{22}e^{j2\tau} + \mathbf{a}_{32}e^{j3\tau} + c.c \tag{7.46}$$

It should be noted that, unlike the Boost converter, the second-order correction $\mathbf{x}_2$ of the Buck converter does not contain the DC component, thus its average value during one switching period is zero, and the duty cycle increment caused by $\mathbf{x}_2$ is also zero, i.e.,$\bar{\mathbf{x}}_2 = 0$ and $d_2 = 0$. Then the solutions of $\mathbf{a}_{22}$ and $\mathbf{a}_{32}$ can be derived by the equation as shown in (7.47).

Table 7.3 Comparison of DC component of the CMC-Buck resulted from ESPM and Pspice5

Variables	Symbolic	Pspice
Inductor current	I_{dc} = 4.1250 A	I_{dc} = 4.1272 A
Output voltage	V_{dc} = 8.5020 V	V_{dc} = 8.3388 V
Duty cycle	d_0 = 0.3542	d_0 = 0.3475

$$G_0(j2\omega)\mathbf{a}_{22} = -b_{20}\mathbf{u} \tag{7.47a}$$

$$G_0(j3\omega)\mathbf{a}_{32} = -b_{30}\mathbf{u} \tag{7.47b}$$

As can be seen in (7.47) that both of the equations are linear, and the coefficients b_{10}, b_{20} and b_{30} can be determined by Eq. (7.17). Then by solving the equations in (7.47) one by one, the symbolic expressions for each element of the column vectors $\mathbf{a}_{22}$ and $\mathbf{a}_{32}$ can be obtained as

$$\begin{cases} V_{22} = \frac{b_{20} \cdot E}{1 + j2\omega L/R - 4\omega^2 LC} \\ I_{22} = (j2\omega C + \frac{1}{R})V_{22} \end{cases} \text{and} \begin{cases} V_{32} = \frac{b_{30} \cdot E}{1 + j3\omega L/R - 9\omega^2 LC} \\ I_{32} = (j3\omega C + \frac{1}{R})V_{32} \end{cases} \tag{7.48}$$

Then the steady-state periodic solution of the state variables, including DC component and the first three harmonics, can be finally obtained as

$$\begin{aligned} i_L &= 4.2510 - 0.6915\cos\tau - 0.4923\cos 2\tau - 0.2684\cos 3\tau \\ v_o &= 8.5020 - 0.4083\cos\tau - 0.0478\cos 2\tau - 0.0013\cos 3\tau \end{aligned} \tag{7.49}$$

where $\tau = \omega t, \omega = 2\pi f_s$.

The DC solutions obtained by the ESPM and Pspice simulation are listed in Table 7.3. Obviously, the difference of the DC values from the two methods is quite small.

As for the ripples of the main circuit state variables during a steady-state period, the comparisons between the two methods are shown in Fig. 7.8, where the solid line is from the Pspice simulation, and the dashed line is from the ESPM, which is the calculation of the sum of the first three harmonic components in the steady-state period solutions shown in (7.49). It can be seen from Fig. 7.8 that the results from the ESPM show good accordance with those from simulations.

7.5 Steady-State Analysis of CMC-Boost in DCM Operation

Discontinuous current mode is a common operating modes in DC/DC converter. When applying traditional State-space Averaging method to analysis PWM converter in DCM, the reduced order model, which do not contain inductor current,

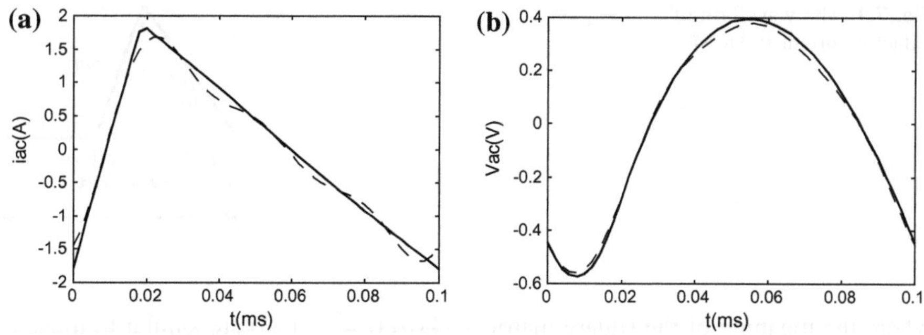

Fig. 7.8 Comparison of ripple of the CMC-Buck between the ESPM and Pspice simulation, **a** inductor current, **b** output voltage

will be obtained. But in some applications, such as PFC (Power Factor Correction) circuit, the inductor current is the ultimate control target, the disappearance of inductor current in the model is not desirable. In the previous section, the ESPM is applied to analyze the current-mode controlled CCM-operated Boost converter. In this section, the analysis method for voltage-mode controlled DCM-operated converters is further extended to the current-mode-controlled converters with DCM operation. Also the obtained steady-state solutions of DC component and ripples of state variables of this type of converter can be expressed in symbolic form.

When using the equivalent-small-parameters method, the analysis for DCM-operated converters of current-mode-control is the same as that of voltage-mode control. The only difference is that the duty ratio is determined differently. Therefore, the related analysis formulas in Chap. 6 will be directly referred to below, and for the convenience of description, some of them are re-listed as follows.

7.5.1 Description of the CMC Converter in DCM

The waveform of inductor current in DCM is shown in Fig. 7.9. According to discussion in Chap. 6, PWM converter operating in DCM can be described with a time-varying differential equation shown in (6.1), here we rewrite it as.

$$\mathbf{G}_0(p)\mathbf{x} + \mathbf{G}_1(p) \cdot \mathbf{f}^{(1)}(\mathbf{x}) + \mathbf{G}_3(p) \cdot \mathbf{f}^{(2)}(\mathbf{x}) = \mathbf{u} \tag{7.50}$$

And the nonlinear vector functions are defined as

$$\mathbf{f}^{(1)} = \delta_1(\mathbf{x} + \mathbf{e}^{(1)}) \tag{7.51}$$

$$\mathbf{f}^{(2)} = \delta_3(\mathbf{x} + \mathbf{e}^{(2)}) \tag{7.52}$$

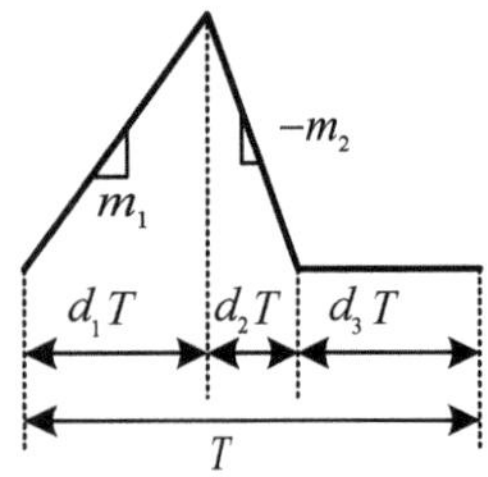

Fig. 7.9 The waveform of inductor current in DCM

where the meaning of the square matrices $\mathbf{G}_i(p)$ (i = 0, 1, 3) are similar to those in Chap. 6, and $\mathbf{e}^{(1)}$, $\mathbf{e}^{(2)}$ are constant vectors. Switching functions δ_1, δ_3 are defined as:

$$\delta_1(t) = \begin{cases} 1 & 0 < t \le d_1(t)T \\ 0 & d_1(t)T < t \le T \end{cases} \tag{7.53}$$

$$\delta_3(t) = \begin{cases} 0 & 0 < t \le [1 - d_3(t)]T \\ 1 & [1 - d_3(t)]T < t \le T \end{cases} \tag{7.54}$$

where δ_1 represents the on-off state of the main switch, d_1 represents the duty cycle when the main switch is on, which is determined by the feedback control algorithm.

According to Chaps. 5 and 6, when using the ESPM to solve Eq. (7.50), the following iterative equations can be obtained.

$$\begin{cases} \mathbf{G}_0(p)\mathbf{x}_0 + \mathbf{G}_1(p)\mathbf{f}_{0m}^{(1)} + \mathbf{G}_3(p)\mathbf{f}_{0m}^{(2)} = \mathbf{u} \\ \mathbf{G}_0(p)\mathbf{x}_1 + \mathbf{G}_1(p)\mathbf{f}_{1m}^{(1)} + \mathbf{G}_3(p)\mathbf{f}_{1m}^{(2)} = -[\mathbf{G}_1(p)\mathbf{R}_1^{(1)} + \mathbf{G}_3(p)\mathbf{R}_1^{(2)}] \\ \mathbf{G}_0(p)\mathbf{x}_2 + \mathbf{G}_1(p)\mathbf{f}_{2m}^{(1)} + \mathbf{G}_3(p)\mathbf{f}_{2m}^{(2)} = -[\mathbf{G}_1(p)\mathbf{R}_2^{(1)} + \mathbf{G}_3(p)\mathbf{R}_2^{(2)}] \\ \cdots \end{cases} \tag{7.55}$$

(1) Determination of the duty cycle d_1 and d_3

The on/off state of switching transistor is determined by feedback compensation network. According to Chap. 6, the duty cycle d_1 can be expressed as a linear function of the system's state variables, namely:

$$d_1(t) = K_0 + \mathbf{K}_1\mathbf{x}(d_1T) \tag{7.56}$$

Let $t_s = d_1T$ represents the moment the switch acts, and d_1 expand into the following series according to Chaps. 5 and 6, i.e.,

$$d_1 = d_{10} + \varepsilon d_{11} + \varepsilon^2 d_{12} + \cdots \varepsilon^i d_{1i} + \cdots \tag{7.57}$$

where each term d_{1i} is determined by

$$\begin{cases} d_{10} = K_0 + \mathbf{K}_1\mathbf{x}_0 \\ d_{11} = \mathbf{K}_1\mathbf{x}_1(t_{s0}) \\ d_{12} = \mathbf{K}_1\frac{\partial \mathbf{x}_1}{\partial t}\Big|_{t=t_{s0}} d_{11}T + \mathbf{K}_1\mathbf{x}_2(t_{s1}) \\ \cdots \end{cases} \tag{7.58}$$

And K_0, $\mathbf{K}_1$ can be determined by the specific circuit, and

$$\begin{aligned} t_{s0} &= d_{10}T \\ t_{s1} &= (d_{10} + \varepsilon d_{11})T \\ t_{s2} &= (d_{10} + \varepsilon d_{11} + \varepsilon^2 d_{12})T \\ &\cdots \end{aligned} \tag{7.59}$$

where t_{si} (i = 0,1,2,...) represent the moment at which the switch acts when the steady-state periodic solution only contains the terms of $\mathbf{x}_0$, $\mathbf{x}_1$, $\mathbf{x}_2$, ...$\mathbf{x}_i$.

Assuming the rising slope and the dropping slop of the inductor current are m_1 and m_2 respectively, then According to Fig. 7.9 and the discussion in Chap. 6, the duty cycle d_3 for DCM operation can be deduced, i.e.,

$$d_3 = 1 - d_1 - \frac{\bar{i}_L - (md_1^2 + I_0 d_1)}{I_0 + md_1} \tag{7.60}$$

where I_0 denotes the initial value of the inductor current at the beginning of each switching cycle T, and $\bar{i}_L$ means the average during one cycle, m = $m_1T/2$.

Generally for most converters, I_0 = 0, and when taking into account the circuit's parasitic parameters, m_1 and m_2 usually are functions of the average input and the output, then d_3 can be expressed as a function of the input, the average state variables $\bar{\mathbf{x}}$ and d_1, i.e.,

$$d_3 = h_1(E, d_1, \bar{\mathbf{x}}) \tag{7.61}$$

According to the series expansion of $\bar{\mathbf{x}}$ and d_1, d_3 is be transformed into the series expansion as follows.

$$d_3 = d_{30} + \varepsilon d_{31} + \varepsilon^2 d_{32} + \cdots \tag{7.62}$$

in which each d_{3i} (i = 1,2,....) can be approximated by

$$
\begin{aligned}
&d_{30} = h_1(E, d_{10}, \bar{\mathbf{x}}_0) \\
&d_{31} = \left(\left. \frac{\partial h_1}{\partial d_1} \right|_{\substack{d_1=d_{10} \\ \bar{\mathbf{x}}=\bar{\mathbf{x}}_0}} \right) \cdot d_{11} + \left(\left. \frac{\partial h_1}{\partial \bar{\mathbf{x}}} \right|_{\substack{d_1=d_{10} \\ \bar{\mathbf{x}}=\bar{\mathbf{x}}_0}} \right) \cdot \bar{\mathbf{x}}_1 = H_{11}^1 \cdot d_{11} + H_{12}^1 \cdot \bar{\mathbf{x}}_1 \\
&d_{32} = \left(\left. \frac{\partial h_1}{\partial d_1} \right|_{\substack{d_1=d_{10}+d_{11} \\ \bar{\mathbf{x}}=\bar{\mathbf{x}}_0+\bar{\mathbf{x}}_1}} \right) \cdot d_{12} + \left(\left. \frac{\partial h_1}{\partial \bar{\mathbf{x}}} \right|_{\substack{d_1=d_{10}+d_{11} \\ \bar{\mathbf{x}}=\bar{\mathbf{x}}_0+\bar{\mathbf{x}}_1}} \right) \cdot \bar{\mathbf{x}}_2 = H_{21}^1 \cdot d_{12} + H_{22}^1 \cdot \bar{\mathbf{x}}_2 \\
&\cdots
\end{aligned}
\tag{7.63}
$$

(2) Determination of Switching functions δ_1 and δ_3

According to the analysis above, d_3 and d_1 have the same form of series expansion. Thus, δ_1 and δ_3 can be transformed into series expansion in the same way.

$$
\delta^{(k)} = \delta_0^{(k)} + \varepsilon\delta_1^{(k)} + \varepsilon^2\delta_2^{(k)} + \cdots \tag{7.64}
$$

Here, k = 1,3. According the Fourier series expansion as

$$
\delta^{(k)} = b_0^{(k)} + \sum_{m=1}^{\infty} [b_m^{(k)} \cdot e^{jm\tau} + \bar{b}_m^{(k)} \cdot e^{-jm\tau}] \tag{7.65}
$$

where $\tau = 2\pi f_s t$. Then each term $\delta_i^{(k)}$ can be chosen as

$$
\begin{aligned}
\delta_0^{(k)} &= d_{k0} + b_{10}^{(k)} e^{j\tau} + c.c \quad , \\
\delta_1^{(k)} &= d_{k1} + b_{11}^{(k)} e^{j\tau} + b_{20}^{(k)} e^{j2\tau} + b_{30}^{(k)} e^{j3\tau} + c.c \quad , \\
\delta_2^{(k)} &= d_{k2} + b_{12}^{(k)} e^{j\tau} + b_{21}^{(k)} e^{j2\tau} + b_{31}^{(k)} e^{j3\tau} + b_{40}^{(k)} e^{j4\tau} + b_{50}^{(k)} e^{j5\tau} + c.c \quad , \\
&\cdots
\end{aligned}
$$

And the coefficient $b_{mi}^{(k)}$ is determined by

$$
\begin{cases}
b_{mi}^{(k)} = \frac{1}{2}(\alpha_{mi}^{(k)} - j\beta_{mi}^{(k)}) \qquad (m = 1, 2, \cdots, i = 0, 1, 2, \cdots) \\
\alpha_{m0}^{(k)} = \frac{\sin 2m\pi d_{k0}}{m\pi}, \ \beta_{m0}^{(k)} = \pm \frac{1 - \cos 2m\pi d_{k0}}{m\pi} \\
\alpha_{mi}^{(k)} \approx 2d_{ki} \cos 2m\pi(d_{k0} + \varepsilon d_{k1} + \cdots + \varepsilon^{i-1} d_{k(i-1)}) \\
\beta_{mi}^{(k)} \approx \pm 2d_i \sin 2m\pi(d_{k0} + \varepsilon d_{k1} + \cdots + \varepsilon^{i-1} d_{k(i-1)}) \\
(i = 1, 2, \cdots)
\end{cases}
\tag{7.66}
$$

7.5.2 Steady-State Solution of the CMC Converter with DCM Operation

For the current-mode control converter shown in Fig. 7.5, when the load resistor is chosen greater than 72.53Ω, i.e. $R > 72.53\Omega$, the circuit will operate in DCM. Choosing the load resistance $R = 112\Omega$, and all the other circuit parameters are the same as those in Table 7.1.

The differential equation to describe the system is shown in Eq. (7.50), in which the constant vectors are $\mathbf{e}^{(1)} = 0$ and $\mathbf{e}^{(3)} = \begin{bmatrix} 0 & -E & 0 \end{bmatrix}^{Tr}$, the square matrices $G_0(p)$ and $G_1(p)$ are the same as those of CCM operation, $G_3(p) = G_1(p)$, they are rewritten as follows.

$$\mathbf{G}_1(p) = \mathbf{G}_3(p) = \begin{bmatrix} p & \frac{1}{L} & 0 \\ -\frac{1}{C} & p+\frac{1}{RC} & 0 \\ 0 & -\frac{1}{R_1 C_a} & p \end{bmatrix} \text{ and } G_2(p) = \begin{bmatrix} 0 & -\frac{1}{L} & 0 \\ \frac{1}{C} & 0 & 0 \\ 0 & 0 & 0 \end{bmatrix} \quad (7.67)$$

The nonlinear vector functions are given by (7.51) and (7.52), and the input vector $\mathbf{u}$ are given by

$$\mathbf{u} = \begin{bmatrix} \frac{E}{L} & 0 & \frac{-1}{C_a}\left(\frac{1}{R_1}+\frac{1}{R_2}\right)V_R \end{bmatrix}^{Tr} \quad (7.68)$$

The duty cycle d_1 is the linear function of state vector, can be expressed as Eq. (7.56), and here it is solved by the simplified method introduced in Sect. 7.4.1, i.e.,

$$d_1(t) = K_0 + \mathbf{K}_1\bar{\mathbf{x}}(d_1 T) \quad (7.69)$$

In which the terms K_0 and $\mathbf{K}_1$ can be determined by the Eq. (7.21), and then d_1 is expanded in series as shown in (7.57), where the term d_{1i} is determined by

$$\begin{cases} d_{10} = K_0 + \mathbf{K}_1\bar{\mathbf{x}}_0 \\ d_{11} = \mathbf{K}_1\bar{\mathbf{x}}_1 \\ d_{12} = \mathbf{K}_1\bar{\mathbf{x}}_2 \\ \cdots \end{cases} \quad (7.70)$$

Then according to Eq. (7.60), d_3 is determined by:

$$d_3 = 1 - \mathbf{K}_{11}\bar{\mathbf{x}}/(md_1) \quad (7.71)$$

where $\mathbf{K}_{11} = \begin{bmatrix} 1 & 0 & 0 \end{bmatrix}, m = (E \cdot T)/(2L)$.

The duty cycle d_3 can be expanded as the series shown in (7.62), in which each term is defined by according to (7.63), i.e.,

$$
\begin{aligned}
d_{30} &= 1 - \mathbf{K}_{11}\bar{\mathbf{x}}_0/(md_{10}) \\
d_{31} &= \frac{\mathbf{K}_{11}\bar{\mathbf{x}}_0}{(md_{10})^2}\cdot d_{11} - \frac{\mathbf{K}_{11}}{(md_{10})}\cdot \bar{\mathbf{x}}_1 = H_{11}^1 \cdot d_{11} + H_{12}^1 \cdot \bar{\mathbf{x}}_1 \\
d_{32} &= \frac{\mathbf{K}_{11}(\bar{\mathbf{x}}_0 + \bar{\mathbf{x}}_0)}{m^2(d_{10}+d_{11})^2}\cdot d_{12} - \frac{\mathbf{K}_{11}}{m(d_{10}+d_{11})}\cdot \bar{\mathbf{x}}_2 = H_{21}^1 \cdot d_{12} + H_{22}^1 \cdot \bar{\mathbf{x}}_2 \\
&\cdots\cdots
\end{aligned}
\tag{7.72}
$$

(1) **Solution of the Main Component**

First, the main wave $\mathbf{x}_0$ to the steady-state value of state variable $\mathbf{x}$ is chosen as

$$
\mathbf{x}_0 = \mathbf{a}_{00} = \begin{bmatrix} I_{00} & V_{00} & V_{f00} \end{bmatrix}^{Tr} \tag{7.73}
$$

It can be obtained by combining the following equations.

$$
\begin{cases}
\mathbf{G}_0(0)\mathbf{a}_{00} + \mathbf{G}_1(0)d_{10}\mathbf{a}_{00} + \mathbf{G}_3(0)d_{30}(\mathbf{a}_{00} + \mathbf{e}^{(3)}) = \mathbf{u} \\
d_{10} = K_0 + \mathbf{K}_1\mathbf{a}_{00} \\
d_{30} = 1 - \mathbf{K}_{11}\mathbf{a}_{00}/(md_{10})
\end{cases}
\tag{7.74}
$$

Obviously, the Eq. (7.73) is nonlinear and need to be solved by symbolic method or numerical method.

(2) **Solution of the First-order Correction**

The first-order correction can be assumed as

$$
\mathbf{x}_1 = \mathbf{a}_{11}e^{j\tau} + \bar{\mathbf{a}}_{11}e^{-j\tau} \tag{7.75}
$$

where the amplitude coefficient $\mathbf{a}_{11}$ can be found by solving the following equation by applying the ESPM.

$$
\begin{aligned}
&[\mathbf{G}_0(j\omega) + \mathbf{G}_1(j\omega)d_{10} + \mathbf{G}_3(j\omega)d_{30}]\mathbf{a}_{11} \\
&= -\mathbf{G}_1(j\omega)(b_{11}^{(1)} + b_{10}^{(1)})\mathbf{a}_{00} - \mathbf{G}_3(j\omega)[(b_{11}^{(3)} + b_{10}^{(3)})(\mathbf{a}_{00} + \mathbf{e}^{(3)})]
\end{aligned}
\tag{7.76}
$$

The Eq. (7.76) is linear. Let $\mathbf{G}_1 = \mathbf{G}_0(j\omega) + \mathbf{G}_1(j\omega)d_{10} + \mathbf{G}_3(j\omega)d_{30}$, the solution for (7.76) can be obtained by

$$
\mathbf{a}_{11} = -\mathbf{G}^{-1}\{\mathbf{G}_1(j\omega)(b_{11}^{(1)} + b_{10}^{(1)})\mathbf{a}_{00} - \mathbf{G}_3(j\omega)[(b_{11}^{(3)} + b_{10}^{(3)})(\mathbf{a}_{00} + \mathbf{e}^{(3)})]\} \tag{7.77}
$$

Here, the superscript "−1" indicates the inversion matrix.

Furthermore, we can define the following symbols as

$$\tau_{s0} = 2\pi d_{10}, \sigma_{s0} = 2\pi d_{30}, \mathbf{A}_{12} = -\mathbf{G}^{-1}\mathbf{G}_1(j\omega), \mathbf{A}_{13} = -\mathbf{G}^{-1}\mathbf{G}_3(j\omega)$$
$$\mathbf{B}_{10} = \mathbf{A}_{12}\mathbf{b}_{10}^{(1)} \cdot e^{j\tau_{s0}}, \mathbf{Q}_{10} = \mathbf{A}_{13}b_{10}^{(3)} \cdot e^{j\sigma_{s0}}, \mathbf{F} = \mathbf{H}_{11}^{1} \cdot e^{j(\tau_{s0}+\sigma_{s0})}$$
$$\mathbf{A}_1 = 1 - \mathbf{K}_1[(\mathbf{A}_{12} + \bar{\mathbf{A}}_{12})\mathbf{a}_{00} + (\mathbf{A}_{13}\mathbf{F} + \bar{\mathbf{A}}_{13}\bar{\mathbf{F}})(\mathbf{a}_{00} + \mathbf{e}^{(3)})]$$
$$\mathbf{B}_1 = \mathbf{K}_1[(\mathbf{B}_{10} + \bar{\mathbf{B}}_{10})\mathbf{a}_{00} + (\mathbf{Q}_{10} + \bar{\mathbf{Q}}_{10})(\mathbf{a}_{00} + \mathbf{e}^{(3)})],$$

Then the first-order correction d_{11} of the duty cycle d_1 can be obtained as

$$d_{11} = \mathbf{B}_1/\mathbf{A}_1 \tag{7.78}$$

(3) **Solution of the Second-order Correction**

Next, according to process of the ESPM, the second-order correction can be assumed as

$$\mathbf{x}_2 = \mathbf{a}_{02} + (\mathbf{a}_{22}e^{j2\tau} + \mathbf{a}_{32}e^{j3\tau} + c.c) \tag{7.79}$$

And then based on the third equation in (7.55), the equations with DC component, the second-order harmonic and the third-order harmonic in the second-order correction can be obtained:

$$\begin{cases} \mathbf{a}_{22} = -\mathbf{G}_2^{-1} \cdot (\mathbf{B}_{22} + \mathbf{Q}_{22}) \\ \mathbf{a}_{32} = -\mathbf{G}_3^{-1} \cdot (\mathbf{B}_{32} + \mathbf{Q}_{32}) \\ \mathbf{G}_0 a_{02} + \mathbf{G}_1(0)\mathbf{a}_{00}d_{12} + \mathbf{G}_3(0)(\mathbf{a}_{00} + e^{(3)})d_{32} = -(\mathbf{B}_{02} + \mathbf{Q}_{02}) \end{cases} \tag{7.80}$$

where the symbolic expressions are defined as follows.

$$\begin{cases} \mathbf{G}_2 = \mathbf{G}_0(j2\omega) + \mathbf{G}_1(j2\omega) \cdot d_{10} + \mathbf{G}_3(j2\omega) \cdot d_{30} \\ \mathbf{B}_{22} = \mathbf{G}_1(j2\omega)[(b_{10}^{(1)} + b_{11}^{(1)})\mathbf{a}_{11} + (b_{20}^{(1)} + b_{21}^{(1)})\mathbf{a}_{00} + b_{30}^{(1)}\bar{\mathbf{a}}_{11}] \\ \mathbf{Q}_{22} = \mathbf{G}_3(j2\omega)[(b_{10}^{(3)} + b_{11}^{(3)})\mathbf{a}_{11} + (b_{20}^{(3)} + b_{21}^{(3)})(\mathbf{a}_{00} + \mathbf{e}^{(3)}) + b_{30}^{(3)}\bar{\mathbf{a}}_{11}] \end{cases}$$

$$\begin{cases} \mathbf{G}_3 = \mathbf{G}_0(j3\omega) + \mathbf{G}_1(j3\omega) \cdot d_{10} + \mathbf{G}_3(j3\omega) \cdot d_{30} \\ \mathbf{B}_{32} = \mathbf{G}_1(j3\omega)[b_{10}^{(1)}a_{22} + b_{20}^{(1)}\mathbf{a}_{11} + (b_{30}^{(1)} + b_{31}^{(1)})\mathbf{a}_{00}] \\ \mathbf{Q}_{32} = \mathbf{G}_3(j3\omega)[b_{10}^{(3)}a_{22} + b_{20}^{(3)}\mathbf{a}_{11} + (b_{30}^{(3)} + b_{31}^{(3)})(a_{00} + \mathbf{e}^{(3)})] \end{cases}$$

$$\begin{cases} \mathbf{G}_0 = G_0(0) + \mathbf{G}_1(0)d_{10} + \mathbf{G}_3(0)d_{30} \\ \mathbf{B}_{02} = \mathbf{G}_1(0)[(b_{10}^{(1)} + b_{11}^{(1)})\bar{\mathbf{a}}_{11} + (\bar{b}_{10}^{(1)} + \bar{b}_{11}^{(1)})\mathbf{a}_{11} + d_{11}\mathbf{a}_{00}] \\ \mathbf{Q}_{02} = \mathbf{G}_3(0)[(b_{10}^{(3)} + b_{11}^{(3)})\bar{\mathbf{a}}_{11} + (\bar{b}_{10}^{(3)} + \bar{b}_{11}^{(3)})\mathbf{a}_{11} + d_{31}(\mathbf{a}_{00} + \mathbf{e}^{(3)})] \end{cases}$$

It can be seen from Eq. (7.80) that the solutions of $\mathbf{a}_{22}$ and $\mathbf{a}_{32}$ are linear, yet the solution of $\mathbf{a}_{02}$ is nonlinear. Assume that

$$\begin{cases} \mathbf{P}_{11} = \mathbf{a}_{11} e^{j\tau_{s0}}, \mathbf{P}_{22} = \mathbf{a}_{22} e^{j\tau_{s1}}, \mathbf{P}_{32} = \mathbf{a}_{32} e^{j\tau_{s1}}, \\ (\tau_{s0} = 2\pi d_{10}, \tau_{s1} = 2\pi(d_{10} + d_{11})) \\ \mathbf{M} = -\mathbf{K}_1 \cdot 4\pi d_{11} \mathrm{Im}(\mathbf{P}_{11}) + 2\mathbf{K}_1 [\mathrm{Re}(\mathbf{P}_{22}) + \mathrm{Re}(\mathbf{P}_{32})] \end{cases} \tag{7.81}$$

in which $\mathrm{Im}(\cdot), \mathrm{Re}(\cdot)$ represent the imaginary part and the real part of conjugate complex respectively.

According to Eq. (7.70) and (7.72), $d_{12} = K_1 \mathbf{a}_{02}$, $d_{32} = H_{21}^1 \cdot d_{12} + H_{22}^1 \cdot \mathbf{a}_{02}$. Then introducing d_{12} and d_{32} into the equation for solving a_{02} in (7.80) gives $\mathbf{a}_{02} = \mathbf{B}_2 / \mathbf{A}_2$,

in which

$$\mathbf{A}_2 = \mathbf{G}_0 + \mathbf{G}_{10}(0)\mathbf{a}_{00}\mathbf{K}_1 - \mathbf{G}_3(0)(\mathbf{a}_{00} + \mathbf{e}^{(3)})(H_{21}^1 \mathbf{K}_1 + H_{22}^1)$$
$$\mathbf{B}_2 = \mathbf{B}_{02} + \mathbf{Q}_{02} - \mathbf{G}_1(0)\mathbf{a}_{00} M - \mathbf{G}_3(0)(\mathbf{a}_{00} + e^{(3)}) H_2 M$$

Finally, with all the circuit parameters being introduced in, the steady-state periodic solution of state variable in main circuit, including DC component and the first three harmonics, can be obtained as

$$\begin{aligned} i_L &= 0.4249 - 0.1939 \cos\tau + 0.4153 \sin\tau - 0.1308 \cos 2\tau \\ &\quad + 0.0852 \sin 2\tau - 0.0616 \cos 3\tau - 0.0195 \sin 3\tau \\ v_o &= 36.0132 - 0.0008 \cos\tau - 0.0013 \sin\tau + 0.0002 \cos 2\tau \\ &\quad - 0.0002 \sin 2\tau + 0.00008 \cos 3\tau - 0.00008 \sin 3\tau \end{aligned} \tag{7.82}$$

The DC components and duty cycles d_1 and d_3 from the symbolic ESPM with simplified duty cycle solution, and that of Pspice simulation are shown in Table 7.4, and the comparison of the ripples of the state variables in the main circuit between the ESPM and the Pspice simulation is shown in Fig. 7.10.

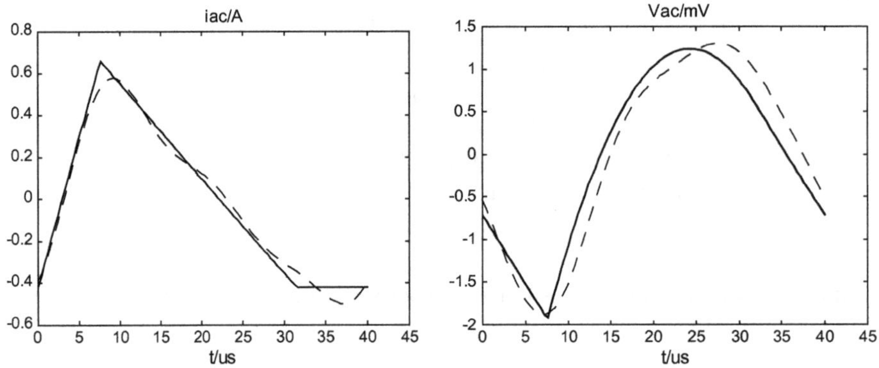

(a) inductor current (b) capacitor voltage

Fig. 7.10 The comparison of ripples of the DCM CMC-Boost from ESPM (dashed line) and Pspice (solid line)

Table 7.4 Comparison of DC component of the DCM CMC-Boost resulted from ESPM and Pspice

Variables	Symbolic	Pspice
Inductor current	$I_{dc} = 0.4249$ A	$I_{dc} = 0.4203$ A
Output voltage	$V_{dc} = 36.0132$ V	$V_{dc} = 36.0031$ V
Duty cycle	$d_1 = 0.1876$ $d_3 = 0.1570$	$d_1 = 0.19$ $d_3 = 0.21$

It can be seen from Fig. 7.10 and Table 7.4 that both the DC components and the AC components from the ESPM are consistent with those from simulation.

7.6 Summary

Current-mode control scheme has been widely used in practice because it has more advantages than duty cycle control scheme. However, due to the double closed loops (current loop and voltage loop) introduced in current-mode control, the duty cycle cannot be expressed explicitly, and thus it is difficult to model and analyze current-mode controlled system. At present, the analysis of current-mode control converter basically adopts the average modelling method based on inductor current waveform analysis. Its premise is that the waveform of inductor current is linear and the assumption of small ripples can be satisfied, and there is no correlation to the analysis of the ripple of the state variable.

For the analysis of closed-loop converter system, the determination of duty cycle is critical. When using the equivalent-small-parameter method (ESPM) to analysis the current-mode controlled converters, two methods for determining the duty cycle are presented in this paper, one is the accurate solution of duty cycle, and the other is the simplified solution of duty cycle. The former would be found wider applicability, as it does not need to satisfy the assumptions of small ripples and linear waveform with inductor current. The results are all analytical expressions, which can help to understand the working mechanism of the circuit. The examples and simulation results confirm that the proposed two methods are of high accuracy.

Two methods have the same manner as the ESPM of the duty-cycle controlled PWM converter mentioned in Chaps. 5 and 6. Thus, current-mode controlled, as well as the duty-cycle controlled converter system, can be analyzed uniformly with the same mathematical model using ESPM.

References

1. Mahabir K et al (1990) Linear averaged and sampled data models for large signal control of power factor AC-DC converters. IEEE PESC Rec:372–381
2. Deisch CW (1978) Simple switching control method changes power converter into a current source. IEEE PESC Rec:300–306
3. Capel A, Ferrante G, O'Sullivan D et al (1978) Application of the injected current model for the dynamics of switching regulators with the new concept of LC3 modulator. IEEE PESC Rec:135–147
4. Lee FC, Carter RA et al (1983) Investigations of stability and dynamic performances of a current-injected regulator. IEEE Trans AES 19(2):126–138
5. Hsu SP, Brown A, Rensink L et al. (1979) Modelling and analysis of switching DC-to-DC converters in constant-frequency current-programmed mode. IEEE PESC Rec:284–301
6. Tan FD, Middlebrook RD (1995) A unified model for current-programmed converters. IEEE Trans PE 10(4):397–408
7. Verghese GC, Bruzos CA, Mahabir KN (1989) Averaged and sampled-data models for current mode control: a re-examination. IEEE PESC Rec:484–491
8. Liu YF, Sen PC (1994) A general unified large signal model for current programmed DC-to-DC converters. IEEE Trans PE 9(4):414–424
9. Liu YF, Sen PC (1996) Large-signal modeling of hysteretic current-programmed converters. IEEE Trans PE 11(3):423–430
10. Bass RM, Heck BS, Khan RA (1994) Average modelling of current-mode controlled converters: instability predictions. Int J Electron 77(5):613–628
11. Tang W, Lee FC, Ridley RB (1993) Small-signal modeling of average current-mode control. IEEE Trans PE 8(2):112–119
12. Wong S-C, Lee Y-S (1993) SPICE modeling and simulation of hysteretic current-controlled Cuk converter. IEEE Trans Power Electron 8:580–587. ISSN 0885-8993
13. Kimhi D, Ben-Yaakov S (1991) A SPICE model for current mode PWM converters operating under continuous inductor current conditions. IEEE Trans PE 6(2):281–286
14. Pavljasevic S, Maksimovic D (1991) Using a discrete-time model for large-signal analysis of a current-programmed boost converter. IEEE PESC Rec:715–721
15. Hong SS, Jo BR, Youn MJ (1996) Duty cycle generator for average model of buck converter with current-mode control—using analog behavioral modeling of Pspice. IEEE Trans PE 11 (6):785–795
16. Middlebrook RD (1985) Topics in multiple-loop regulators and current-mode programming. IEEE PESC Rec:716–732
17. Amran Y, Huliehel F, Ben-Yaakov (1991) A unified SPICE compatible average model of PWM converters. IEEE Trans PE 11(6):585–594
18. Rodriguez FD, Chen JE (1991) A refined nonlinear averaged model for constant frequency current mode controlled PWM converters. IEEE Trans PE 11(6):656–664
19. Ben-Yaakov S (1993) Averaged simulation of PWM converters by direct implementation of behavioral relationships. APEC Proc:510–516
20. Nanov ID (1994) Large signal analysis of switching regulators employing adaptive gain current injected control. IEEE Trans IE 41(3):339–343

Chapter 8
Analysis of PFM Quasi-resonant DC/DC Converters Based on ESPM

8.1 Introduction

Conventional PWM switching technology forces the switching device to turn off when the voltage is not zero or when the current is not zero [1, 2]. Ideally, the switching device voltage and current waveforms are square waves. However, due to the non-ideal characteristics of the switching devices and the parasitic parameters of the circuits, the voltage and current waveforms of the switching devices are not completely that ideal, and the switching loss occurs during the switching process. This loss increases when the switching frequency rises. Therefore, the traditional PWM switching power converter has large switching losses, and the switching stress is considerable. As a result, the electromagnetic interference (EMI) is severe in conventional PWM switching regulators and the switching frequency is difficult to be increased [3]. In order to reduce the switching losses and stresses, and to eliminate or mitigate the effects of parasitic parameters in the circuit and achieve the switching converter miniaturization, different kinds of technologies such as high operating frequency, series resonant, parallel resonant, quasi-resonant, class-E resonance, multi-resonance, nonlinear resonance and quasi-square wave resonant, etc., are applied [4], and they are collectively referred to as resonant converters. The Quasi Resonant Converter (QRC) has been extensively used because of its relatively simple circuit topology [3–8]. The quasi-resonant technology is developed by adding a resonant inductor and a resonant capacitor on the basis of PWM switching converter. The resonant inductor and capacitor form the concept of "resonant switch" together with the original PWM switch, wherein the resonant working mode can be attained, thus the switching voltage and the current waveform are improved, while the switching losses are reduced. During the running procedures, the circuits sustain resonant state only in part of the whole switching cycle, and in the rest of the cycle, they will be running in non-resonant state, so this kind of topology is called as "quasi-resonant converter". Additionally, this kind of topology is a mix type of PWM converter and resonant converter. As a result, advantages of

Y. Chen and B. Zhang, *Equivalent-Small-Parameter Analysis of DC/DC Switched-Mode Converter*, CPSS Power Electronics Series,
https://doi.org/10.1007/978-981-13-2574-8_8

both the simple control of PWM technology and small switching loss of the resonant circuit are absorbed.

The common point of all kinds of resonant converters is to use the resonant principle to make the voltage or current waveforms of the switch change in a sinusoidal or quasi-sinusoidal fashion, by which the zero voltage or zero current switching condition can be achieved, while at the same time, the switching losses can be reduced. However, due to the existence of resonant components, the increase of the switching topologies and the use of variable frequency control technology [9, 10], etc., the difficulty of analysis and design for resonant converters has been increased. Additionally, since the existence of resonant elements, the time constant of such topologies is often the same as or less than the switching period, which is clearly contrary to the linear ripple assumption, so that the traditional state space averaging method is no longer applicable [5].

The state space averaging method is one main method for the analysis of switching power converters, which has been widely used in practice with its distinctive physical meaning and simple processing [11, 12]. So it is a natural idea that the state space averaging method is expected to be applied to the analysis of resonant converters, and Witulski [5], Ninomiya [6] and Lin Botao [13] put forward different approaches based on this idea. Essentially speaking, these approaches can be regarded as a generalized average modeling method, and the common point is the case that, the resonant element and the filter element in the circuit are processed separately to obtain the order-reduced average model, which describes the dynamic characteristics of the filtering elements, and by which the small signal frequency domain model can be obtained by small signal disturbance. More specifically, in Witulski's work, by considering the advantages and disadvantages of the QRC technology and PWM technology, the equivalent duty cycle μ is introduced to replace the duty cycle d of the conventional PWM converters to establish a state space averaging model suitable for the resonant converters. Honestly speaking, the execution procedures of this method is somewhat a little complicated, but the introduction of equivalent PWM switch has a distinct physical meaning, which can help to understand the working mechanism of converters. Ninomiya and Lin divide the circuit into fast-changing parts (high-frequency networks) and slow-changing parts (low-frequency networks), wherein the former one is usually made up by the switching elements and resonant components, and the latter one generally consists of filtering capacitors, input and output components. From a mathematical point of view [6], the modeling procedures can be divided into two steps:

(1) The state differential equations of fast varying systems and slow varying systems are listed separately. The slow variable is treated as constant, and the response of the fast variable in a switching period is calculated when the initial value is known.
(2) The fast variables are substituted into the slow variable equations, and the newly generated slow variable equations are averaged over a switching period to obtain the time-invariant model that describes the dynamic response characteristics of the slow variables.

The high frequency network averaging method [13] is a generalized method that extended based on the equivalent circuit averaging method. Specifically, the high frequency components (switches and resonators) are replaced by controlled sources, and the values of the controlled sources are averaged in a switching cycle. By utilizing the simplification principle for power supplies, the number of controlled sources in the circuit can be reduced, and the QRC simplified order-orduced equivalent circuit model can be obtained. Generally, there are only low-frequency filter elements in dynamic components of the model, and the resonant elements are replaced by time-invariant controlled source. The model is consistent with the equivalent circuit model of the corresponding PWM master converter. Thus, the physical meaning of the high frequency network averaging method is more distinct than that of [6].

The order-reduced averaging method [12] needs to know the analytical solution of the fast variable system and assumes that the solution is periodic. In order to facilitate the analysis and calculation, the fast variable system application is as simple as possible, which means that the influence of most parasitic parameters must be ignored. On the other hand, only when the fast variable system does not contain continuous energy storage components, the solution is periodic. Otherwise, the fast variable system will show effective low frequency dynamic characteristics, its impact on the slow change system cannot be ignored, since the average of the fast variables in a switching cycle only effectively retains its DC component, significant errors are introduced in the final average model. Therefore, it is necessary to establish a full-order nonlinear large-signal model of the resonant class converter.

In [14], circuit averaging method for PWM converters are applied to the analysis of quasi-resonant converters. More specifically, the switching elements are replaced by time-invariant control sources, hence the full-order equivalent circuit model consistent with the original circuit topology can be obtained, in which the values of the controlled sources are usually the average of the current or voltage of the switching elements within a switching period. Both the order-reduced averaging and the full-order averaging model deal the circuits with an averaged process, and thus they cannot fully reflect the dynamic characteristics of the circuits. In this chapter, from the perspective of nonlinear analysis, the order-reduced and the full-order nonlinear circuit model of the quasi-resonant converter are established based on the switching function, and the equivalent small parameter method is used to analyze the ripples of state variables belong to the resonant elements and the filtering elements, in which, the equivalent small parametric analysis method of the full-order circuit model has the same form as that of the PWM converter. Moreover, it is shown that the non-linear modeling and equivalent small parameter analysis of the switching function can be used to analyze the PWM and resonant converters.

8.2 Classification of Quasi-resonant Converters

The quasi-resonant converter (QRC) has two types, the zero-current switching QRC and the zero-voltage switching QRC.

8.2.1 *Zero-Current Switch*

The zero-current-switch (ZCS) has two forms as shown in Fig. 8.1. The characteristic of the zero-current switch is that the resonant inductor and the resonant capacitor resonate during the on period of the switch S_T. When the resonant inductor current reaches zero according to sinusoidal law, the transistor is naturally turned off, which ensures that the current flowing through the running switch has dropped to zero before the turn-off signal arrives. The main advantage of the ZCS is that the turn-off loss of the switching device can be greatly reduced, but the turn-on loss cannot be eliminated, which is its main disadvantage. This means that the energy stored in the switching transistor during the off state is converted to the internal loss of the switch when it is turned on. This loss is proportional to the switching frequency, thus limiting the further increase of the switching frequency. In general, switching frequency of ZCS can be up to 1–2 MHz [4].

According to whether the current through the switch is bidirectional, the ZCS can be divided into two modes of operation, i.e. the full-wave zero-current-switch (FW-ZCS) and the half-wave zero-current-switch (HW-ZCS). Since the current flowing through the switch can be reversed, the FW-ZCS is often represented by the switching transistor in reverse parallel with a diode. While in the half-wave mode, current can only flow through the transistor switch in one direction, the HW-ZCS is usually represented by a transistor in series with a forward diode. In both cases, the transistor is considered an ideal switch.

8.2.2 *Zero-Voltage-Switch*

Zero voltage switches are also available in two basic forms, as shown in Fig. 8.2, which is characterized by the capacitor and the inductor operating in resonant mode during the off period of the switch S_T, so that the voltage over the switch changes in

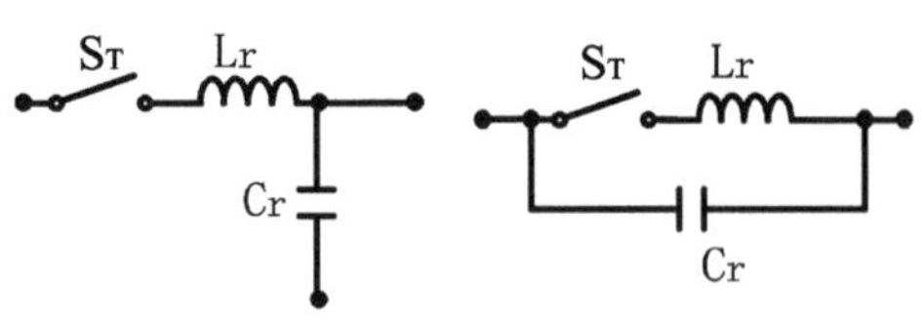

Fig. 8.1 Zero-current-switch

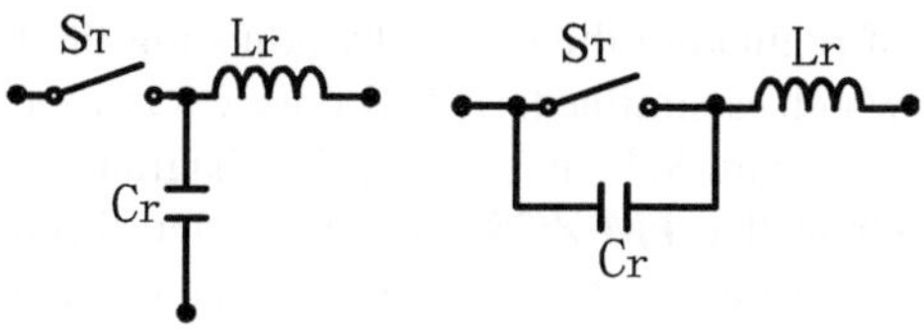

Fig. 8.2 Zero-voltage-switch

a sinusoidal law to fulfill the zero voltage conditions. The zero voltage switches eliminate the opening loss associated with the parasitic capacitor of the switch, and the operating frequency can reach above 10 MHz [4].

Similarly, according to whether the voltage over the switching transistor is bidirectional, the ZVS can also be divided into two modes, i.e. the HW-ZVS and the FW-ZVS. The former can be represented by the transistor in reverse parallel with a diode, since the voltage over the switching transistor cannot be reversed, and the later is usually represented by a transistor in series with a forward diode. We can see that for FW and HW mode, the diode connection is just the opposite of the zero-current-switch.

8.2.3 Duality Between ZVS and ZCS

The ZCS technology causes the switch to be turned off in a zero current state. With the help of the auxiliary LC resonant elements, the current flowing during the on time of the switching device is changed according to a sinusoidal law, thereby creating conditions for the switch to be turned off in a zero current state. Similarly, the ZVS technology causes the switch to be turned on in a zero voltage state. With the aid of the auxiliary LC resonant elements, the voltage across the switching device in the off state changes sinusoidally, creating a zero voltage condition for the device to be turned on. Obviously, the two types of ZCS and ZVS converters have duality, which is strictly established in [12]. The duality means that for any ZCS quasi-resonant converter, there is always a ZVS quasi-resonant converter that is its dual circuit, and vice versa. Therefore, this chapter focuses on the analysis of closed-loop ZCS converter systems, while the characteristics of closed-loop ZVS systems can be obtained by duality.

8.3 Modulation Principle of the Quasi-resonant Converter

By means of the steady-state analysis of the resonant converter [7], it can be found that the voltage transmission gain of the quasi-resonant converter is generally related to the ratio of the switching frequency to the resonant angular frequency, and thus this type of converter is often frequency-modulated, which is also called pulse-frequency-modulation (PFM). The PFM control usually achieves the purpose of regulating the output by adjusting the time of the converter in a constant phase, which is similar to PWM duty cycle modulation with constant on-time [10].

Figure 8.3 is a block diagram of a typical single-voltage feedback controlled FM-ZCS Buck converter. Similar to the PWM control, in order to satisfy the closed-loop characteristics, the PFM control also needs a suitable compensation correction circuit, usually using the circuit with two poles and two zero points, which can make the whole system have a higher low-frequency gain and a sufficient phase margin at its corner frequency [10].

In Fig. 8.3, the output of the error amplifier (EA) is inputted as a control voltage signal to a Voltage Controlled Oscillator (VCO) such that the frequency of the VCO's output rectangular wave signal is linear with this control voltage. The DSP (Digital Signal Processor) is typically composed of a switch driver circuit and a logic circuit that converts the VCO output signal into a constant on-time pulse signal. The common effect of the VCO and the DSP is to make the cut-off time of the pulse change with the output voltage of the converter, that is, the time of the converter in the constant phase is adjustable, and the frequency of the switching transistor is generally adjusted, so that the converter system can achieve a stable output.

Single loop control is sensitive to changes in the main circuit parameters of the converter. At the same time, since the voltage controlled oscillator has high sensitivity to noise, it is difficult for the converter to have good stability for all operating conditions and component changes. Therefore, multi-loop control mode is often used in quasi-resonant converters [9, 10], that is, in addition to voltage feedback, current feedback is also added in. However, from the analysis in Chaps. 5–7 of this book, it can be observed that the analysis of the multi-loop controlled converter is similar to that of the single-loop controlled converter when using the equivalent small-parameters method. Therefore, we only discuss the analysis of a frequency-modulated quasi-resonant converter system with single closed-loop control.

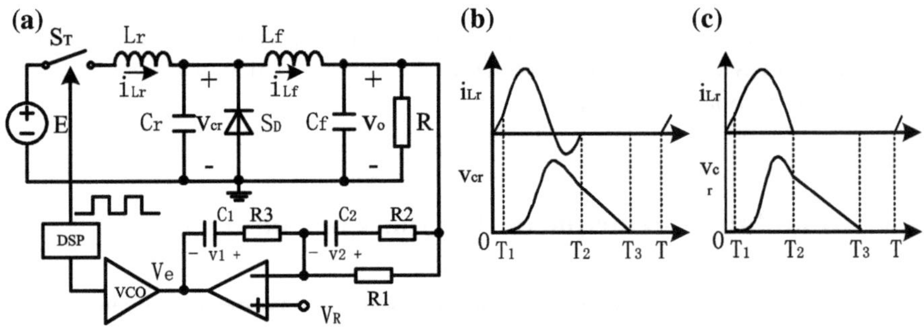

Fig. 8.3 FM ZCS QR buck regulator: **a** Circuit diagram, **b** waveforms of i_{Lr} and v_{Cr} in FW mode, **c** waveforms of i_{Lr} and v_{Cr} in HW mode

8.4 Symbolic Analysis of ZCS PFM Quasi-resonant Buck Converter System

The Zero-Current-Switching (ZCS) quasi-resonant Buck converter is taken as an example for analysis. For other types of resonant converters, the similar method can be used. Figure 8.3a shows the circuit diagram of a frequency-modulated quasi-resonant ZCS-Buck converter. Wherein the low-frequency filter elements L_f, C_f and the load resistance R constitute the output stage of the circuit, and the zero-current resonant switch consists of the high-frequency resonant components L_r, C_r together with the switches S_T and S_D. If the switch S_T is current bi-directional, then the circuit works in full-wave mode with the typical waveforms of resonant inductor current i_{Lr} and resonant voltage v_{Cr} shown in Fig. 8.3b; otherwise, the converter operates in half-wave mode with typical waveforms shown in Fig. 8.3c. The feedback network is used to adjust the operating frequency of the main circuit transistor. It consists of a feedback correction circuit with dual zero points and two poles, a voltage controlled oscillator (VCO) module, and a digital signal processor (DSP) module. The role of the feedback network is detailed in the previous section.

The parameters of the whole converter system are as follows, where the parameters of the main circuits are given as: $E = 12\text{V}$, $L_r = 30\text{nH}$, $C_r = 0.3\mu\text{F}$, $L_f = 5.7\mu\text{H}$, $C_f = 0.63\mu\text{F}$, $R = 0.5\ \Omega$; and the parameters of the control loop are given as: $C_1 = 800$ pF, $C_2 = 6300$ pF, $R_1 = 3.1$ kΩ, $R_2 = 1.0$ kΩ, $R_3 = 10$ kΩ, the gain of the VCO module is chosen as $f_g = 0.32$ MHz/V, and the reference voltage $V_R = 5$ V.

8.4.1 Circuit Operating Principle of the Converter

The following assumptions are usually made for steady-state analysis of quasi-resonant converters, i.e.,

(i) The values of filter elements is much larger than those of resonant elements, i.e., $L_f \gg L_r$ and $C_f \gg C_r$. Therefore the inductor L_f can be regarded as a constant current source, and C_f as a constant voltage source when analyzing the operating modes of the converter. In this example, in order to show that the analysis method is still applicable when the output voltage has a large ripple, we deliberately select the case where the output filter capacitor has a small value.

(ii) All components have desirable characteristics, i.e., their parasitic parameters are not taken into account. In fact, this analysis method still applies when considering the loss resistance of the inductor and capacitor components.

Let f_s and f_n denote the switching frequency and the resonant frequency respectively, and then set the following symbol expressions as

$$Z_n = \sqrt{L_r/C_r}, a = Z_n i_{Lf}/E, f_n = 1/\left(2\pi\sqrt{L_r C_r}\right), \omega_n = 2\pi f_n, T = 1/f_s$$

We consider the case where the filter inductor current is continuous. In steady state operation, each cycle of the ZCS-QRC can be divided into four working stages, that is, the power-converter-stage (PCS) has four topology modes during one switching cycle, as shown in Fig. 8.4. Assume that at the moment of $t = 0^-$, the switch S_T is off, then the current flowing through the diode S_D is i_{Lf}, and the resonant inductor current and capacitor voltage keep to be zero, i.e., $i_{Lr} = 0$ and $v_{Cr} = 0$. And at the moment of $t = 0$ the switch S_T is turned on by an external control signal, then the working process of the ZCS-QR Buck converter can be analyzed as follows.

(1) **Stage-1: Linear Charging Stage of Resonant Inductor** $L_r (0 \le t \le T_1)$

During this stage, the switches S_T and S_D are all in on state, and the converter can be equivalent as the topology shown in Fig. 8.4a. It can be seen that the voltage across the inductor L_r equals to the input voltage E, and then L_r is charged. If the voltage E keeps constant during this period, the inductor current i_{Lr} would increase linearly with the time t. While the capacitor voltage v_{Cr} remains at zero since the diode in parallel with the capacitor C_r is conductive. The forward current flowing through the diode S_D is $i_d = i_{Lf} - i_{Lr}$. And at the moment of $t = T_1$ when the condition of $i_{Lf} = i_{Lr}$ is satisfied, the diode S_D is cut off naturally, as the forward

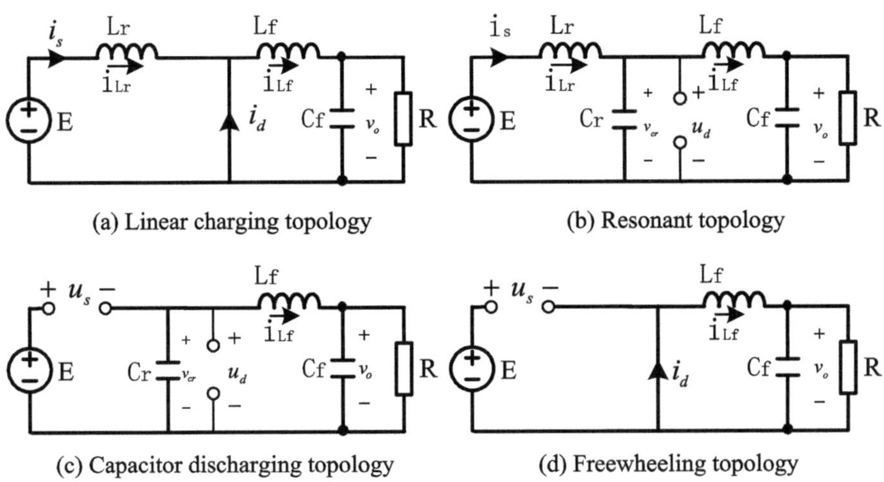

Fig. 8.4 Operation topologies of ZCS quasi-resonant buck converter

current through the diode is zero. Then the circuit would go into the next stage after $t = T_1$. From Fig. 8.4a, the following equations can be obtained as:

$$\begin{cases} i_{Lr} = i_{Lf}\omega_n t/a \\ v_{Cr} = 0 \end{cases}$$

Then the duration T_1 of stage-1 can be obtained as $T_1 = a/\omega_n$ by letting $i_{Lr}(T_1) = i_{Lf}$.

(2) **Stage-2: Resonant Stage** $(T_1 \le t \le T_2)$

During this stage, the switch S_T is on and the diode S_D is off, and the inductor L_r and the capacitor C_r form a series resonant circuit. The equivalent topology during this stage is shown in Fig. 8.4b. With the initial conditions of $i_{Lr}(T_1) = i_{Lf}$ and $v_{Cr}(T_1) = 0$ one can get the following equations according to the Fig. 8.4b, i.e.,

$$\begin{cases} i_{Lr} = i_{Lf} + (i_{Lf}/a)\sin\omega_n(t - T_1) \\ v_{Cr} = E[1 - \cos\omega_n(t - T_1)] \end{cases}$$

If the converter is operating in full-wave mode, that is, S_T is a current bidirectional switch, the inductor current i_{Lr} continues to flow through the internal body diode of the switch after the first time it reaches zero, so that i_{Lr} becomes a negative value. During the period when i_{Lr} is negative, if the drive signal of S_T is kept to be a low level, S_T will be turned off in the zero current state at the second time of i_{Lr} to be zero value, then the next stage comes.

And if the converter is operating in half-wave mode, that is, S_T is current unidirectional, then the switch turns off naturally at the first time when the inductor current i_{Lr} reaches zero. Thus, in the two cases, the end time of the stage-2 can be deduced as

$$\begin{aligned} T_2 &= (2\pi - \sin^{-1} a)/\omega_n + T_1 \quad &(\text{for full} - \text{wave mode}) \\ T_2 &= (\pi + \sin^{-1} a)/\omega_n + T_1 \quad &(\text{for half} - \text{wave mode}) \end{aligned}$$

(3) **Stage-3: Capacitor discharge stage** $(T_2 \le t \le T_3)$

The equivalent topology during stage-3 is shown in Fig. 8.4c. Both S_T and S_D are turned off, the resonant capacitor is discharged, and its voltage v_{Cr} declines linearly with the time. This stage continues until the moment that the value of v_{Cr} goes to zero, and thus the diode switch S_D is turned on naturally. From Fig. 8.4c, the following equations can be derived as:

$$\begin{cases} i_{Lr} = 0 \\ v_{Cr} = E[1 - \cos\omega_n(T_2 - T_1)] - E\cdot a\omega_n(t - T_2) \end{cases}$$

And at the moment of $t = T_3$, there is $v_{Cr} = 0$, then T_3 can be determined by solving the above equation as

$$T_3 = T_2 + [1 - \cos\omega_n(T_2 - T_1)]/(a\omega_n)$$

(4) **Stage-4: Constant Stage** $(T_3 \le t \le T)$

The equivalent topology during stage-4 is shown in Fig. 8.4d. The switch S_T is turned off while the diode S_D is on, the resonant inductor current and the resonant capacitor voltage are kept at zero, until the turn-on signal of S_T comes, and then a new cycle begins. During this stage, there exists the following equation as

$$\begin{cases} i_{Lr} = 0 \\ v_{Cr} = 0 \end{cases}$$

To sum up, the functions of i_{Lr} and v_{Cr} during different stages in a switching cycle can be expressed as follows:

$$i_{Lr} = \begin{cases} i_{Lr}\omega_n t/a & (0 \le t \le T_1) \\ i_{Lf} + (i_{Lf}/a)\sin\omega_n(t - T_1) & (T_1 \le t \le T_2) \\ 0 & (T_2 \le t \le T_3) \\ 0 & (T_3 \le t \le T) \end{cases} \tag{8.1a}$$

$$v_{Cr} = \begin{cases} 0 & (0 \le t \le T_1) \\ E[1 - \cos\omega_n(t_2 - T_1)] & (T_1 \le t \le T_2) \\ E[1 - \cos\omega_n(T_2 - T_1) - a\omega_n(t - T_2)] & (T_2 \le t \le T_3) \\ 0 & (T_3 \le t \le T) \end{cases} \tag{8.1b}$$

In which the time T_i (i = 1, 2, 3) is determined by

$$\begin{cases} T_1 = a/\omega_n \\ T_2 = (2\pi - \sin^{-1} a)/\omega_n + T_1 \\ T_3 = T_2 + [1 - \cos\omega_n(T_2 - T_1)]/(a\omega_n) \end{cases} \tag{8.1c}$$

for full-wave mode, and

$$\begin{cases} T_1 = a/\omega_n \\ T_2 = (\pi + \sin^{-1} a)/\omega_n + T_1 \\ T_3 = T_2 + [1 - \cos\omega_n(T_2 - T_1)]/(a\omega_n) \end{cases} \tag{8.1d}$$

for half-wave mode.

8.4.2 *Nonlinear Model of Quasi-resonant Converter and Its Solution by ESPM*

The non-linear modeling method of ZCS-QRC is illustrated by the converter working in full-wave mode. According to the operating process of the converter discussed in Sect. 8.4.1, we define the duty cycle and the switching functions corresponding to each topology as follows:

$$\begin{cases} d_1 = T_1/T = \frac{f_s}{\omega_n} a \\ d_2 = (T_2 - T_1)/T = \frac{f_s}{\omega_n}(2\pi - \sin^{-1} a) \\ d_3 = (T_3 - T_2)/T = \frac{f_s}{\omega_n}\left(\frac{1-\sqrt{1-a^2}}{a}\right) \\ d_4 = 1 - d_1 - d_2 - d_3 \end{cases} \tag{8.2}$$

$$\begin{matrix} \delta^{(1)} = \begin{cases} 1 & 0 \le t \le T_1 \\ 0 & others \end{cases}, & \delta^{(2)} = \begin{cases} 1 & T_1 \le t \le T_2 \\ 0 & others \end{cases} \\ \delta^{(3)} = \begin{cases} 1 & T_2 \le t \le T_3 \\ 0 & others \end{cases}, & \delta^{(4)} = \begin{cases} 1 & T_3 \le t \le T \\ 0 & others \end{cases} \end{matrix} \tag{8.3}$$

Assume that the current flowing through the switch S_T is i_s, the voltage across S_T is u_s; the current flowing through the diode S_D is i_d, and the reverse voltage is u_d, and their direction is shown in Fig. 8.4. According to Fig. 8.4 and the modeling method discussed in Chap. 2, we can express i_s, u_s, i_d and u_d during one cycle by the switching functions as follows, i.e.,

$$\begin{cases} i_s = \delta^{(1)} i_{Lr} + \delta^{(2)} i_{Lr} \\ u_d = \delta^{(2)} v_{cr} + \delta^{(3)} v_{cr} \end{cases} \tag{8.4}$$

$$\begin{cases} i_d = \delta^{(1)}(i_{Lf} - i_{Lr}) + \delta^{(4)} i_{Lf} = -\delta^{(1)} i_{Lr} + (1 - \delta^{(2)} - \delta^{(3)}) i_{Lf} \\ u_s = \delta^{(3)}(E - v_{cr}) + \delta^{(4)} E = (1 - \delta^{(1)} - \delta^{(2)}) E - \delta^{(3)} v_{cr} \end{cases} \tag{8.5}$$

8.4.2.1 The Order-Reduced Model of QRC Main Circuit

By replacing the switches S_T and S_D of the QRC with a time-varying controlled current source i_s and a time-varying controlled voltage source u_d respectively, the nonlinear equivalent circuit model can be obtained as shown in Fig. 8.5a. According to the analysis of operating process of the QRC, we can find that during a switching cycle, the current i_s flowing through S_T is just the resonant inductor current i_{Lr}. And the voltage u_d across the diode equals to the voltage v_{Cr} across the resonant capacitor. Thus Fig. 8.5a can be represented by its order-reduced model shown as Fig. 8.5b.

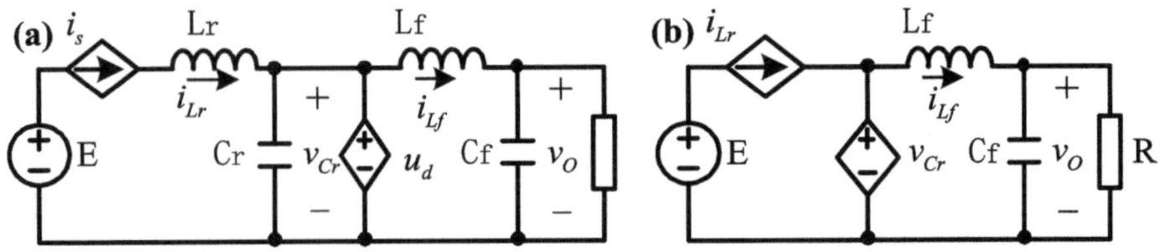

Fig. 8.5 Circuit models of the PCS **a** An equivalent circuit model; **b** the order-reduced model

Regarding the order-reduced model, there are some explanations as follows:

(1) From a circuit perspective, in Fig. 8.5a, there is a cut set consisting of a pure current source and an inductor, and a loop composed of a pure capacitor and a voltage source. So actually the circuit in Fig. 8.5a is 2-ordered, which can be replaced by its order-reduced model in Fig. 8.5b, where the waveforms of i_{Lr} and v_{Cr} are obtained by means of steady-state analysis of the converter.
(2) From a non-linear point of view, the inductor L_{r} and the capacitor C_{r} are not linear components in the ordinary sense due to the action of the switch. Since they are time-divisionally present in the circuit topology during one switching cycle, the inductance L_{r} and the capacitance C_{r} themselves can be regarded as nonlinear time-varying components.

According to Fig. 8.5b we can get the lower order differential equation of the main circuit of the converter. Similar to the analysis of the PWM feedback controlled system, the state variables of the equivalent circuit model of the main circuit are combined with the state variables of the feedback network, that is, the state variable vector **x** is chosen as

$$\mathbf{x} = \begin{bmatrix} i_{Lf} & v_O & v_1 & v_2 \end{bmatrix}^{Tr} \tag{8.6}$$

In which i_{Lf} is the current flowing through the filter inductor, v_{O} is the output voltage, they are state variables belong to the main circuit; and the state variables v_1 and v_2 belong to the control loop, they are the voltages across the capacitors C_1 and C_2, respectively.

Then the order-reduced time-varying differential equations of the whole closed-loop ZCS-QRC system can be obtained according to the Figs. 8.5b and 8.3a as follows:

$$\mathbf{G}_0(p)\mathbf{x} + b \cdot f = \mathbf{u} \tag{8.7}$$

In which the square matrix is determined by:

$$G_0(p) = \begin{bmatrix} p & 1/L_f & 0 & 0 \\ -1/C_f & p+1/RC_f & 0 & 0 \\ 0 & -(1/R_1C_1+1/R_2C_1) & p & 1/R_2C_1 \\ 0 & -1/R_2C_2 & 0 & p+1/R_2C_2 \end{bmatrix}$$

And the coefficient column vectors of the input, and the coefficient are given as

$$u = \begin{bmatrix} 0 \\ 0 \\ -V_R(1/R_1C_1+1/R_2C_1) \\ -V_R/R_2C_2 \end{bmatrix}, \quad b = \begin{bmatrix} -1/L_f \\ 0 \\ 0 \\ 0 \end{bmatrix}$$

It is worth noting that the nonlinear function f is a scalar function, which is defined as

$$f = v_{Cr}$$

The switching frequency can be expressed as a linear combination of system state variables, that is:

$$f_s = K_0 + \mathbf{K}_1\mathbf{x} \tag{8.8}$$

In which the coefficient K_0 and the row vector K_1 are related to the circuit parameters, they are determined by:

$$\begin{aligned} K_0 &= f_g(1+R_3/R_1+R_3/R_2)V_R \\ \mathbf{K}_1 &= [0 \quad -f_g(R_3/R_1+R_3/R_2) \quad -f_g \quad f_gR_3/R_2] \end{aligned}$$

where f_g is the gain of the VCO module.

Using a processing method similar to the switching function δ(t) in PWM controlled converters, the nonlinear function f is expanded into a Fourier series over a switching period as follows:

$$f = f_0 + \varepsilon f_1 + \varepsilon^2 f_2 \tag{8.9}$$

Here, the $\varepsilon = 1$ is only an introduced small amount mark. And the term f_0 and f_m ($m = 1, 2, \ldots$) in Eq. (8.9) are determined according to the principle of Fourier transformation, i.e.,

$$\begin{aligned} f_0 &= \frac{1}{T}\int_0^T v_{cr}(t)dt \\ f_m &= b_m e^{jmt} + c.c \end{aligned} \tag{8.10}$$

In which the expression of $v_{cr}(t)$ is given in Eq. (8.1b), and the coefficient b_m is given as

$$b_m = \frac{1}{T}\left(\int_0^T v_{cr} \cdot \cos m\omega t \cdot dt - j\int_0^T v_{cr} \cdot \sin m\omega t \cdot dt\right) \tag{8.11}$$

In this way and similar to the solving process for PWM closed-loop systems, one can obtain DC component and ripple components in filtering inductor current and capacitor voltage of QRC closed-loop systems. Since the operating frequency of the quasi-resonant converter is usually very high, and the feedback control circuit has a compensation correction network, thus the feedback output ripple is smoothed, and the ripple component of the output has little effect on the switching frequency. As a result, it can be considered that, the switching frequency is actually determined by the DC component of the feedback, i.e.,

$$f_s = K_0 + \mathbf{K}_1\mathbf{x}_0 \tag{8.12}$$

According to the expression for each f_m of the nonlinear function f in Eq. (8.10), we can set:

$$\mathbf{x} = a_0 + \varepsilon \cdot \mathbf{a}_1 e^{j\tau} + \varepsilon^2 \cdot \mathbf{a}_2 e^{j2\tau} + \varepsilon^3 \cdot \mathbf{a}_3 e^{j3\tau} + \cdots c.c \tag{8.13}$$

Then according to the equivalent parameter method, the harmonic components of the steady-state periodic solution can be obtained, wherein f_0 is the average value of $v_{cr}(t)$ in a switching cycle, and the main oscillation component $a_0 = \begin{bmatrix} I_{Lf0} & V_{o0} & V_{10} & V_{20} \end{bmatrix}^{Tr}$ can be solved by the following equations when the QRC operating in full-wave mode:

$$\begin{cases} \mathbf{G}_0(0) \cdot \mathbf{a}_0 + \mathbf{b} \cdot f_0 = \mathbf{u} \\ f_0 = \frac{f_s}{\omega_n} \cdot \left(\frac{a}{2} + 2\pi - \sin^{-1} a + \frac{1-\sqrt{1-a^2}}{a}\right) \\ f_s = K_0 + K_1 \cdot a_0 \\ a = \frac{Z_n I_{Lf0}}{E} \end{cases} \tag{8.14}$$

It can be seen that the solution of the zero-order component is nonlinear, which needs to be solved by numerical or symbolic method. In this chapter, the numerical method is used to solve the Eq. (8.14), and it can be obtained that f_s = 700 kHz, and the main component (DC component) of the filter inductor current and the output voltage can be got as

$$I_{Lf0} = 10\,\text{A}, \quad V_{O0} = 5\,\text{V}$$

According to the previous discussion, we believe that the switching frequency of the PFM-QRC closed-loop system can be obtained when the zero-order component is solved. So the harmonic component can be deduced by solving the following linear equation, that is

$$\mathbf{a}_m = -G_0^{-1}(jm\omega) \cdot b_m \tag{8.15}$$

in which b_m is determined by (8.10).

Based on the ESPM, the approximated steady-state periodic solution the filtering inductor current and the output voltage, which is only calculated to the first three harmonics according the ESPM, can be obtained finally by the following equation:

$$\begin{aligned} i_{Lf} &= 10.0000 - 0.3555\cos\tau + 0.0563\sin\tau - 0.0371\cos 2\tau - 0.1194\sin 2\tau \\ &\quad + 0.0383\cos 3\tau - 0.0185\sin 3\tau \\ v_O &= 5.0000 - 0.0742\cos\tau - 0.0747\sin\tau + 0.0169\cos 2\tau - 0.0128\sin 2\tau \\ &\quad + 0.0032\cos 3\tau + 0.0038\sin 3\tau \end{aligned} \tag{8.16}$$

Numerical simulation results show that, the values of DC components are $I_{\mathrm{Lf0}} = 9.9632$ A, $V_{\mathrm{o0}} = 4.9789$ V which are quite close to those from ESPM. Figure 8.6 shows the comparison of ripples obtained by different methods. The dash line in the figure is the summation of the 1st, 2nd and 3rd harmonics in Eq. (8.15). The solid line represents those obtained by numerical simulation. Obviously, the results meet well with each other.

8.4.2.2 The Full-Order Model of QRC Main Circuit

If switch S_T of the QRC power main circuit is replaced by a time-varying controlled voltage source, while the diode is replaced by a time-varying controlled current source, we can get an equivalent full-order circuit model for the QRC power level main circuit, as shown in Fig. 8.7. The time-dependent controlled source in Fig. 8.7 is represented by Eq. (8.5).

According to the previous analysis, for the QRC feedback network, the switching frequency is mainly determined by the DC component of the output feedback, so we only analyze the state variables of the main circuit. The role of the

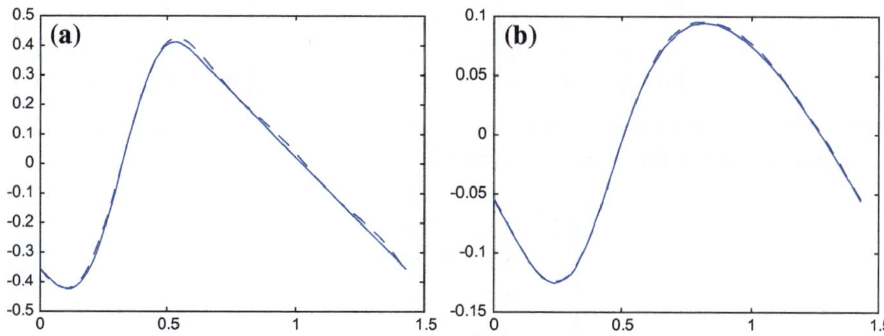

Fig. 8.6 Comparison of ripple waveforms between the ESPM and numerical simulation **a** the filter inductor current **b** the output voltage

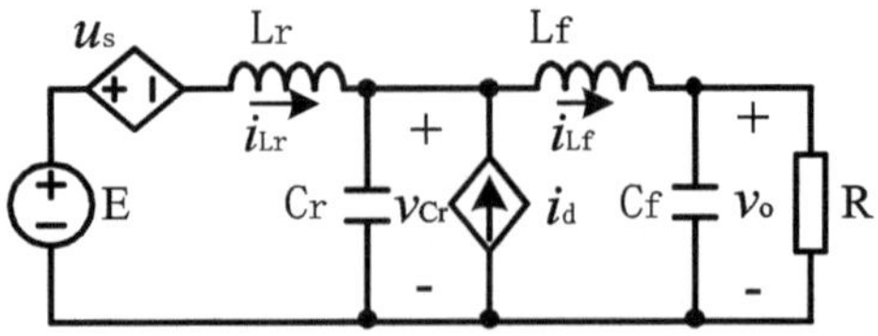

Fig. 8.7 Full-order equivalent circuit model of power stage

control circuit is to determine the circuit switching frequency during the calculating processes for the zero-order component. So in the following analysis, the system's mathematical model is independent with state variables of the feedback circuit. Actually, if the state variables of the feedback circuit are combined with the state variables of the main circuit as in the previous analysis method, the full-order mathematical model of the whole closed-loop system can be established, but at this situation, the order of the system matrix is higher. However, from the programming point of view, for example in MATLAB, the difficulty for solving does not increase with the increase of matrix orders. Thus the state variable vector consists of only the state variables of the main circuit, that is,

$$\mathbf{x} = \left[i_{Lr}\ v_{cr}\ i_{Lf}\ v_o\right]^{Tr}$$

And by Fig. 8.7, we can establish the time-varying differential equation as follows:

$$G_0(p) + G_1(p)\delta^{(1)}\mathbf{x} + G_2(p)\delta^{(2)}\mathbf{x} + G_3(p)\delta^{(3)}\mathbf{x} = (\delta^{(1)} + \delta^{(2)}) \cdot \mathbf{e} \tag{8.17}$$

in which the constant vector is $\mathbf{e} = [E/L_r\ 0\ 0\ 0]^{Tr}$, and the square matrices are defined by

$$G_0(p) = \begin{bmatrix} p & 1/L_r & 0 & 0 \\ -1/C_r & p & 0 & 0 \\ 0 & -1/L_f & p & 1/L_f \\ 0 & 0 & -1/C_f & p+1/RC_f \end{bmatrix}, \quad G_1(p) = \begin{bmatrix} 0 & 0 & 0 & 0 \\ 1/C_r & 0 & 0 & 0 \\ 0 & 0 & 0 & 0 \\ 0 & 0 & 0 & 0 \end{bmatrix},$$

$$G_2(p) = \begin{bmatrix} 0 & 0 & 0 & 0 \\ 0 & 0 & 1/C_r & 0 \\ 0 & 0 & 0 & 0 \\ 0 & 0 & 0 & 0 \end{bmatrix}, \quad G_3(p) = \begin{bmatrix} 0 & -1/L_r & 0 & 0 \\ 0 & 0 & 1/C_r & 0 \\ 0 & 0 & 0 & 0 \\ 0 & 0 & 0 & 0 \end{bmatrix}$$

Let the nonlinear functions be defined as :

$$\begin{cases} f^{(1)}(\mathbf{x}) = \delta^{(1)}\mathbf{x}, \\ f^{(2)}(\mathbf{x}) = \delta^{(2)}\mathbf{x} \\ f^{(3)}(\mathbf{x}) = \delta^{(3)}\mathbf{x} \\ f^{(4)}(\mathbf{x}) = (\delta^{(1)} + \delta^{(2)})\mathbf{e} \end{cases}$$

According to the definition of the switching function and the duty cycle of each stage, together with the general method of analyzing the switching power converter in Chap. 4, the periodic switching function can be developed into the Fourier series:

$$\delta^{(n)} = b_0^{(n)} + \sum_{m=1}^{\infty} [b_m^{(n)} \exp(jm\tau) + \bar{b}_m^{(n)} \exp(-jm\tau)] \tag{8.18}$$

In which the coefficients can be determined by

$$\begin{cases} b_0^{(n)} = d_n \\ b_m^{(n)} = \frac{1}{2}(\alpha_m^{(n)} - j\beta_m^{(n)}), \quad m = 1, 2 \cdots \end{cases} \tag{8.19}$$

where

$$\begin{cases} \alpha_m^{(n)} = [\sin 2(\sum_{k=0}^{n} d_k) m\pi - \sin 2(\sum_{k=0}^{n-1} d_k) m\pi]/m\pi \\ \beta_m^{(n)} = [\cos 2(\sum_{k=0}^{n-1} d_k) m\pi - \cos 2(\sum_{k=0}^{n} d_k) m\pi]/m\pi \end{cases}$$

in which, $n = 1, 2, 3, 4$. With the equivalent small parameter method, we turn the switch function into the sum of the main and the small corrections, i.e.,

$$\delta^{(n)} = \delta_{.0}^{(n)} + \varepsilon \delta_1^{(n)} + \varepsilon^2 \delta_2^{(n)} + \cdots \tag{8.20}$$

in which:

$$\begin{aligned} \delta_0^{(n)} &= d_n + b_1^{(n)} e^{j\tau} + c.c \\ \delta_i^{(n)} &= b_{2i}^{(n)} e^{j2i\tau} + b_{(2i+1)}^{(n)} e^{j(2i+1)\tau} + c.c \end{aligned}$$

Then the ESP method can be used to solve the steady-state periodic solution of the system, in which the main oscillation components and the first three harmonic components are as follows:

(1) **Main oscillation component**

Based on the principle of the ESPM, the main term (DC component) can be assumed as

$$\mathbf{x}_0 = \mathbf{a}_{00} = \left[I_{Lr0} \; V_{Cr0} \; I_{Lf0} \; V_{o0}\right]^{Tr}$$

According to (3.46), the following equation for solutions of $\mathbf{a}_{00}$ can be obtained, i.e.,

$$[G_0(0) + G_1(0)d_1 + G_2(p)d_2 + G_3(0)d_3]a_{00} = (d_1 + d_2) \cdot e \tag{8.21}$$

According to Eq. (8.8) and Fig. 8.3, one can deduce that $V_{o0} = V_R$, then the switching frequency can be determined by:

$$f_s = f_g(V_R - V_1) \tag{8.22}$$

Combining the equations of (8.20) (8.21) and (8.2), the calculation formula of the duty ratio d_{i} (i = 1,2,3,4), we can obtain the symbolic expression of the main component $\mathbf{a}_{00}$ and the duty ratio d_{i} of each stage. It should be noted that the solution of the main component is still nonlinear, just like those obtained by the order-reduced method, thus it should be solved by numerical method or symbolic method.

(2) **Solution of the first-order correction**

The first-order correction term can be assumed as

$$\mathbf{x}_1 = a_{11}e^{j\tau} + c.c$$

In which $\tau = \omega t, \omega = 2\pi f_s$ is switching angular frequency, and the amplitude coefficient vector $\mathbf{a}_{11}$ can be calculated by:

$$[G_0(j\omega) + \sum_{n=1}^{3} G_n(j\omega)d_n]\mathbf{a}_{11} = -\sum_{n=1}^{3} G_n(j\omega)b_1^{(n)}\mathbf{a}_{00} + (b_1^{(1)} + b_1^{(2)})\times\mathbf{e} \tag{8.23}$$

(3) **Solution of the second-order correction**

Based on the ESPM, the second-order correction term can be assumed as

$$\mathbf{x}_2 = \mathbf{a}_{02} + \mathbf{a}_{22}e^{j2\tau} + \mathbf{a}_{32}e^{j3\tau} + c.c,$$

where the DC component and the coefficients of the harmonics are calculated by the following equations:

$$\begin{aligned}
&[G_0(0) + \sum_{n=1}^{3} G_n(0)d_n]\mathbf{a}_{02} = -\sum_{n=1}^{3} G_n(0)(b_1^{(n)}\bar{\mathbf{a}}_{11} + \bar{b}_1^{(n)}\mathbf{a}_{11})\\
&[G_0(j2\omega) + \sum_{n=1}^{3} G_n(j2\omega)d_n]\mathbf{a}_{22} = -\sum_{n=1}^{3} G_n(j2\omega)(b_2^{(n)}\mathbf{a}_{00} + b_1^{(n)}\mathbf{a}_{11} + b_3^{(n)}\bar{\mathbf{a}}_{11}) + (b_2^{(1)} + b_2^{(2)})\mathbf{e}\\
&[G_0(j3\omega) + \sum_{n=1}^{3} G_n(j3\omega)d_n]\mathbf{a}_{32} = -\sum_{n=1}^{3} G_n(j3\omega)(b_3^{(n)}\mathbf{a}_{00} + b_2^{(n)}\mathbf{a}_{11} + b_1^{(n)}\mathbf{a}_{22}) + (b_3^{(1)} + b_3^{(2)})\mathbf{e}
\end{aligned} \tag{8.24}$$

The coefficients $b_m^{(n)}$ in (8.23) and (8.24) are determined by Eq. (8.19). As can be seen from (8.23) and (8.24), when the duty cycles of each stage are determined, the three equations in (8.24) are all linear. This is because we assume that the switching frequency is determined by the DC component of the output feedback based on the actual situation of the circuit operation, so that it is obtained when solving the main oscillation component. And the duty cycle at each stage is a function of the switching frequency, which is determined by the steady-state analysis of the Eq. (8.2). Then according to (8.2), the duty ratio d_i can be derived when solving the main oscillation term after the switching frequencies are determined. Therefore, the correction of each order is linear, as in the case of open loop.

In Fig. 8.3 circuit, when the reference voltage of the error amplifier is taken as $V_R = 2.85\,\text{V}$, the switching frequency $f_s = 400\,\text{kHz}$ is obtained by the solution of the main term. When $V_R = 5\,\text{V}$, we can get $f_s = 700\,\text{kHz}$. In both cases, the steady-state periodic solutions of state variables are as follows, which are the sum of the DC component and the first three harmonics.

$$\begin{aligned} i_{Lf} &= 5.7383 - 0.2724\cos\tau + 0.2836\sin\tau - 0.1743\cos 2\tau - 0.0009\sin 2\tau \\ &\quad -0.0666\cos 3\tau - 0.0684\sin 3\tau \\ v_o &= 2.8692 - 0.1527\cos\tau + 0.0209\sin\tau - 0.2046\cos 2\tau - 0.0395\sin 2\tau \\ &\quad +0.0072\cos 3\tau - 0.0171\sin 3\tau \end{aligned} \qquad (f_s = 400\,\text{kHz})$$

$$\begin{aligned} i_{Lf} &= 10.0083 - 0.3619\cos\tau + 0.0521\sin\tau - 0.0379\cos 2\tau - 0.1206\sin 2\tau \\ &\quad +0.0408\cos 3\tau - 0.0163\sin 3\tau \\ v_o &= 5.0042 - 0.0744\cos\tau - 0.0769\sin\tau + 0.0171\cos 2\tau - 0.0130\sin 2\tau \\ &\quad +0.0030\cos 3\tau + 0.0042\sin 3\tau \end{aligned} \qquad (f_s = 700\,\text{kHz})$$

The DC components obtained from the ESPM and numerical simulations are shown in Table 8.1. It can be seen that results from the two methods are very close.

As for the ripples of the filter state variables and the resonant state variables during a steady-state period, the comparisons between the ESPM (dashed line) and the numerical simulation (solid line) under different conditions are shown in Figs. 8.8, 8.9, and 8.10. Figures 8.8 and 8.9 show the ripple comparison of state variables under the condition of switching frequency fs = 400 kHz, where in Fig. 8.8, the dashed lines from the ESPM are calculated by summing the first three harmonic components in the steady-state period solutions, while in Fig. 8.9, the dashed lines from the ESPM are calculated by summing the first five harmonics in the steady-state periodic solutions.

Table 8.1 Comparison of DC components obtained by ESPM and simulation

	$f_s = 400\,\text{kHz}$				$f_s = 700\,\text{kHz}$			
	I_{Lr}(A)	V_{Cr}(V)	I_{Lf}(A)	V_o(V)	I_{Lr}(A)	V_{Cr}(V)	I_{Lf}(A)	V_o(V)
Simulation	1.2618	2.8359	5.7027	2.8462	3.9049	4.9570	9.9631	4.9787
ESPM	1.3931	2.8692	5.7383	2.8692	4.1736	5.0042	10.0083	5.0042

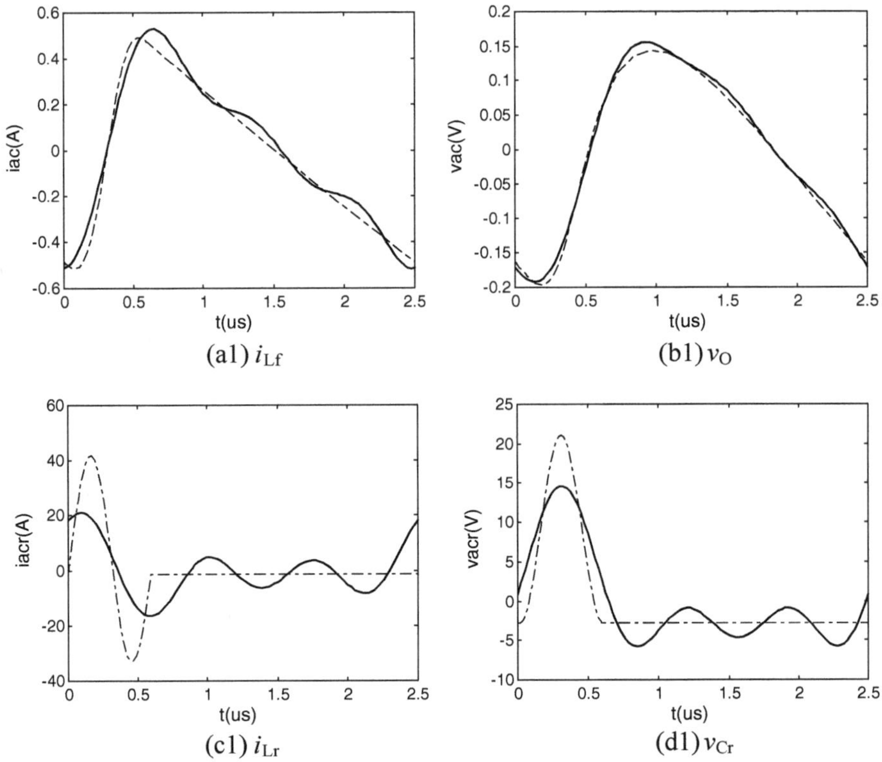

Fig. 8.8 Comparison of ripple waveforms between ESPM (solid line, the first three harmonics are considered) and simulation (dashed line) with fs = 400 kHz

Comparing Figs. 8.8 and 8.9, the higher harmonics (such as the 5th harmonic) in the periodic solution have little effect on the ripple of the output inductor current and capacitor voltage, which is mainly due to the low-pass filter characteristics of the output LC network. So that for the output state variable, when using the ESPM, we only needs to calculate the first three harmonics of the state variable to have sufficient accuracy. However, the more the numbers of harmonics are calculated, the more precise the solutions for resonant variables are, since the resonant variable contains more harmonic components. From another point, this also shows that although the resonant variables contain more harmonic components, but the impact on the output is not great, that is, whether for PWM converters or resonant converters, it is the advantage of this algorithm for the output state variables that the ESPM has enough accuracy even if only the first three harmonics are calculated.

Figures 8.10 shows the ripple comparison of state variables under the condition of switching frequency fs = 700 kHz. It can be found from Fig. 8.10 that when the switching frequency is high, whether for the resonant state variables or for the filter state variables, the ESPM only needs to calculate the first three harmonics of the

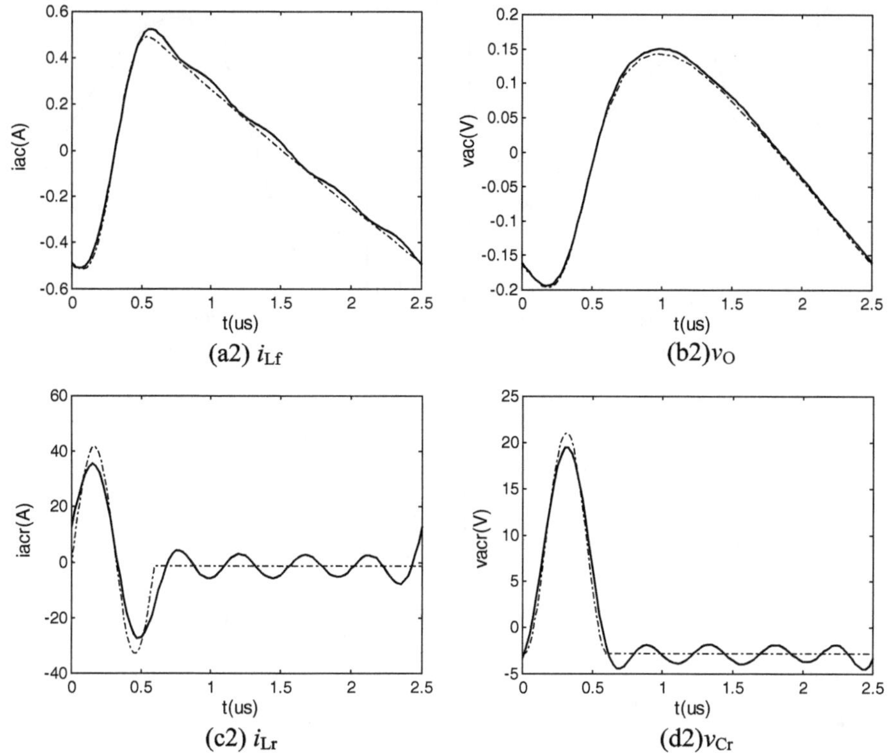

Fig. 8.9 Comparison of ripple waveforms between ESPM (solid line, the first five harmonics are considered) and simulation (dashed line) with fs = 400 kHz

state variables with high accuracy. This is because when the switching frequency is sufficiently high to approach the resonant frequency, the resonant state variable can be considered to exhibit more low frequency characteristics.

The symbolic algorithm of the full-order model has the following advantages compared with that of the order-reduced model, i.e.:

(1) The proposed method avoids the complex operation of the resonant inductor current or the capacitor voltage waveform during Fourier decomposition, since the Fourier series decomposition of the binary (0-1) switching function is obviously much easier.
(2) It can be seen from the analysis process that the full-order analytical model can make the analysis of QRCs have the same form as that of PWM converters, so that the analysis of PWM and resonant converters can be unified based on the ESPM.

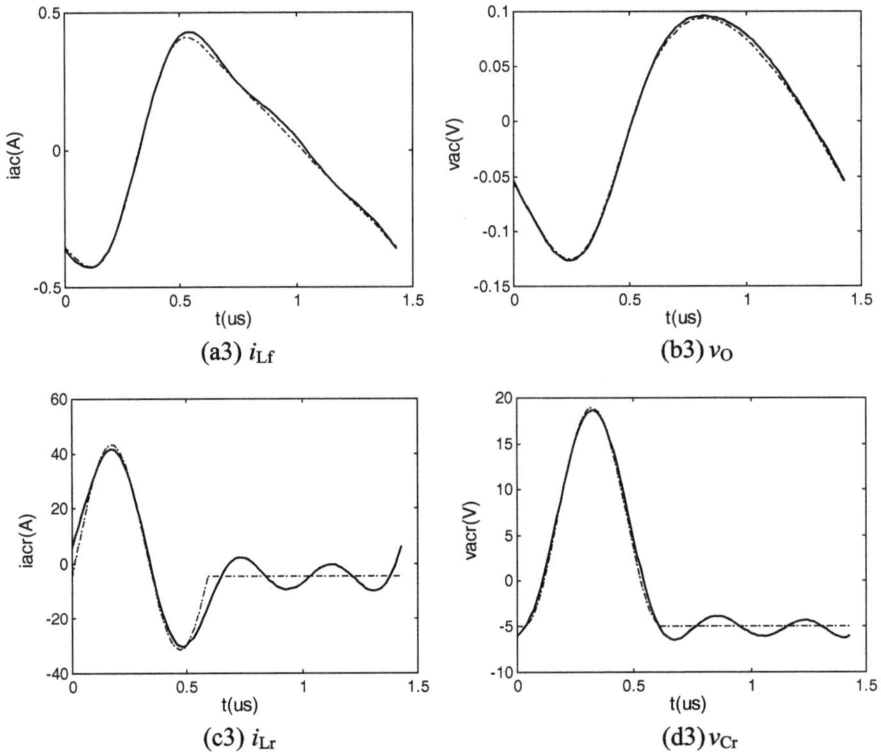

Fig. 8.10 Comparison of ripple waveforms between ESPM (solid line, the first three harmonics are considered) and simulation (dashed line) with fs = 400 kHz

8.5 Summary

In this chapter, according to the nonlinear modeling method based on the switching function proposed in Chap. 2, the order-reduced and full-order equivalent circuit model of the quasi-resonant converter (QRC) are obtained. The full-order model has the same form as the analysis of the PWM converters, so that the analysis of the PWM and the resonant converter can be unified.

At the same time, the full-order model analysis method can more intuitively explain that the higher harmonic components in the resonant variables in the QRC circuit have no significant influence on the output filter state variables. Therefore, it can be explained that the order-reduced model that ignores the dynamic characteristics of the resonant variables and analyzes only the low-frequency state variables of the circuit is reasonable. It is worth noting that the output filter components in engineering design are calculated according to I_{Lf} and Vo, and Table 8.1 shows that the values of these two variables obtained by ESPM have higher accuracy.

References

1. Wu KC (1997) Pulse width modulated DC-DC converters. Chapman & Hall, New York
2. Griffio A, Wang J (2012) Large signal stability analysis of 'more electric' aircraft power systems with constant power loads. IEEE Trans Aeros Electron Syst 48(1):477–489
3. Wong LK et al (1997) A simple large-signal nonlinear modeling approach for fast simulation of zero-current-switch quasi-resonant converters. IEEE Trans PE 12(3):437–442
4. Jonvanovic MM (1994) Resonant, quasi-resonant, multi-resonant and soft-switching techniques—merits and limitations. Int J Electronics 77(5):537–554
5. Witulski AF, Erickson RW (1990) Extension of state-space averaging to resonant switches and beyond. IEEE Trans PE 5(1):98–109
6. Ninomiya T, Nakahara M, Higashi T et al (1991) A unified analysis of resonant converters. IEEE Trans PE 6(2):260–270
7. Bhat AKS (1991) A unified approach for the steady-state analysis of resonant converters. IEEE Trans Indus Electron 38:251–259. ISSN 0278-0046
8. Sabate JA, Farrington RW, Jovanovic MM, Lee FC (1992) Effect of switch capacitance on zero-voltage switching of resonant converters. In: PESC '92 Record. 23rd Annual IEEE Power Electronics Specialists Conference 1992, vol 1, pp 213–220
9. Ridley RB, Cho BH, Lee FC (1988) Analysis and interpretation of loop gains of multi-loop-controlled switching regulators. IEEE Trans Power Electron 3(4):489–498 (Oct 1988)
10. Ridley RB, Tabisz WA, Lee FC et al (1991) Multi-loop control for quasi-resonant converters. IEEE Trans PE 6(1):28–37
11. Middlebrook RD, Cuk S (1976) A general unified approach to modelling switching-converters power stages. IEEE PESC Rec:18–34
12. Sun J, Grotstollen H (1997) Symbolic analysis methods for averaged modeling of switching power converters. IEEE Trans PE 12(3):537–546
13. Lin BT, Qiu SS (1994) A unified equivalent circuit analysis of quasi resonant converters: high frequency network averaging. Proceedings of IEEE International Conference on Industrial Technology-ICIT'94, pp 490–494
14. Xu JP, Lee CQ (1998) A unified averaging technique for the modeling of quasi-resonant converters. IEEE Trans PE 13(3):556–563

Chapter 9
Stability Analysis of PWM Power Switching Converters Based on ESPM

9.1 Stability Analysis of Equilibrium Points

Closed-loop DC/DC power converters are highly nonlinear, which makes it difficult to analyze the system's dynamics. In particular, when the system suffers from large signal disturbances, the small-signal frequency domain model obtained by the State-Space-Averaging (SSA) method [1] cannot analyze the stability of the system in this case [2]. The average large-signal model proposed in [2] retains the assumption of linear ripple of state-space averaging method, and can analyze multiple equilibrium points of the system and its stability, but the obtained model has nothing to do with the switching frequency, which limits its further applications in practice. Yet the frequency dependent averaged model proposed in [3] cannot find multiple solutions that may exist in the closed-loop system. The ESPM (Equivalent-Small-Parameters Method) [4–7] is a simple symbolic analysis method with higher accuracy and relatively simpler analysis process, which is suitable for high-order strongly nonlinear systems. Chapters 3 and 4 systematically illustrate the basic principles of ESPM and its applications in steady-state and transient analysis of open-loop switching power converters. Based on the steady-state periodic solution obtained by ESPM, the characteristic equation analysis method (CEAM) is used in the stability analysis of open-loop converters.

From Chaps. 5 to 7, the ESPM is extended to the steady-state analysis of the closed-loop DC/DC converters. It is shown that the equivalent small-parameter method is used for the analysis of the closed-loop system of PWM converter without losing the advantages of simplicity and accuracy. At the same time, the main oscillation equation can be transformed into a single-variable high-order equation, which makes the analysis of multiple solutions and stability more convenient. In this chapter, we extend CEAM to the stability analysis of closed-loop systems. The basic principle is described in detail, and an example is analyzed to show that it is possible for more than one equilibrium point existing in some the systems, which can lead to instability under large transient conditions.

Y. Chen and B. Zhang, *Equivalent-Small-Parameter Analysis of DC/DC Switched-Mode Converter*, CPSS Power Electronics Series,
https://doi.org/10.1007/978-981-13-2574-8_9

9.1.1 Mathematical Model of PWM Close-Loop System

For the sake of analysis, this section briefly introduces the analysis method of the Chap. 5 above. According to the analysis in Chap. 5, the closed-loop system of PWM DC/DC switching power converter can be uniformly represented by the time-varying state differential equation as follows.

$$\begin{aligned} &\mathbf{G}_0(p)\mathbf{x} + G_1(p)\mathbf{f} = \mathbf{u} \\ &\mathbf{f} = \delta(t)(\mathbf{x} + \mathbf{e}) \end{aligned} \tag{9.1}$$

where the differential operator $p = d/dt$. The terms $\mathbf{G}_0(p)$ and $\mathbf{G}_1(p)$ are corresponding square matrices of the differential operator *p*, which are related to the circuit configurations and parameters of the system. And the term $\mathbf{x}$ is vector state variables of either the power conversion stage (PCS) or the feedback network (FN). The symbol $\mathbf{e}$ represents the constant vector, and $\mathbf{u}$ denotes the input source vector of either PCS or FN. The switching function $\delta(t) = 1\,(0)$ indicate the ON (OFF) state of the main switch (transistor). Given the converter going through two topologies during a switching period, that is the converter operates in CCM state, then:

$$\delta(t) = \begin{cases} 1 & 0 < t \leq d(t)T \\ 0 & d(t)T < t \leq T \end{cases} \tag{9.2}$$

where *T* is the switching period, $d(t)$ is the duty cycle, which can be expressed as a linear function of state variable $\mathbf{x}$. For linear feedback control laws, such as proportional-integral-derivative (PID) control, it can be expressed as

$$d(t) = K_0 + \mathbf{K}_1\mathbf{x}(t)|_{t=t_s} \tag{9.3}$$

where K_0 is a constant, and $\mathbf{K}_1$ is a line vector, the time symbol $t_s = dT$ stands for the switching moment of the switch from ON to OFF state. Letting $\tau = \omega t$ (here $\omega = 2\pi/T$), then the switching function δ(t) can be expanded in terms of its Fourier series as

$$\delta(t) = b_0 + \sum_{m=1}^{\infty} [b_m \exp(jm\tau) + \bar{b}_m \exp(-jm\tau)] \tag{9.4}$$

where

$$b_0 = d(t),\ b_m = \frac{1}{2}(\alpha_m - j\beta_m),\ \alpha_m = \frac{\sin 2md(t)\pi}{m\pi},\ \beta_m = \frac{1 - \cos 2md(t)\pi}{m\pi}$$

Here the integer $m = 1, 2, 3\ldots$.

It should be noted that the DC component and the coefficients of high-order harmonic terms in (9.4) are functions of duty cycle in closed-loop power systems. Based on the above discussion, the discrete PWM switching converter system in (9.1) can be described by a periodic time-continuous system by using Eq. (9.4). Thus the characteristic equation analysis method [4] (CEAM) is applicable.

9.1.2 Stability Analysis Method of Equilibrium Points: Characteristic Equation Analysis Method

The characteristic equation analysis method is an approximation algorithm for stability analysis of periodic solutions in nonlinear systems [4, 8–11], it has higher accuracy since the effects of higher harmonics on the stability have been taken into account. The periodic steady-state solution of the closed-loop system is obtained by ESPM as follows (Referring to Chap. 5 for detail):

$$\mathbf{x} = \mathbf{a}_0 + \mathbf{a}_1 \cos\tau + \mathbf{b}_1 \sin\tau + \mathbf{a}_2 \cos 2\tau + \mathbf{b}_2 \sin 2\tau + \mathbf{a}_3 \cos 3\tau + \mathbf{b}_3 \sin 3\tau + \cdots \tag{9.5}$$

When the switching frequency is large enough, the main oscillation component of the state variables can be viewed as a DC component, i.e.

$$\mathbf{x} \approx \mathbf{a}_0 \approx \mathbf{x}_0 \tag{9.6}$$

Substituting (9.6) in (9.1), one can get

$$\mathbf{G}_0(p)\mathbf{x}_0 + \mathbf{G}_1(p)\mathbf{f} = \mathbf{u} \tag{9.7}$$

where $d_0 = K_0 + \mathbf{K}_1\mathbf{x}_0$, the quasi-linearized function of $\mathbf{f}$ is:

$$\mathbf{f} \approx d_0(\mathbf{x}_0 + \mathbf{e}) \tag{9.8}$$

Defining

$$\mathbf{x}_0 = \mathbf{a}_{00} + \Delta\mathbf{a}_0 \tag{9.9}$$

then one can get

$$d_0 = K_0 + \mathbf{K}_1\mathbf{a}_{00} + \mathbf{K}_1\Delta\mathbf{a}_0 = d_{00} + \mathbf{K}_1\Delta\mathbf{a}_0 = d_{00} + \Delta d_0 \tag{9.10}$$

Here, $\mathbf{a}_{00}$ and $\Delta\mathbf{a}_0$ are the steady-state value and the disturbance of $\mathbf{x}_0$, the term d_{00} stands for the steady-state value of d_0. Then the steady-state equation can be obtained from (9.7), that is,

$$\begin{cases} \mathbf{G}_0(0)\mathbf{a}_{00} + \mathbf{G}_1(0)(\mathbf{a}_{00} + \mathbf{e})d_{00} = \mathbf{u} \\ d_{00} = K_0 + K_1\mathbf{a}_{00} \end{cases} \tag{9.11}$$

To get a_{00} and d_{00}, the nonlinear equations obtained from (9.11) can be solved applying the numerical method (depends on specific PCS and FN).

The linearized incremental function of **f** can be obtained from (9.8):

$$\Delta\mathbf{f} = \frac{\partial\mathbf{f}}{\partial d_0}\Delta d_0 + \frac{\partial\mathbf{f}}{\partial\mathbf{x}_0}\Delta\mathbf{a}_0 = (\mathbf{a}_{00} + \mathbf{e})\mathbf{K}_1\Delta\mathbf{a}_0 + d_{00}\Delta\mathbf{a}_0 \tag{9.12}$$

Introducing (9.8)–(9.10) and (9.12) to (9.7), and combining steady state equation (9.11), lead to the incremental equation as follows

$$\mathbf{G}_0(p)\Delta\mathbf{a}_0 + \mathbf{G}_1(p)\Delta\mathbf{f} = 0 \tag{9.13}$$

which can be written as

$$\{\mathbf{G}_0(p) + \mathbf{G}_1(p)[(\mathbf{a}_{00} + \mathbf{e})\mathbf{K}_1 + d_{00}\mathbf{I}]\}\Delta\mathbf{a}_0 = 0 \tag{9.14}$$

where the symbol "**I**" represents a unit matrix, and from which the characteristic equation can be resulted as

$$\det\{\mathbf{G}_0(p) + \mathbf{G}_1(0)[(\mathbf{a}_{00} + \mathbf{e})\mathbf{K}_1 + d_{00}\mathbf{I}]\} = 0 \tag{9.15}$$

Then the stability properties of equilibrium points depend on the eigenvalues of Eq. (9.15). If all the eigenvalues have negative real parts, the equilibrium points are stable, otherwise they are unstable. In addition, we can plot the state-plane trajectories according to (9.7) under different initial values, by which the stability properties can be determined directly (see the following example for detail).

When the circuit operates at a lower switching frequency, the effect on stability of higher harmonic terms should be considered according to (9.4). The detailed analysis method similar to that mentioned above can be found in [4, 9]. We note that the analysis method in [2] preserves the linear-ripple assumption in SSA, and considers that the output of the system only contains the averaged values (i.e., DC component). In fact, it can be regarded as the zero-order approximation of the method proposed in this section.

9.1.3 Example

A PWM Boost regulator with simple feedback control circuit [12], which is the same as the example in Chap. 5, is given again in Fig. 9.1, with the circuit parameters being set as follows, i.e., E = 4 V, L = 5.24 μH, C = 0.2 μF, R = 16 Ω, Vref = 0.48 V, $g_1 = -0.1$, $g_2 = 0.01$, $V_L = 0$, $V_u = 1$ V, $f_s = 5$ MHz.

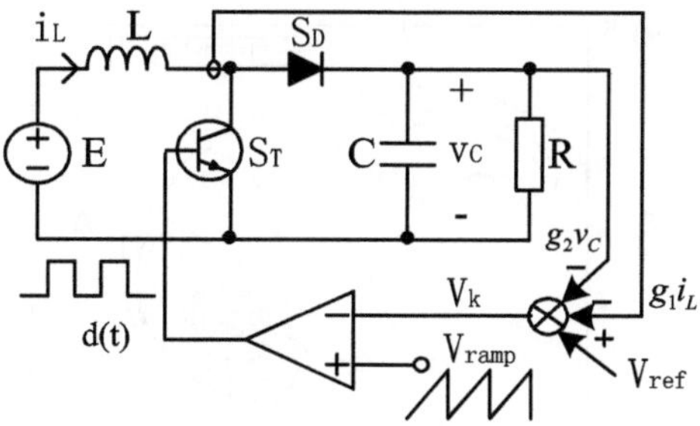

Fig. 9.1 PWM controlled boost regulator

In Fig. 9.1, the symbol V_{ramp} denotes a periodic ramp function. The system's differential equation and duty cycle equation are the same as Eqs. (9.1) and (9.3) respectively, where the square matrices, the state vector, the input vector and constant vector, and the nonlinear vector function are given as follows.

$$\mathbf{G}_0(p) = \begin{bmatrix} p & \frac{1}{L} \\ \frac{-1}{C} & p + \frac{1}{RC} \end{bmatrix}, \ \mathbf{G}_1(p) = \begin{bmatrix} 0 & \frac{-1}{L} \\ \frac{1}{C} & 0 \end{bmatrix}, \ \mathbf{x} = \begin{bmatrix} i_L \\ v_C \end{bmatrix}, \ \mathbf{u} = \begin{bmatrix} E/L \\ 0 \end{bmatrix}, \ \mathbf{f} = \delta\mathbf{x}, \ \mathbf{e} = \mathbf{0}$$

And the coefficients related to the feedback control circuit are determined by:

$$K_0 = \frac{V_{ref} - V_L}{V_u - V_L} \text{ and } \mathbf{K}_1 = \begin{bmatrix} \frac{-g_1}{V_u - V_L} & \frac{-g_2}{V_u - V_L} \end{bmatrix}$$

Here the symbols V_u and V_L stands for the maximum and minimum values of the ramp signal V_{ramp}, respectively.

There are three equilibrium points of the system according to Eq. (9.11) (We call this system a multiple-solution system), i.e.,

$$\begin{aligned}
&\mathrm{A}: d_{00} = 0.7662 \text{ and } \mathbf{a}_{00} = [4.5735\mathbf{A} \quad 17.1086\mathrm{V}]^{Tr}; \\
&\mathrm{B}: d_{00} = 1.2138 \text{ and } \mathbf{a}_{00} = [5.4692\mathrm{A} \quad -18.7091\mathrm{V}]^{Tr}; \\
&\mathrm{C}: d_{00} = 0.5 \text{ and } \mathbf{a}_{00} = [1\mathbf{A} \quad 8\mathbf{V}]^{Tr}.
\end{aligned}$$

The second solution is a virtual equilibrium point that does not exist for the actual system, as it is impossible for the duty ratio greater than 1.

Substituting the first and the third condition into Eq. (9.15) respectively, we have negative real parts of all the eigenvalues for the third equilibrium point, while there is a characteristic root with positive real part for the first equilibrium point. Obviously the point C is stable while point A is unstable.

Figures 9.2 and 9.3 give some numerical simulation results of the system under different conditions. The state-plane trajectories with different values of K_1 are shown in Fig. 9.2, where Fig. 9.2a corresponds the condition of $K_1 = [0.1, -0.01]$, from which it can be seen that the system is locally stable since only part of the solutions converge to the stable point C, and part of the solutions converge to the

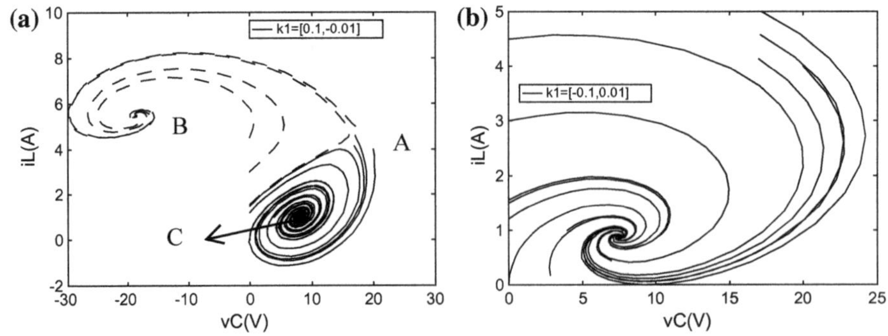

Fig. 9.2 State-plane trajectories under different feedback coefficients

virtual stable point B, yet no solutions trends to the point A, which can lead to system's instability under large-signal disturbances.

Figures 9.3a, b show the simulation curves of state variable transient response when the system is unstable. Since point B doesn't exist physically, that the state-plane trajectories converge to point B means that the output voltage is approaching towards zero, while the inductor current is becoming infinite when the system operates unstably. This will lead to practical switch-device damage or operate in non-regular regime (Referring to the waveforms in Figs. 9.3a, b). The transient responses of state variables, when the system has an initial value close to the unstable equilibrium point A, is shown in Figs. 9.3c, d, from which it can be seen that the system remains in point A for some time, then tends towards the stable point C. Figure 9.2b shows the system's state-plane trajectories when $K_1 = [-0.1, 0.01]$ (We call the system in this case a mono-solution system). As a comparison, Figs. 9.3e, f show the simulated transient waveforms under the same initial conditions as Figs. 9.3c, d. It can be seen from Figs. 9.2b, and 9.3e, f, that the system returns to stable operating point at once even under large-signal disturbance, since the system has only one stable equilibrium point and is globally stable.

9.2 Large-Signal Stability Analysis Methods of Buck Regulator

A Buck regulator circuit is given in Fig. 9.4, and its parameters (which are the same as in [13]) are set as follows, i.e., $E = 30$ V, $L = 0.9$ mH, $r_L = 0.25\ \Omega$, $C = 22\ \mu$F, $R = 20\ \Omega$, $R_1 = R_2 = 7.5$ kΩ, $R_3 = 60$ kΩ, $C_2 = 0.4\ \mu$F, $Vr = 10$ V, $v_l = 2.8$ V, $v_u = 8.2$ V, $f_s = 15$ kHz.

For the sake of analysis simplicity, the parasitic resistance of capacitors has not been taken into account. The system is in stable operation when E = 30 V and has large-signal oscillation when E = 50 V, which can be seen from the waveforms in Figs. 9.5 and 9.6, respectively. Figure 9.5 shows the steady-state waveforms of the

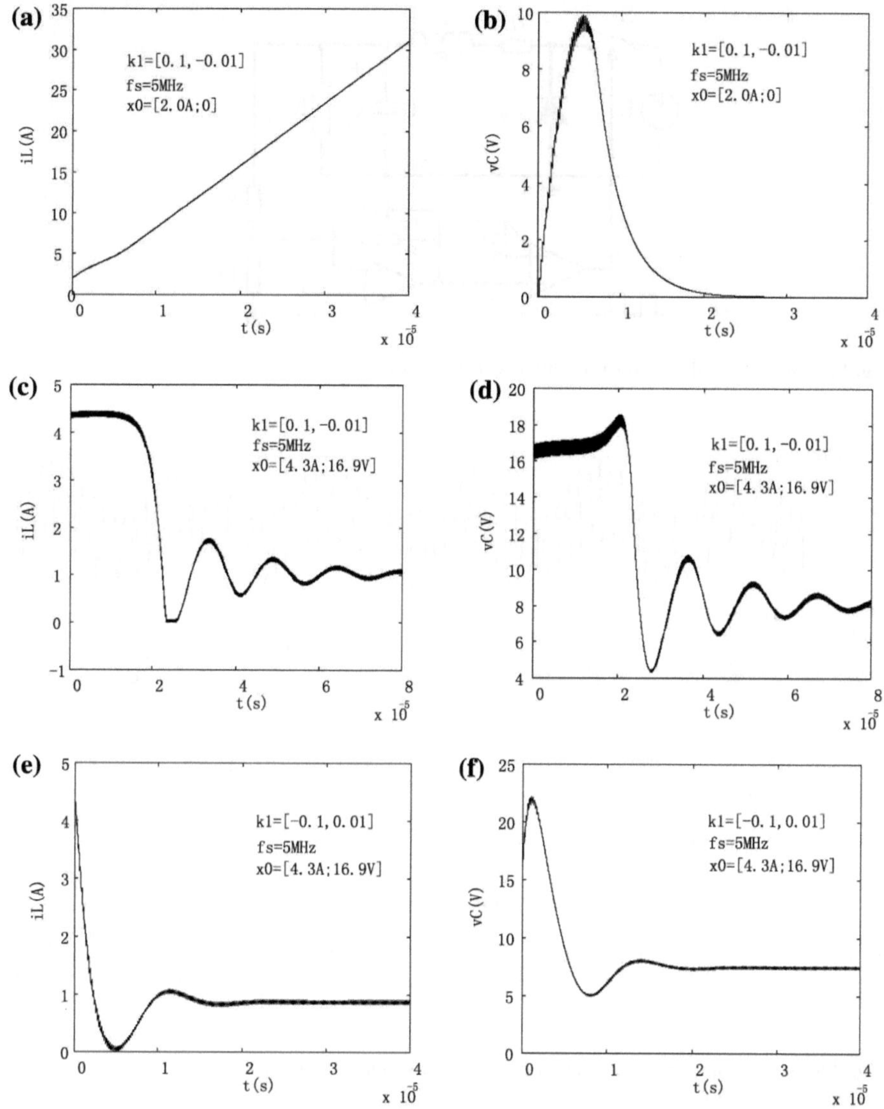

Fig. 9.3 State-variables transient response waveforms under different conditions

buck regulator with E = 50 V, in which $i(l)$ and $v(4)$ stand for the inductor current and output capacitor voltage of the main circuit respectively, and $v(52,7)$ is the capacitor voltage in the feedback loop. It can be seen obviously from Fig. 9.5 that Low-frequency Oscillations (LFO) with large amplitudes appear in the state variables.

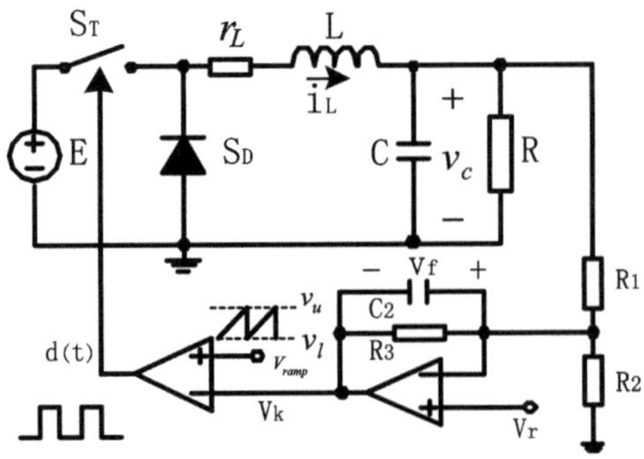

Fig. 9.4 Schematic diagram of a regulated Buck converter

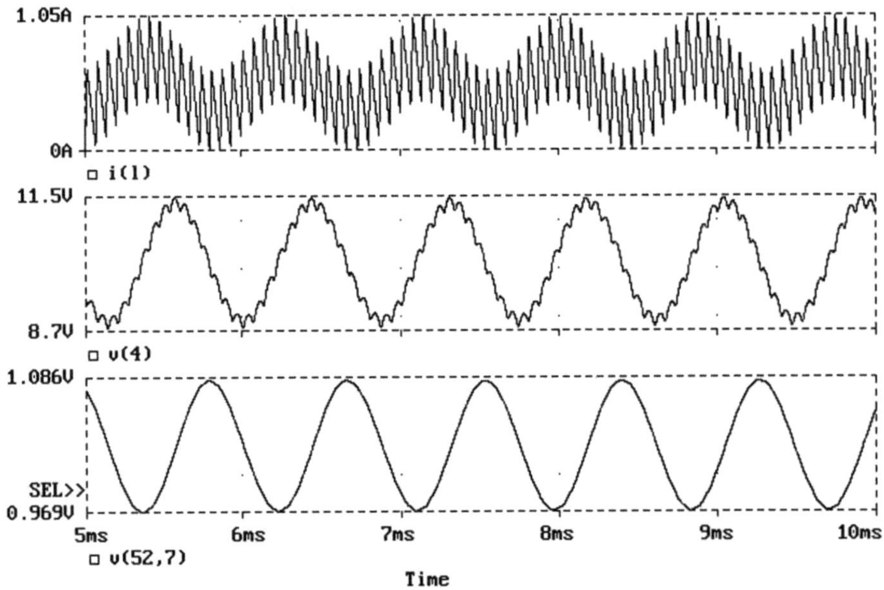

Fig. 9.5 Stead-state simulated waveforms of state variables when E = 50 V

Suppose that the inductor current and capacitor voltage are initially zero, with the converter turning on and large-signal disturbance is then applied, the system is going into the discontinuous conduction mode. The simulated transient waveforms when there is a sudden change of the input voltage from 30 to 50 V are shown in Fig. 9.6. The simulated waveforms in Figs. 9.5 and 9.6 show good agreement with those experimental results in [14]. Then, can the frequency, amplitude and phase of the low frequency oscillation signal be solved by analytic method? What is their relationship with the parameters of the circuit? There is no current literature to analyze such phenomena from an analytical perspective.

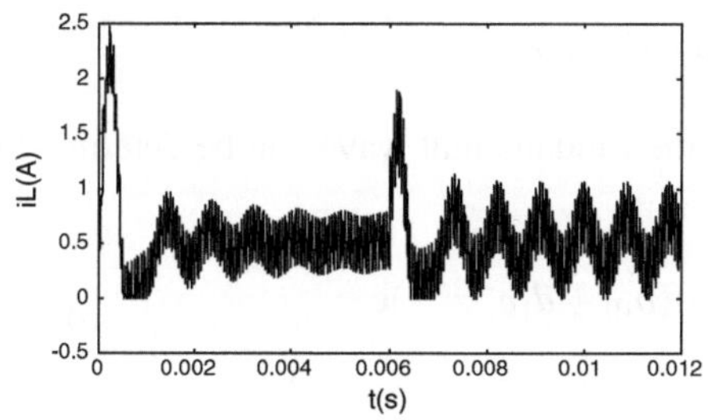

(a) Inductor current

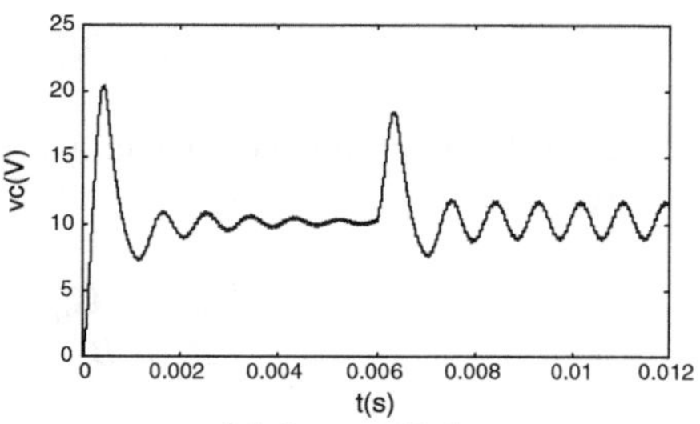

(b) Output Voltage

Fig. 9.6 Transient response of the Buck regulator when input is subjected to a sudden change from 30 to 50 V

9.2.1 Steady-State Analysis Method-1 Based on ESPM

The vector state variable differential equation and the duty cycle equation of the Buck regulator (as shown in Fig. 9.4) with CCM operation can be described as follows:

$$\begin{cases} \mathbf{G}_0(p)\mathbf{x} - \delta\mathbf{e} = \mathbf{u} \\ d(t) = K_0 + \mathbf{K}_1\mathbf{x} \end{cases} \tag{9.16}$$

where

$$G_0(p) = \begin{bmatrix} p+\frac{r_L}{L} & \frac{1}{L} & 0 \\ -\frac{1}{C} & p+\frac{1}{RC} & 0 \\ 0 & -\frac{1}{R_1C_2} & p+\frac{1}{R_3C_2} \end{bmatrix},\ e = \begin{bmatrix} \frac{E}{L} \\ 0 \\ 0 \end{bmatrix},\ u = \begin{bmatrix} 0 \\ 0 \\ -\frac{V_r}{C_2}\left(\frac{1}{R_1}+\frac{1}{R_2}\right) \end{bmatrix},\ x = \begin{bmatrix} i_L \\ v_c \\ v_f \end{bmatrix},$$
$$\delta = \begin{cases} 1 & 0 \le t \le d(t)T \\ 0 & d(t)T \le t \le T \end{cases},\ K_0 = \frac{V_r - v_l}{v_u - v_l},\ K_1 = \begin{bmatrix} 0 & 0 & \frac{-1}{v_u - v_l} \end{bmatrix}.$$

Based on the ESP method discussed in Chap. 5, the main wave and correction of the steady-state periodic solution can be derived as follows.

(1) The main wave is assumed to be

$$\mathbf{x}_0 = \mathbf{a}_{00} = [I_{00} \quad V_{00} \quad V_{f0}]$$

It can be obtained by the following equations as

$$\begin{cases} \mathbf{a}_{00} = [\mathbf{G}_0(0) - \mathbf{e}\mathbf{K}_1]^{-1}(\mathbf{u} + \mathbf{e}K_0) \\ d_0 = K_0 + K_1\mathbf{a}_{00} \end{cases} \tag{9.17}$$

(2) The first-order correction can be chosen as

$$\mathbf{x}_1 = \mathbf{a}_{11}e^{j\tau} + c.c$$

Then the amplitude coefficient $\mathbf{a}_{11}$ of the fundamental wave can be determined by

$$\begin{cases} \mathbf{a}_{11} = \mathbf{G}_0^{-1}(j\omega)(b_{10} + d_1 e^{-j2\pi d_0})\mathbf{e} \\ d_1 = B/A \end{cases} \tag{9.18}$$

And the symbols in (9.18) are defined as

$$A = 1 - \mathbf{K}_1\left[\mathbf{G}_0^{-1}(j\omega) + \overline{\mathbf{G}_0^{-1}(j\omega)}\right]\mathbf{e}$$

$$B = \left[\mathbf{G}_0^{-1}(j\omega)\cdot b_{10}e^{j2\pi d_0} + \overline{\mathbf{G}_0^{-1}(j\omega)b_{10}}\cdot e^{-j2\pi d_0}\right]\mathbf{e}$$

(3) The second-order correction can be chosen as

$$\mathbf{x}_2 = \mathbf{a}_{02} + \mathbf{a}_{22}e^{j2\tau} + \mathbf{a}_{32}e^{j3\tau} + c.c$$

Then the amplitude coefficients $\mathbf{a}_{22}$, $\mathbf{a}_{32}$ and $\mathbf{a}_{02}$ of the second-order, the third-order harmonics and the DC component, can be determined respectively by the following equations as

$$\mathbf{a}_{22} = \mathbf{G}_0^{-1}(j2\omega)(b_{20} + b_{21})\mathbf{e} \tag{9.19a}$$

$$\mathbf{a}_{32} = \mathbf{G}_0^{-1}(j3\omega)(b_{30} + b_{31})\mathbf{e} \tag{9.19b}$$

$$\mathbf{a}_{02} = [\mathbf{G}_0^{-1}(0) - \mathbf{e}K_1]^{-1}d_1\mathbf{e} \tag{9.19c}$$

Based on the discussion in Chap. 5, for a regulator with appropriate compensation in FN, the ripple (harmonic component) has little effect on the duty cycle, thus a simplified approximation is made in Eq. (9.19c), i.e., $d_2 = \mathbf{K}_1\mathbf{a}_{02}$. And the coefficient b_{mi} is defined by:

$$b_{mi} = \frac{1}{2}(\alpha_{mi} - j\beta_{mi})\ (m = 1,2,\ldots,i = 0,1,2,\ldots)$$

$$\alpha_{m0} = \frac{\sin 2m\pi d_0}{m\pi},\ \beta_{m0} = \frac{1 - \cos 2m\pi d_0}{m\pi}$$

$$\alpha_{mi} \approx 2d_i \cos 2m\pi(d_0 + \varepsilon d_1 + \cdots + \varepsilon^{i-1}d_{i-1}) \quad (i = 1,2,\ldots)$$

$$\beta_{mi} \approx 2d_i \sin 2m\pi(d_0 + \varepsilon d_1 + \cdots + \varepsilon^{i-1}d_{i-1}) \quad (i = 1,2,\ldots)$$

It can be seen that the main wave and corrections can be solved through the above linear equations for Buck regulator. For the input E = 30 V, the resulted main wave are

$$\begin{cases} \mathbf{a}_{00} = [0.5022 \quad 10.0441 \quad 0.3532] \\ d_0 = 0.3420 \end{cases}$$

And for the input E = 50 V, the resulted main wave are

$$\begin{cases} \mathbf{a}_{00} = [0.5067 \quad 10.1352 \quad 1.0819] \\ d_0 = 0.2071 \end{cases}$$

According to the CEAM method introduced in Sect. 9.1, the equilibrium point a_{00} is stable for E = 30 V, while unstable for E = 50 V. Though the stability criteriais given, characteristic parameters (such as phase, amplitude and frequency) of the LFO signal in a unstable system still can't be obtained.

9.2.2 Steady-State Analysis Method-2 Based on ESPM

According to the simulated waveforms in Fig. 9.5, the system has low frequency oscillation (LFO) when E = 50 V, thus the LFO component should be included when solving the main wave of the steady-state periodic solution by ESP.

To facilitate analysis, the parasitic parameters have not been considered, i.e., $r_L = 0$ and $R_3 = \infty$. Waveforms simulated by Pspice show that the LFO still remains when E = 50 V, which is the same as the situation when the parasitic parameters are taken into account. Then the vector state variable differential equation of the Buck regulator in CCM operation can be simplified as follows:

$$\begin{cases} di_L/dt = \frac{-1}{L}v_c + \delta \cdot \frac{E}{L} & \text{(a)} \\ dv_c/dt = \frac{1}{C}i_L - \frac{1}{RC}v_c & \text{(b)} \\ dv_f/dt = \frac{1}{R_1C_2}v_c - V_r\left(\frac{1}{R_1C_2} + \frac{1}{R_2C_2}\right) & \text{(c)} \end{cases} \tag{9.20}$$

The higher-order differential operator function of v_f can be deduced from Eq. (9.20) as

$$(p^3 + a_2p^2 + a_1p)v_f = B_1\delta - B_0 \tag{9.21a}$$

Then the main oscillation equation can be written as:

$$(p^3 + a_2p^2 + a_1p)v_{f0} = B_1 \cdot d - B_0 \tag{9.21b}$$

where d is the duty cycle, $p = d/dt$, is the differential operator, the polynomial coefficients are determined by:

$$a_2 = 1/RC, \quad a_1 = 1/LC, \quad B_1 = E/(LCR_1C_2), \quad B_0 = V_r\left(1+\frac{R_1}{R_2}\right)/(LCR_1C_2).$$

Thus Eq. (9.21a) can be solved by ESPM. Similar to Eq. (5.31) in Chap. 5, the nonlinear function δ can be expanded into series. According to Eq. (9.16), the duty cycle can be described as

$$d = k_0 + k_1 v_f$$

where $k_0 = (V_r - v_l)/(v_u - v_l)$, $k_1 = -1/(v_u - v_l)$, then Eq. (9.21b) can be rewritten as:

$$(p^3 + a_2p^2 + a_1p + a_0)v_{f0} = B_1k_0 - B_0 \tag{9.22}$$

where $a_0 = -B_1k_1$. The main wave of v_f can be described as:

$$v_{f0} = V_f + A\cos\omega_0 t \tag{9.23}$$

where ω_0, A are the angular frequency and amplitude of the LFO component respectively. Substituting (9.23) into Eq. (9.22) and with the help of harmonic balance method, one can get the following equation:

$$\begin{cases} V_f = (B_1k_0 - B_0)/a_0 \\ (-j\omega_0^3 - a_2\omega_0^2 + a_1\cdot j\omega_0 + a_0)A = 0 \end{cases} \tag{9.24}$$

It is clear the A should be zero, which means the LFO component can not be solved by this method. The main reason is that the mathematical model of closed-loop system may not be correct. It can be seen from (9.22) that the main wave is a linear time-invariant function, obviously it cannot represent the nonlinear phenomena in the system.

9.2.3 Analysis Method Considering the Saturation of Duty-Ratio

For the function of duty cycle in Eq. (9.16), only the linear part was considered, while the nonlinear part was ignored. Is this the reason for a linear wave in Eq. (9.22)?

Because in practical closed-loop system, duty cycle has saturation nonlinearity, that is, duty cycle is bounded, i.e., $d_{min} \le d \le d_{max}$, for this example, $d_{min} = 0$, $d_{max} = 1$. In the linear region, the duty cycle is determined by

$$d = \frac{2.2 - v_f}{5.4}$$

It can be seen from the above equation that when $v_f \geq 2.2V$, $d = 0$, and when $v_f \leq -3.2V$, $d = 1$. Then a third-order fitted polynomial of d can be obtained by MATLAB:

$$d = 0.4114 - 0.1995v_f + 0.0038v_f^2 + 0.004v_f^3 \tag{9.25}$$

The real curve (thick line) and fitted curve (thin line) of the duty cycle are shown in Fig. 9.7.

Then the nonlinear function of main wave can be solved based on the ESPM by introducing (9.25) into (9.21b). Suppose the main wave component still has the same expression as Eq. (9.23), and $\omega_0 = \frac{1}{\sqrt{LC}}$, then the period of the LFO component can be expressed as $T_0 = 2\pi/\omega_0$, and for this example, it can be obtained as $T_0 = 0.8841\,\text{ms}$. However, the solution of the amplitude of the LFO signal is complicated and there may be no solution.

When the power switching converter system operates in a steady state, its duty cycle is linearly determined by the state variables. However, when the system starts up or is disturbed by large signals, the duty cycle may become non-linear. Therefore, the method considering the saturation nonlinearity of the duty ratio is useful for analyzing the stability of the system, since the duty cycle is in this case for the actual PWM converter system.

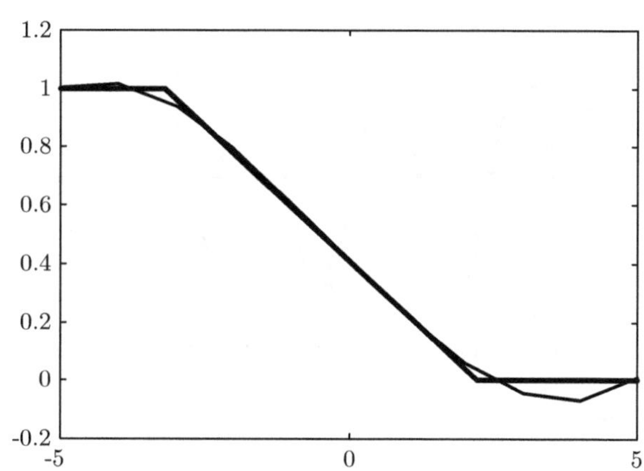

Fig. 9.7 d(t) and its 3-order fitted polynomial

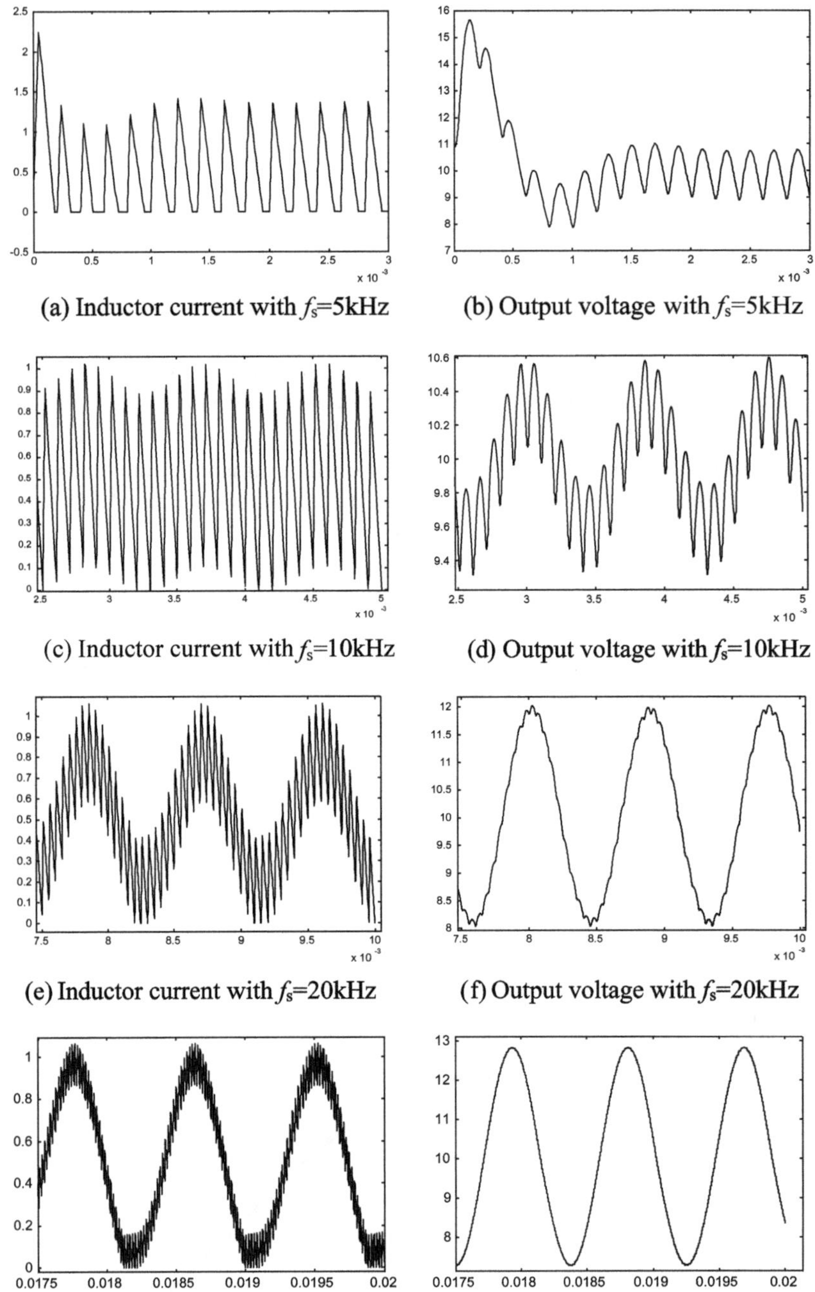

(a) Inductor current with f_s=5kHz (b) Output voltage with f_s=5kHz

(c) Inductor current with f_s=10kHz (d) Output voltage with f_s=10kHz

(e) Inductor current with f_s=20kHz (f) Output voltage with f_s=20kHz

(g) Inductor current with f_s=50kHz (h) Output voltage with f_s=50kHz

Fig. 9.8 Simulated ripple waveforms of state variables with different switching frequency

9.2.4 Stability Analysis by Numerical Simulation

9.2.4.1 LFO Under Different Switching Frequency

Figure 9.8 shows the numerical simulation waveforms of the state variables with the main circuit in steady state at different switching frequencies under E = 50 V and other parameters unchanged. It can be seen that: (1) the regulator can work stably when operates only with DCM in steady state (see Figs. 9.8a, b), or only with CCM (see Figs. 9.1 with E = 30 V); and (2) when the circuit exhibits stable low-frequency oscillation, each low-frequency oscillation period consists of several switching cycles. In some of the switching cycles, the circuit enters DCM mode, which means that the Buck regulator that generates low-frequency oscillations does not work in a single state mode, that is, periodically switching between CCM and DCM; (3) the amplitude of the low-frequency oscillation component contained in the output voltage is related to the switching frequency, the higher the frequency, the larger the amplitude of the oscillation; however, the frequency of the oscillation is independent of the switching frequency. As seen from Figs. 9.8c, d, the period T_0 of the LFO is about 0.9 ms, which is close to the value (0.8841 ms) that obtained by ESPM in Sect. 9.2.3. Yet the amplitude of the inductor current (about 1A) is almost independent of the switching frequency.

Taking the switching frequency f_s = 15 kHz as an example, it can be seen from Fig. 9.5 that the period of the low-frequency oscillation for the circuit operating in steady-state is 13 times of the switching period, that is, a low-frequency oscillation period consists of 13 switching cycles, of which there are 2 switching cycles that the regulator works in DCM. Yet according to the judgment criterion of the DCM and CCM that the regulator will always operates with CCM at the condition of $k_{crit} = 2Lf_s/R = 1.35 > 1$, which means that this criterion does not apply when the transient or system is in an unstable state of operation.

9.2.4.2 Simulations Under Different Feedback Capacitor

Figure 9.9 shows the simulation waveforms of the main circuit state variable when the feedback capacitor takes different values under the conditions of f_s = 15 kHz and E = 50 V.

When C_2 = 0.2 μF, there are some typical discontinuous switching periods of the inductor current at steady state, and the output contains a large amplitude of low frequency oscillation.

When C_2 = 0.8 μF, the converter operates in the DCM state in some switching cycles during the start-up process, while when entering the steady state, there is a small amplitude of low-frequency oscillation component, and finally the converter can work stably. From Fig. 9.9, we can see that the low frequency oscillation component in the output can be suppressed by selecting a suitable feedback capacitor of the compensation network.

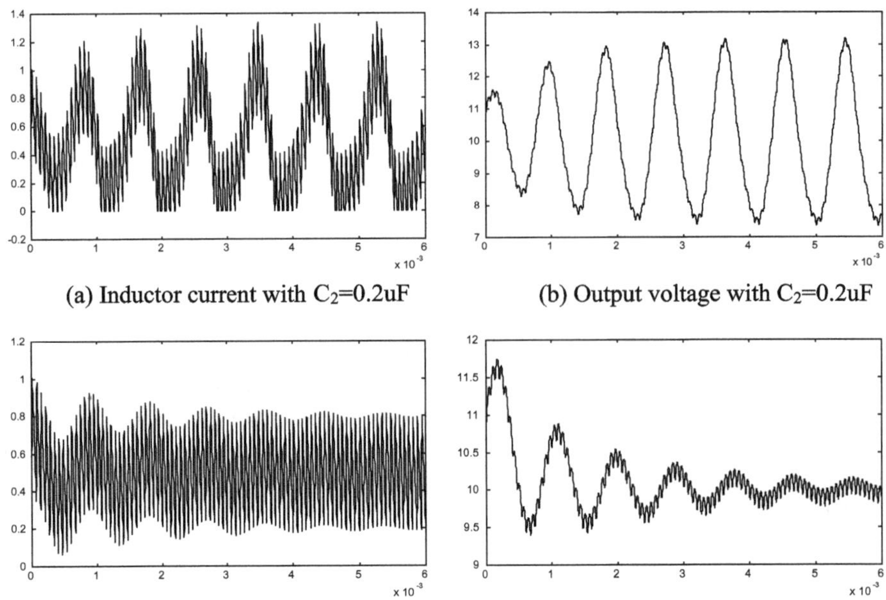

(a) Inductor current with C_2=0.2uF (b) Output voltage with C_2=0.2uF

(c) Inductor current with C_2=0.8uF (d) Output voltage with C_2=0.8uF

Fig. 9.9 Simulated ripple waveforms of state variables with different feedback capacitor

9.3 Summary

Based on the resulted periodic solutions and main oscillation equation obtained by ESP method. the stability of equilibrium points of closed-loop PWM power switching converter systems has been studied by CEAM. The example shows that the method has significant practicality and offers the advantage of directness and validness when combining with state-plane trajectories. When the system operates in a higher frequency, the results from this method is identical with that from SSA. When the switching frequency is lower, the final steady-state DC solutions is different from that from SSA due to the high-order terms resulted from ESP have corrections for DC component, and the analysis method proposed in this paper is still applicable. In addition, when switching frequency is lower and the ripple is larger, the effect of higher harmonic terms of the periodic solutions can be considered and this method is still valid and the solutions can be obtained as well. Therefore, the analysis method proposed in this paper is better than the method based on SSA in [2].

For a Buck regulator, it is difficult to analyze its large signal characteristics analytically because of the complexity of its operation: (1) The closed-loop system state differential equation is time-varying nonlinear, while the main wave equation obtained by ESP is linear; (2) The frequency of LFO signal is independent of the switching frequency, but the amplitude is related to the switching frequency;

(3) Taking fs = 15 kHz as an example, when E = 50 V, the output voltage contains an oscillation signal with a peak-to-peak value of about 3 V. Each LFO period contains 13 switching cycles, in which the system is in DCM state with 1∼2 switching cycles. It can be seen that the system does not work in a single mode during steady state, but periodically switches between CCM and DCM modes, which can be verified by simulation, and it causes the difficulty in analyzing the circuit by the analytical method. Therefore, both the DCM and CCM operating modes should be taken into account in the system's state differential equations. That is, the key is to establish an appropriate mathematical model which is suitable for large signal stability analysis.

References

1. Middlebrook RD, Cuk S (1976) A general unified approach to modelling switching-converters power stages. IEEE PESC Rec 18–34
2. Takashi Nabeshima, Kovsuke H (1982) Large-signal transient response of switching regulators. IEEE Trans AES 18(5):545–551
3. Lehman B, Bass RM (1996) Switching frequency dependant averaged models for PWM DC/DC converters. IEEE Trans PE 11(1):89–98
4. Qiu SS, Filanovsky IM (2000) Harmonic analysis of PWM converters. IEEE Trans Circuits Syst I: Fundam Theory Appl 47:1340–1349, ISSN 1057-7122
5. Qiu SS, Filanovsky IM, Stromsmoe KA (1985) An equivalent small parameter method for strongly nonlinear systems. In: Proceedings 28th Midwest symposium on circuits and systems, Louisille, pp 61–64
6. Qiu SS, Filanovsky IM (1989) Calculation of steady-state oscillation in nonlinear circuits. Int J Electronics 67(3):403–414
7. Qiu SS, Filanovsky IM, Stromsmoe KA (1987) On one combination of the harmonic balance method and perturbation techniques. Control-theory Adv Technol 3:189–195
8. Czarkowski D, Pujara LR, Kazimierczuk MK (1995) Robust stability of state-feedback control of PWM DCDC push-pull converter. IEEE Trans Ind Electron 42(1):108–111
9. Kabalan M, Singh P, Niebur D Large signal Lyapunov-based stability studies in microgrids: a review. IEEE Trans Smart Grid 99:1–9
10. Kharitonov VL (1978) Asymptotic stability of an equilibrium position of a family of systems of linear differential equations. Differ Equa 14:2086–2088
11. Uidyasager M (1976) Nonlinear systems analysis. Prentice-Hall, Inc., Englewood Cliffs, NJ
12. Lehman B, Bass RM (1996) Extensions of averaging theory for power electronic systems. IEEE Trans PE 11(4):542–553
13. Tse KK, Chung HS (1998) Decoupled technique for the simulation of PWM switching regulators using second-order-output extrapolations. IEEE Trans PE 13(2):222–234
14. Tse KK, Chung HS et al (1999) Stepwise quadratic state-space modeling technique for simulation of power electronics circuits. IEEE Trans IE 46(1):91–99

Chapter 10
Extension of ESPM to Fractional-Order DC/DC Converters

10.1 Induction

Concept from fractional-order circuits and systems has recently attracted much attention from the electrical engineering community [1]. Many novel ideas have been generated by exploiting concepts of fractional circuits. For example, fractional-order models of capacitors, inductors, memristors, and CMOS metamaterial transmission lines are built [2–11]. Practical fractional-order elements are fabricated, such as those reported in [12, 13]. In addition, new topologies of circuits based on fractional-order elements are constructed [14], though the underlying characteristics of these circuits remain elusive.

In general, systems include energy storage components could be described by integral or differential equations. However, because of their long memory characteristic, fractional-order calculus equations do not have simple analytical solutions. Therefore, numerical algorithms are widely applied to the analysis of fractional-order systems, which may cause an exponential increasing of computational efforts [15, 16]. Besides, unlike integer-order systems, fractional-order systems do not permit fractional derivatives to be defined uniformly. Thus, existing numerical methods for finding derivatives are ineffective [17]. Therefore, taking the fractional-order DC/DC Boost converter as an example, this chapter presents a time-domain modeling and analysis scheme, which is different from the methods reported in [18] and [19]. And the main theoretical basis of the scheme is derived from the formula proposed in [20], that is:

$$\frac{d^{\mu} e^{\omega t}}{dt^{\mu}} = \omega^{\mu} e^{\omega t} \tag{10.1}$$

in which μ is the order of the differential operation, it can be an integer or non-integer, and ω is the angular frequency, it could be any real number. As it can be seen from (10.1) that, the differential operation affects only the amplitude of the exponential function, but has no effect on the angular frequency. Thus, the principle

Y. Chen and B. Zhang, *Equivalent-Small-Parameter Analysis of DC/DC Switched-Mode Converter*, CPSS Power Electronics Series,
https://doi.org/10.1007/978-981-13-2574-8_10

of harmonic balance could be used, and a scheme based on the ESP method [21] would be put forward. According to the discussion in the previous chapters, the Equivalent Small Parameter (ESP) method is designed for obtaining periodic steady-state solutions of strong nonlinear systems, and it could be used for the analysis of conventional integer-order nonlinear systems, including integer-order DC/DC converters. And in this chapter, the conventional ESP method is further extended to describe the fractional-order DC/DC converters and acquire their steady-state solutions. By using this scheme, the approximate analytical time-domain solution can be derived, which is expressed as the summations of harmonic contents. Also, the steady-state performance of the converter can be analyzed.

The following contents in this chapter are organized as follows. Section 10.2 deals with the modeling of the fractional-order DC/DC Boost converter with continuous-conduction-mode (CCM) operation. In Sect. 10.2, a general form of vector state variable differential equation is presented, and its equivalent linear system is obtained. Subsequently, proper parameters for CCM operating are selected; thereby numerical simulations and comparisons are done. Following this, the rational approximation method is discussed, and then corresponding modeling simulations and experimental works are provided for a further verification. Section 10.3 further extends this method to the fractional-order converters with DCM operation. Finally, concluding remarks are given in Sect. 10.4.

10.2 Mathematical Model of Fractional-Order Boost Converter Operating in CCM

The fractional-order DC/DC Boost converter is shown in Fig. 10.1, in which both the inductor L and capacitor C are fractional-order energy-storage components, besides, the switch S_T and diode S_D are ideal.

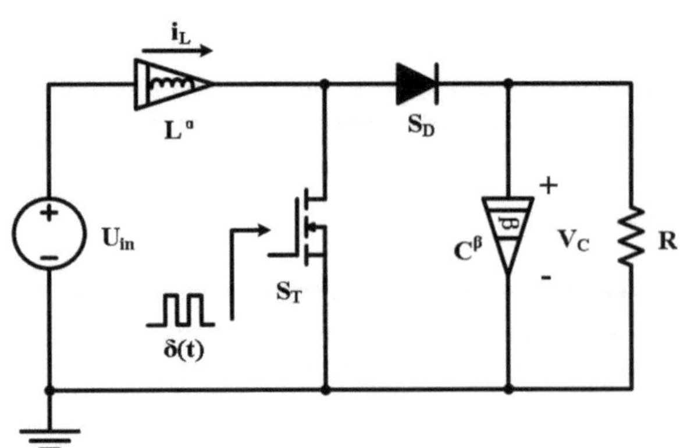

Fig. 10.1 Fractional-order DC/DC boost converter

The converter could be regarded as a system with a variable structure toggles its topologies according to the states of the switch S_T. Typically, when operating in CCM, two switch states can be identified.

State 1: switch S_T on and diode S_D off.
State 2: switch S_T off and diode S_D on.

The two switch states toggle periodically in the steady state. We assume that the converter operates in state 1 for $n\ T_s < t \leq (n + D)\ T_s$, and state 2 for $(n + D)\ T_s < t \leq (n + 1)T_s$, where n is an integer, T_s is the switching period and D is the duty cycle, which is defined as the ratio of the turn-on time of switch S_T to the switching period Ts. Then a scalar function $\delta(t)$ is introduced to represent the switching state, which is defined as $\delta(t) = 1$ and $\delta(t) = 0$ for state 1 and state 2 respectively.

According to Westerlund's model [2], relationships between voltage and current of the fractional-order components are redefined as:

$$\begin{cases} v_L(t) = L\frac{d^\alpha i_L}{dt^\alpha} \\ i_C(t) = C\frac{d^\beta v_C}{dt^\beta} \end{cases} \tag{10.2}$$

Here, α and β are orders of derivations. When $\alpha = 1$ and $\beta = 1$, the inductor and capacitor are conventional integer-order components, and this could be considered as a special case which has been discussed widely discussed in the previous chapters. Choosing i_L and v_C as state variables, the fractional-order Boost converter in state 1 can be described by the following equations as:

$$\begin{cases} \frac{d^\alpha i_L}{dt^\alpha} = \frac{1}{L}U_{in} \\ \frac{d^\beta v_C}{dt^\beta} + \frac{v_C}{RC} = 0 \end{cases} \tag{10.3}$$

And when operating in state 2, it could be described by:

$$\begin{cases} \frac{d^\alpha i_L}{dt^\alpha} + \frac{1}{L}v_C = \frac{1}{L}U_{in} \\ \frac{d^\beta v_C}{dt^\beta} - \frac{1}{C}i_L + \frac{v_C}{RC} = 0 \end{cases} \tag{10.4}$$

When we use a symbol p to represent the differential operator d/dt, then the differential operator d^α/dt^α and d^β/dt^β could be replaced by p^α and p^β separately. By combining (10.3) and (10.4) with $\delta(t)$, the fractional-order DC/DC Boost converter can be described by the following general vector state variable differential equation as

$$G_1\left(p^\alpha, p^\beta\right)\mathbf{x} + G_2\mathbf{f}(\mathbf{x}) = \mathbf{U} \tag{10.5}$$

in which the nonlinear function f(**x**) = $\delta(t)\mathbf{x}$, the state vector $\mathbf{x} = [i_L\ v_C]^{Tr}$ and the constant vector $\mathbf{U} = [U_{in}/L\ 0]^{Tr}$, where the subscript T_r means the transpose of a matrix. And the square matrices $\mathbf{G}_1(p^\alpha, p^\beta)$ and G_2 are given by:

$$\mathbf{G}_1(p^\alpha, p^\beta) = \begin{bmatrix} p^\alpha & \frac{1}{L} \\ -\frac{1}{C} & p^\beta + \frac{1}{RC} \end{bmatrix}, \ \mathbf{G}_2 = \begin{bmatrix} 0 & -\frac{1}{L} \\ \frac{1}{C} & 0 \end{bmatrix} \tag{10.6}$$

It could be seen from the Eq. (10.6), as to Boost converter, the orders of derivations can only affect the matrix G_1, yet have no effects on the matrix G_2. Similarly, other topologies of fractional-order DC/DC converter operates in CCM could be modeled by the same form of matrix Eq. (10.5), yet the matrix G_1 and G_2 may have different situations.

10.2.1 An Equivalent Model Based on the ESPM

As being discussed above, the fractional-order DC/DC converters can be described by the matrix vector differential equation (10.5). According to the principle of perturbation method, we can assume that the state vector **x** could be expanded into the following series

$$\mathbf{x} = \mathbf{x}_0 + \sum_{i=1}^{n} \varepsilon^i \mathbf{x}_i \tag{10.7}$$

where the zero order approximation $\mathbf{x}_0$ and the *i*th order approximation $\mathbf{x}_i$ are called as the main wave and corrections, and represent the main components and harmonics of state vector **x** respectively [21]. Using the same processing method as the integer-order DC/DC converter system, the switching function $\delta(t)$ is expressed as the series of the sum of main wave δ_0 and the corrections δ_i too, as shown in (10.8).

$$\delta(t) = \delta_0 + \sum_{i=1}^{\infty} \varepsilon^i \delta_i \tag{10.8}$$

Note that, here the parameter $\varepsilon = 1$, which can also be called as "intrinsic small parameter", is introduced temporarily to indicate that each term $\mathbf{x}_i$ would have different order of magnitude. For example, the term $\varepsilon^i \mathbf{x}_i$ means that the correction $\mathbf{x}_i$ is much smaller than the main wave $\mathbf{x}_0$ [22], and the inequality equation $\|\mathbf{x}_i\| < \|\mathbf{x}_{i-1}\| \ll \|\mathbf{x}_0\|$ is satisfied.

Substituting (10.7) and (10.8) into **f**(**x**) = δ(t)**x**, and combining terms which have the same order of indicator parameter ε^i, one can get the expression for f(**x**), that is in the same way with state vector **x**, i.e.,

$$\mathbf{f}(\mathbf{x}) = \mathbf{f}_0 + \varepsilon \mathbf{f}_1 + \varepsilon^2 \mathbf{f}_2 + \ldots \tag{10.9}$$

in which:

$$\begin{cases} \mathbf{f}_0 = \delta_0 \mathbf{x}_0 \\ \mathbf{f}_1 = \delta_0 \mathbf{x}_1 + \delta_1 \mathbf{x}_0 \\ \mathbf{f}_2 = \delta_0 \mathbf{x}_2 + \delta_1 \mathbf{x}_1 + \delta_2 \mathbf{x}_0 \\ \quad \ldots \end{cases} \tag{10.10}$$

Similarly, $\mathbf{f}_0$ and $\mathbf{f}_i$ represent the main components and corrections of nonlinear function $\mathbf{f}(\mathbf{x})$ respectively.

Based on (10.1), the order of the fractional differential operation has no effect on the angular frequency of the exponential function. Then the principle of harmonic balance could be available to solve non-integer order differential equations too. Thus, similar to the method for integer-order DC/DC converters, the correction $\mathbf{x}_i$ in (10.7) can also be chosen as

$$\mathbf{x}_i = \sum_{k \in \{E_{ir}\}} \mathbf{x}_{ik} = \mathbf{a}_{i0} + \sum \left(\mathbf{a}_{ik} e^{jk\tau} + c.c \right) \tag{10.11}$$

in which *c.c* denotes conjugate items, *k* is an integer, and $\tau = \omega t$ (where $\omega = 2\pi/T_s$) is the normalized time. The spectral content $\{E_0\}$ of vector $\mathbf{x}_0$, which consists of a set of numbers representing relative frequencies of harmonics, is dictated by the physical sense of the study object. For example, as DC/DC converters have low-pass filtering properties, we usually assume that the main wave $\mathbf{x}_0$ includes DC components only, i.e. $\mathbf{x}_0 = a_{00} = [I_{00}\ V_{00}]^{Tr}$, thus the set $\{E_0\}$ is $\{0\}$. Similarly, the set $\{E_{ir}\}$ contains the spectral contents for each $\mathbf{x}_i$, yet it is not known in advance and is determined in the process of iterated calculations, i.e. it is determined by $\mathbf{x}_0$, $\mathbf{x}_1$, …, and $\mathbf{x}_{i-1}$ during the iteration. The term $\mathbf{a}_{i0}$ and $\mathbf{a}_{ik}$ represent the DC components and the *k*th harmonic magnitude that belong to $\mathbf{x}_i$ respectively.

Performing the same procedure, δ(t) is expanded into its Fourier series form, i.e., the sum of a constant and some sinusoidal functions:

$$\delta(t) = b_0 + \sum_{k=1}^{\infty} \left(b_k e^{jk\tau} + \bar{b}_k e^{-jk\tau} \right) \tag{10.12}$$

Here, $b_0 = \frac{1}{T}\int_0^T \delta dt = D$, $b_k = \frac{1}{2}(\alpha_k - j\beta_k)$ and $\bar{b}_k$ is the conjugate complex of b_k, in which:

$$\alpha_k = \frac{2}{T}\int_0^{T_s} \delta(t)\cos(k\omega t)dt = \frac{\sin(2Dk\pi)}{k\pi} \tag{10.13}$$

and

$$\beta_k = \frac{2}{T}\int_0^{T_s} \delta(t)\sin(k\omega t)dt = \frac{1-\cos(2Dk\pi)}{k\pi} \tag{10.14}$$

In general, δ_0 and δ_i can be chosen as:

$$\begin{cases} \delta_0 = b_0 + b_1 e^{j\tau} + c.c \\ \delta_i = b_{2i}e^{j2i\tau} + b_{2i+1}e^{j(2i+1)\tau} + c.c \end{cases} \tag{10.15}$$

Because of the nonlinear characteristics of $\mathbf{f}(\mathbf{x})$, when the related $\mathbf{x}_i$ and δ_i are substituted into Eq. (10.10), the spectral content of the term $\mathbf{f}_i$ becomes wider than the spectral content of the term $\mathbf{x}_i$, and thus each $\mathbf{f}_i$ in Eq. (10.10) can be represented as:

$$\mathbf{f}_i = \mathbf{f}_{im} + \varepsilon\mathbf{R}_{i+1} \quad (i = 0, 1, 2\cdots) \tag{10.16}$$

In (10.16), the main term $\mathbf{f}_{0m}$ contains the terms that have the same spectral contents with $\mathbf{x}_0$, i.e., both $\mathbf{f}_{0m}$ and $\mathbf{x}_0$ have the same harmonics. The remaining term $\mathbf{R}_1$ includes the rest contents which are different with $\mathbf{x}_0$. Generally, the magnitude of $\mathbf{R}_1$ is smaller than $\mathbf{f}_{0m}$, so the indicator parameter ε is used here again. These newly generated harmonics in $\mathbf{f}_0$ determine the spectral content in the following spectral content set $\{E_{1r}\}$, that is to say, the term $\mathbf{f}_{1m}$ has the same spectral content set $\{E_{1r}\}$ with the correction term $\mathbf{x}_1$. Similarly, the content of harmonics in the term $\mathbf{R}_2$ determines the following term $\mathbf{x}_2$, etc. Then substituting (10.7) and (10.10) into the vector state variable differential Eq. (10.5), it could be represented as:

$$G_1\left(p^\alpha, p^\beta\right)\left(\mathbf{x}_0 + \sum_{i=1}^{n}\varepsilon^i\mathbf{x}_i\right) + G_2\left(\mathbf{f}_0 + \sum_{i=1}^{n}\varepsilon^i\mathbf{f}_i\right) = \mathbf{U} \tag{10.17}$$

Then introducing Eq. (10.11) and Eq. (10.16) into Eq. (10.17), and equating the terms with the same ε^i on the right- and left-hand sides, we can get the equivalent linear non-integer order differential equations as in Eq. (10.18):

$$\begin{cases} G_1\left(p^\alpha, p^\beta\right)\mathbf{x}_0 + G_2\mathbf{f}_{0m} = \mathbf{U} \\ G_1\left(p^\alpha, p^\beta\right)\mathbf{x}_1 + G_2(\mathbf{f}_{1m} + \mathbf{R}_1) = 0 \\ G_1\left(p^\alpha, p^\beta\right)\mathbf{x}_2 + G_2(\mathbf{f}_{2m} + \mathbf{R}_2) = 0 \\ \cdots \end{cases} \tag{10.18}$$

Equations in Eq. (10.18) are all linear and they could be solved step by step by using the method of harmonic balance, where the first equation is used to obtain the main wave $\mathbf{x}_0$, and following equations are used to solve the corrections $\mathbf{x}_1$, $\mathbf{x}_2$, …, etc. If the harmonic amplitude in the *k*th correction $\mathbf{x}_k$ is much smaller than those in the $(k - 1)$th correction $\mathbf{x}_{k-1}$, the calculation process is terminated. Hence, according to discussions above, the steady-state solution for the vector state variable Eq. (10.5) can be approximated by $\mathbf{x} \approx \mathbf{x}_0 + \mathbf{x}_1 + \mathbf{x}_2 + \ldots$.

In theory, under the criterion of minimum mean-square error, the Fourier series is the best approximation of the periodic signal. From the discussion above, we can see that the nonlinear switching function δ(t) is expanded into Fourier series, and in fact the expansion (10.7) of the state vector $\mathbf{x}$ is based on the principle that the periodic function can be represented as Fourier series. Actually, for the nonlinear system with low-pass filtering property, the magnitudes of harmonics with order higher than 3 are small. It means that, the steady-state solution can be approximated by finite numbers of corrections $\mathbf{x}_i$. Thus based on the principle in [23], a tolerance error index is introduced here as a termination criterion of the iterated calculation of the correction $\mathbf{x}_i$, which is defined as the ratio between 2-norm of $\mathbf{a}_{iK}$ and $\mathbf{a}_{00}$, i.e.:

$$Tolerance = \frac{\|\mathbf{a}_{iK}\|_2}{\|\mathbf{a}_{00}\|_2} \times 100\% \tag{10.19}$$

in which $\mathbf{a}_{iK}$ represents the magnitude of the highest Kth harmonic in the correction $\mathbf{x}_i$. In this chapter the tolerance error index is set to be 1%, which is sufficient for most engineering applications. And the correcting process is terminated once this index is smaller than 1%, which means the harmonic magnitude in $\mathbf{x}_i$ is much smaller than the DC component, and the approximate solution obtained under this condition is accurate enough for most DC/DC converters.

Usually, as the low-pass filtering property of DC/DC converters, the magnitudes of harmonics with much higher frequencies are small, so they are neglected, and only the first three equations in Eq. (10.18) need to be solved, these solutions suffice for most technical applications.

10.2.2 Periodic Steady-State Solutions of Fractional-Order Boost Converter in CCM

As it is known that DC/DC converters have low-pass properties, the main wave $\mathbf{x}_0$ to the steady-state value of state variable $\mathbf{x}$ could be chosen as $\mathbf{x}_0 = a_{00} = [I_{00}\ V_{00}]^T$, here I_{00} and V_{00} are DC values. According to the basic principle of ESP, $\mathbf{f}_{0m}$ should include the same harmonics as $\mathbf{x}_0$, and the rest harmonics in $\mathbf{f}_0$ belong to $\mathbf{R}_1$, hence it can be deduced that

$$\mathbf{f}_{0m} = \mathbf{a}_{00} b_0 = \mathbf{a}_{00} D \tag{10.20a}$$

$$\mathbf{R}_1 = \mathbf{a}_{00}(b_1 e^{j\tau} + \bar{b}_1 e^{-j\tau}) \tag{10.20b}$$

then introducing $\mathbf{x}_0$ and f_{0m} into the first equation of Eq. (10.18) gives:

$$(G_{10} + G_2 D)\mathbf{a}_{00} = \mathbf{U} \tag{10.21}$$

The coefficient matrix $G_{10} = G_1(0^{\alpha},0^{\beta})$, therefore, Eq. (10.21) can be rewritten in the matrix form as:

$$\left(\begin{bmatrix} 0 & \frac{1}{L} \\ -\frac{1}{C} & 0 + \frac{1}{RC} \end{bmatrix} + D\begin{bmatrix} 0 & -\frac{1}{L} \\ \frac{1}{C} & 0 \end{bmatrix}\right)\begin{bmatrix} I_{00} \\ V_{00} \end{bmatrix} = \begin{bmatrix} \frac{U_{in}}{L} \\ 0 \end{bmatrix} \tag{10.22}$$

Thereby the term $\mathbf{a}_{00}$ could be obtained. Then according to the content of harmonics in the term $\mathbf{R}_1$ (there only exits the first-harmonic), we can deduce that the spectral content of vector $\mathbf{x}_1$ is $\{E_{1r}\} = \{1\}$, and then the first order correction $\mathbf{x}_1$ can be assumed to be:

$$\mathbf{x}_1 = \mathbf{a}_{11}e^{j\tau} + \bar{\mathbf{a}}_{11}e^{-j\tau} \tag{10.23}$$

in which $\bar{\mathbf{a}}_{11}$ means complex-conjugate, and $\mathbf{a}_{11} = [I_{11}\ V_{11}]^{Tr}$. Substituting $\mathbf{x}_0$, $\mathbf{x}_1$, δ_0 and δ_1 into Eq. (10.10), the following the expressions for $\mathbf{f}_{1m}$ and $\mathbf{R}_2$ can be derived as:

$$\begin{cases} \mathbf{f}_{1m} = b_0\mathbf{a}_{11}e^{j\tau} + c.c \\ \mathbf{R}_2 = (b_1\bar{\mathbf{a}}_{11} + \bar{b}_1\mathbf{a}_{11}) + (b_1\mathbf{a}_{11} + b_2\mathbf{a}_{00})e^{j2\tau} + b_3\mathbf{a}_{00}e^{j3\tau} + c.c \end{cases} \tag{10.24}$$

where $\mathbf{f}_{1m}$ should consist of all the terms in $\mathbf{f}_1$ with the same harmonics as $\mathbf{x}_1$, and $\mathbf{R}_2$ includes the rest. Introducing $\mathbf{x}_1$, $\mathbf{R}_1$ and $\mathbf{f}_{1m}$ into the second equation of Eq. (10.18) leads to the following equation as

$$G_1(p^{\alpha}, p^{\beta})\mathbf{a}_{11}e^{j\tau} + G_2\left(b_0\mathbf{a}_{11}e^{j\tau} + b_1\mathbf{a}_{00}e^{j\tau}\right) = 0 \tag{10.25}$$

According to Eq. (10.1), we have $p^{\alpha}e^{j\omega t} = (j\omega)^{\alpha}e^{j\omega t}$ and $p^{\beta}e^{j\omega t} = (j\omega)^{\beta}e^{j\omega t}$, then the vector $\mathbf{a}_{11}$ could be found by solving the following equation

$$(G_{11} + G_2 b_0)\mathbf{a}_{11} = -G_2 b_1\mathbf{a}_{00} \tag{10.26}$$

in which the coefficient matrix $G_{11} = G_1[(j\omega)^{\alpha}, (j\omega)^{\beta}]$ is obtained by replacing the operator p with $j\omega$ in Eq. (10.6). Hence Eq. (10.26) can be rewritten in the matrix form as:

$$\left(\begin{bmatrix} (j\omega)^{\alpha} & \frac{1}{L} \\ -\frac{1}{C} & (j\omega)^{\beta} + \frac{1}{RC} \end{bmatrix} + \begin{bmatrix} 0 & -\frac{D}{L} \\ \frac{D}{C} & 0 \end{bmatrix}\right)\begin{bmatrix} I_{11} \\ V_{11} \end{bmatrix} = \begin{bmatrix} 0 & \frac{1}{L} \\ \frac{-1}{C} & 0 \end{bmatrix}\begin{bmatrix} b_1 I_{00} \\ b_1 V_{00} \end{bmatrix} \tag{10.27}$$

Obviously, due to the effects of $(j\omega)^{\alpha}$ and $(j\omega)^{\beta}$ in the matrix G_{11}, the harmonic amplitude in the correction $\mathbf{x}_1$ has a complex nonlinear relationship with the order α and β.

Next, based on harmonics of $\mathbf{R}_2$, the spectral content of $\mathbf{x}_2$ is determined, that is $E_{2r} = \{0,2,3\}$. Therefore the second-order correction $\mathbf{x}_2$ can be assumed to be:

$$\mathbf{x}_2 = \mathbf{a}_{20} + \mathbf{a}_{22}e^{j2\tau} + \bar{\mathbf{a}}_{22}e^{-j2\tau} + \mathbf{a}_{23}e^{j3\tau} + \bar{\mathbf{a}}_{23}e^{-j3\tau} \tag{10.28}$$

Here $\mathbf{a}_{20} = [I_{20}\ V_{20}]^{Tr}$ is the vector of DC values, it will give the corrections of DC components in $\mathbf{a}_{00}$, $\mathbf{a}_{22} = [I_{22}\ V_{22}]^{Tr}$ and $\mathbf{a}_{23} = [I_{23}\ V_{23}]^{Tr}$ are the magnitudes of the second- and third-harmonics of the state variable $\mathbf{x}$ respectively.

Similarly, substituting $\mathbf{x}_0$, $\mathbf{x}_1$, $\mathbf{x}_2$, and δ_0, δ_1, δ_2 into $\mathbf{f}_2$, we can get the expression for $\mathbf{f}_{2m}$, which should include the same harmonics as $\mathbf{x}_2$.

$$\mathbf{f}_{2m} = b_0\mathbf{a}_{20} + (b_0\mathbf{a}_{22} + \bar{b}_1\mathbf{a}_{23} + b_3\bar{\mathbf{a}}_{11})e^{j2\tau} + (b_0\mathbf{a}_{23} + b_1\mathbf{a}_{22} + b_2\mathbf{a}_{11})e^{j3\tau} + c.c \tag{10.29}$$

Next, substituting $\mathbf{x}_2$, $\mathbf{R}_2$ and $\mathbf{f}_{2m}$ into the third equation of Eq. (10.18), and still using the method of harmonic balance, one can obtain the following three equations as:

$$(G_{10} + G_2b_0)\mathbf{a}_{02} = -G_2(b_1\bar{\mathbf{a}}_{11} + \bar{b}_1\mathbf{a}_{11}) \tag{10.30}$$

$$(G_{12} + G_2b_0)\mathbf{a}_{22} = -G_2(b_1\mathbf{a}_{11} + b_2\mathbf{a}_{00} + b_3\bar{\mathbf{a}}_{11}) \tag{10.31}$$

$$(G_{13} + G_2b_0)\mathbf{a}_{23} = -G_2(b_1\mathbf{a}_{22} + b_2\mathbf{a}_{11} + b_3\mathbf{a}_{00}) \tag{10.32}$$

Likewise, the coefficient matrix $\mathbf{G}_{1k} = \mathbf{G}_1[(jk\omega)^{\alpha}, (jk\omega)^{\beta}]$ (k = 0, 2, 3) can be obtained by replacing the operator p with $jk\omega$ in Eq. (10.6). It should be noticed that during the derivation process, the term $\bar{b}_1\mathbf{a}_{23}$ in Eq. (10.29) is omitted with respect to the term $(b_0\mathbf{a}_{22} + b_3\bar{\mathbf{a}}_{11})$, as according to the principle of ESP, the magnitudes of harmonics with higher frequencies are smaller than those with lower frequencies. Thus Eq. (10.30) can be rewritten in matrix form as:

$$\left(\begin{bmatrix} 0 & \frac{1}{L} \\ -\frac{1}{C} & \frac{1}{RC} \end{bmatrix} + \begin{bmatrix} 0 & -\frac{D}{L} \\ \frac{D}{C} & 0 \end{bmatrix}\right)\begin{bmatrix} I_{02} \\ V_{02} \end{bmatrix} = \begin{bmatrix} 0 & \frac{1}{L} \\ \frac{-1}{C} & 0 \end{bmatrix}\begin{bmatrix} \bar{b}_1 I_{11} + b_1\bar{I}_{11} \\ \bar{b}_1 V_{11} + b_1\bar{V}_{11} \end{bmatrix} \tag{10.33}$$

where $b_0 = D$ and $b_1 = [\sin 2D\pi - j(1 - \cos 2D\pi)]/2\pi$.

And the Eqs. (10.31) and (10.32) can be rewritten in matrix form as shown in (10.34) and (10.35) respectively, where the coefficient $b_k = [\sin 2\,kD\pi - j(1 - \cos 2kD\pi)]/2k\pi$ (k = 1,2,3).

$$\begin{aligned} &\left(\begin{bmatrix} (j2\omega)^{\alpha} & \frac{1}{L} \\ -\frac{1}{C} & (j2\omega)^{\beta} + \frac{1}{RC} \end{bmatrix} + \begin{bmatrix} 0 & -\frac{D}{L} \\ \frac{D}{C} & 0 \end{bmatrix}\right)\begin{bmatrix} I_{22} \\ V_{22} \end{bmatrix} \\ &= \begin{bmatrix} 0 & \frac{1}{L} \\ \frac{-1}{C} & 0 \end{bmatrix}\begin{bmatrix} b_1 I_{11} + b_2 I_{00} + b_3\bar{I}_{11} \\ b_1 V_{11} + b_2 V_{00} + b_3\bar{V}_{11} \end{bmatrix} \end{aligned} \tag{10.34}$$

$$\left(\begin{bmatrix} (j3\omega)^{\alpha} & \frac{1}{L} \\ -\frac{1}{C} & (j3\omega)^{\beta} + \frac{1}{RC} \end{bmatrix} + \begin{bmatrix} 0 & -\frac{D}{L} \\ \frac{D}{C} & 0 \end{bmatrix}\right)\begin{bmatrix} I_{23} \\ V_{23} \end{bmatrix} = \begin{bmatrix} 0 & \frac{1}{L} \\ \frac{-1}{C} & 0 \end{bmatrix}\begin{bmatrix} b_1 I_{22} + b_2 I_{11} + b_3 I_{00} \\ b_1 V_{22} + b_2 V_{11} + b_3 V_{00} \end{bmatrix} \tag{10.35}$$

Following this, the term $\mathbf{a}_{ik} = [I_{ik}, V_{ik}]^{Tr}$ could be solved step by step, and the detailed information is listed in Table 10.1, in which, the $\Delta(jk\omega)$ is defined as:

$$\Delta(jk\omega) = (1 - D)^2 + (jk\omega)^{\alpha+\beta} LC + (jk\omega)^{\alpha} L/R \tag{10.36}$$

Additionally, it should be noted that there is a DC term $\mathbf{a}_{20}$ in $\mathbf{x}_2$ which is related to the term $\mathbf{a}_{11}$. Despite there is no α and β in expressions of G_{10} and G_2, but the term $\mathbf{a}_{11}$ is determined by the order dependent matrix G_{11}, so the DC term $\mathbf{a}_{20}$ is also order dependent. That is to say, orders of derivative have effects on the values of DC component of state variables in non-integer order Boost converter.

Hence, according to the discussions above, with the main wave $\mathbf{x}_0$, the corrections $\mathbf{x}_1$ and $\mathbf{x}_2$ being obtained, the approximate periodic solutions for non-integer order Boost converter could be expressed as

$$\begin{aligned} \mathbf{x} &= \begin{bmatrix} i_L & v_C \end{bmatrix}^{Tr} \\ &= (\mathbf{a}_{00} + \mathbf{a}_{02}) + \left(\mathbf{a}_{11} e^{j\tau} + \mathbf{a}_{22} e^{j2\tau} + \mathbf{a}_{32} e^{j3\tau} + c.c\right) \end{aligned} \tag{10.37}$$

The components of this vector are

$$\begin{aligned} i_L \approx\ & (I_{00} + I_{02}) + 2(\operatorname{Re} I_{11} \cos \omega t - \operatorname{Im} I_{11} \sin \omega t \\ & + \operatorname{Re} I_{22} \cos 2\omega t - \operatorname{Im} I_{22} \sin 2\omega t + \operatorname{Re} I_{32} \cos 3\omega t - \operatorname{Im} I_{32} \sin 3\omega t) \end{aligned} \tag{10.38}$$

in amperes, and

$$\begin{aligned} v_C \approx\ & (V_{00} + V_{02}) + 2(\operatorname{Re} V_{11} \cos \omega t - \operatorname{Im} V_{11} \sin \omega t \\ & + \operatorname{Re} V_{22} \cos 2\omega t - \operatorname{Im} V_{22} \sin 2\omega t + \operatorname{Re} V_{32} \cos 3\omega t - \operatorname{Im} V_{32} \sin 3\omega t) \end{aligned} \tag{10.39}$$

Table 10.1 Values of $\mathbf{a}_{ik}$

i	k	$\mathbf{a}_{ik}$	
		V_{ik}	I_{ik}
0	0	$\frac{U_{in}}{(1-D)}$	$\frac{U_{in}}{R(1-D)^2}$
1	1	$\frac{(1-D)b_1 V_{00} - (j\omega)^{\alpha} L b_1 I_{00}}{\Delta(j\omega)}$	$\frac{b_1 V_{00} - (1-D)V_{11}}{(j\omega)^{\alpha} L}$
2	0	$\frac{\bar{b}_1 V_{11} + b_1 \overline{V}_{11}}{(1-D)}$	$\frac{1}{(1-D)}\left(\frac{V_{02}}{R} + \bar{b}_1 I_{11} + b_1 \overline{I}_{11}\right)$
	2	$\frac{(1-D)(b_1 V_{11} + b_2 V_{00} + b_3 \overline{V}_{11}) - (j2\omega)^{\alpha} L(b_1 I_{11} + b_2 I_{00} + b_3 \overline{I}_{11})}{\Delta(j2\omega)}$	$\frac{(b_1 V_{11} + b_2 V_{00} + b_3 \overline{V}_{11}) - (1-D)V_{22}}{(j2\omega)^{\alpha} L}$
	3	$\frac{(1-D)(b_1 V_{22} + b_2 V_{11} + b_3 V_{00}) - (j3\omega)^{\alpha} L(b_1 I_{22} + b_2 I_{11} + b_3 I_{00})}{\Delta(j3\omega)}$	$\frac{(b_1 V_{22} + b_2 V_{11} + b_3 V_{00}) - (1-D)V_{23}}{(j3\omega)^{\alpha} L}$

in volts. In Eq. (10.38) and Eq. (10.39), the symbol **Re** $\mathbf{x}_{ik}$ and **Im** $\mathbf{x}_{ik}$ represent the real- and imaginary-part of the complex term $\mathbf{x}_{ik}$ separately. Obviously, the DC values and harmonic components of state variables can be obtained from Eqs. (10.38) and (10.39), and we can find that they are both order dependent.

10.2.3 Discussion of the Order-Related Phenomena

According to the discussion on the boundary between modes of the conventional Boost converter, if the converter operates in CCM, the average value of the inductor current i_L must be greater than its peak-peak ripple [24]. Here, we use $\bar{I}_L$ and Δi_L to represent the average value and the ripple respectively. Then $\bar{I}_L$ can be approximated by the summation of I_{00} and I_{20}, and the ripple Δi_L can be derived by the principle of inductor volt-second balance. Hence, the CCM-operating criterion of the fractional-order Boost converter can be calculated by the Eq. (10.3) under Caputo's definition of fractional calculus [17], that is

$$\Phi = \bar{I}_L - \Delta i_L \approx I_{00} + I_{20} - \frac{U_{in}(DT)^{\alpha}}{2\alpha L\Gamma(\alpha)} \tag{10.40}$$

in which $\Gamma(\alpha)$ can be obtained by the "Gamma function" in MATLAB. According to (10.40), if $\Phi > 0$, the Boost converter would be in CCM, otherwise, it would be in DCM operation.

For the circuit in Fig. 10.1, parameters are chosen as list in the Table 10.2.

Note the unit of fractional-order capacitor, it is μF/(second)$^{1-\beta}$, instead of μF [2]. However, since there is no unified definition of the unit of the fractional-order inductor now, the fractional inductance still takes μH as the unit. Then, a contour plot of the CCM-operating boundary can be obtained, as shown in Fig. 10.2, in which the area filled with color represents the CCM-operating region.

Additionally, according to the solutions obtained by the ESPM, a statistical analysis of harmonics under different fractional orders is made, as shown in Fig. 10.3.

It could be seen from Fig. 10.3 that, the RMS values of harmonics increase with the decrease of order α and β, which will lead to the increasing of the ripples with

Table 10.2 Circuit parameters of the fractional-order boost converter in CCM

Parameters	Values
Input voltage U_{in}	12 V
Switching frequency f_s	25 kHz
Fractional inductance L	477 μH
Fractional capacitance C	10 μF/(s)$^{1-\beta}$
Load resistance R	40 Ω
Duty ratio D	0.4

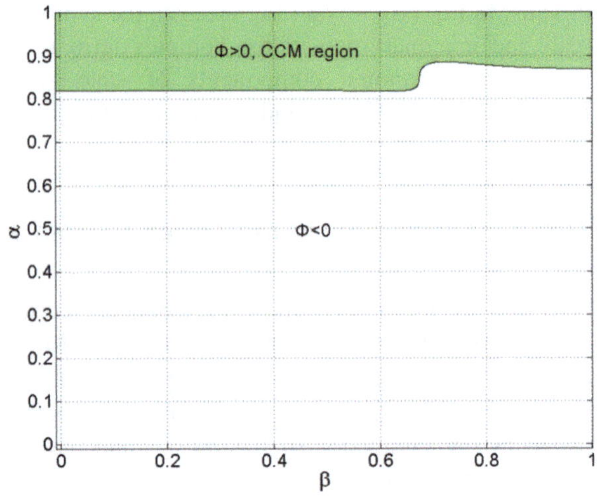

Fig. 10.2 The CCM-operating boundary

state variables, and is usually undesirable for DC/DC converters. Besides, orders also affect the terminating criterion of the ESPM, as shown in Fig. 10.4.

As can be seen in Fig. 10.4 that, for the condition of $(\alpha, \beta) = (1, 1)$, the tolerance error index for the term $\mathbf{a}_{11}$ is less than 1% when just one time of correction is done, though one more time of correction is still accomplished. When $(\alpha, \beta) = (1, 0.9)$, the index for a_{11} is greater than 1%, which means more than one time of corrections are required, thus the second correction is performed. One can also see that the tolerance error index increases with the decrease of α and β, which means that more correction times, should be done for smaller fractional orders.

10.2.4 Numerical Simulations and Comparisons

Using the circuit parameters shown in Table 10.2, the order α and β are chosen to be no less than 0.8 in following works. The comparison is made among the results from different methods. An Adams-Bashforth-Moulton typed predictor-corrector method (abbreviated as PECE-ABM) is applied to perform numerical simulations [15]. Equations (10.3) and (10.4) can be solved in each switching state. Then a cycle-by-cycle numerical simulation is carried out. And the results obtained in each switching state would be collected in one matrix, in which values of the inductor current i_L and the capacitor voltage v_C are included. The final matrix obtained contains the full information about state variables from the zero initial state to the steady-state. Also, by using Oustaloups' approximation method [25], a MATLAB/Simulink based simulation is offered. Then results from these two methods can be used as the criterion. The DC components obtained from the two above mentioned numerical methods and the ESPM are listed together in Table 10.3.

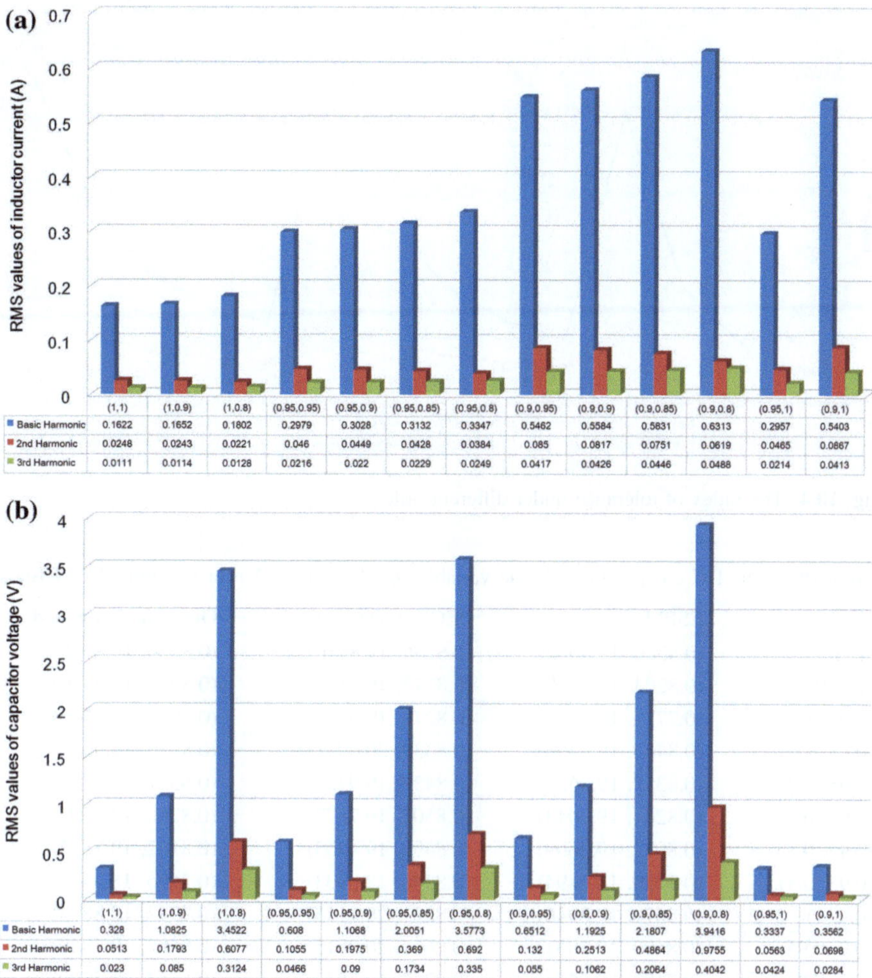

(a)

	(1,1)	(1,0.9)	(1,0.8)	(0.95,0.95)	(0.95,0.9)	(0.95,0.85)	(0.95,0.8)	(0.9,0.95)	(0.9,0.9)	(0.9,0.85)	(0.9,0.8)	(0.95,1)	(0.9,1)
Basic Harmonic	0.1622	0.1652	0.1802	0.2979	0.3028	0.3132	0.3347	0.5462	0.5584	0.5831	0.6313	0.2957	0.5403
2nd Harmonic	0.0248	0.0243	0.0221	0.046	0.0449	0.0428	0.0384	0.085	0.0817	0.0751	0.0619	0.0465	0.0867
3rd Harmonic	0.0111	0.0114	0.0128	0.0216	0.022	0.0229	0.0249	0.0417	0.0426	0.0446	0.0488	0.0214	0.0413

(b)

	(1,1)	(1,0.9)	(1,0.8)	(0.95,0.95)	(0.95,0.9)	(0.95,0.85)	(0.95,0.8)	(0.9,0.95)	(0.9,0.9)	(0.9,0.85)	(0.9,0.8)	(0.95,1)	(0.9,1)
Basic Harmonic	0.328	1.0825	3.4522	0.608	1.1068	2.0051	3.5773	0.6512	1.1925	2.1807	3.9416	0.3337	0.3562
2nd Harmonic	0.0513	0.1793	0.6077	0.1055	0.1975	0.369	0.692	0.132	0.2513	0.4864	0.9755	0.0563	0.0698
3rd Harmonic	0.023	0.085	0.3124	0.0466	0.09	0.1734	0.335	0.055	0.1062	0.2064	0.4042	0.0424	0.0284

Fig. 10.3 RMS values of harmonics in state variables: **a** for i_L, **b** for v_C

It can be seen from Table 10.3 that the DC components from the three methods agree well with each other, and are obviously dependent on the orders of the fractional-order elements. Furthermore, steady-state ripples from these three methods are put together in Figs. 10.5 and 10.6, where blue straight lines, black straight lines and red dash-and-dot lines represent results from the PECE-ABM method, Oustaloup's method and the ESPM respectively.

As been shown in Figs. 10.5 and 10.6, the magnitudes of harmonics are also order dependent. More specifically, when the order α decreases, the magnitudes of

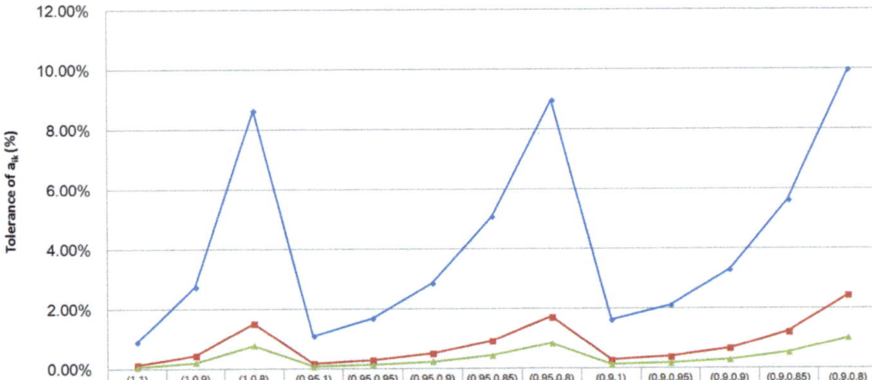

	(1,1)	(1,0.9)	(1,0.8)	(0.95,1)	(0.95,0.95)	(0.95,0.9)	(0.95,0.85)	(0.95,0.8)	(0.9,1)	(0.9,0.95)	(0.9,0.9)	(0.9,0.85)	(0.9,0.8)
Tolerance of a11	0.9100%	2.7400%	8.6300%	1.1100%	1.6900%	2.8700%	5.0700%	8.9700%	1.6200%	2.1200%	3.2900%	5.6400%	9.9700%
Tolerance of a22	0.1400%	0.4500%	1.5200%	0.1800%	0.2900%	0.5100%	0.9300%	1.7300%	0.2800%	0.3900%	0.6600%	1.2300%	2.4400%
Tolerance of a23	0.0637%	0.2100%	0.7800%	0.0807%	0.1300%	0.2300%	0.4400%	0.8400%	0.1300%	0.1700%	0.2900%	0.5300%	1.0200%

Fig. 10.4 The index of tolerance under different orders

Table 10.3 The DC components of state variables i_L (A) and v_C (V) for fractional CCM-Boost

(α, β)	ESPM	PECE-ABM method	Oustaloup's method
(1, 1)	(0.8306, 19.9662)	(0.8236, 19.8810)	(0.8720, 19.8628)
(1, 0.9)	(0.8204, 19.7870)	(0.8377, 19.7376)	(0.8379, 19.6360)
(1, 0.8)	(0.7771, 18.9121)	(0.8270, 19.0599)	(0.8001, 18.5520)
(0.95, 1)	(0.8403, 19.9406)	(0.8339, 20.0031)	(0.8335, 19.9315)
(0.95, 0.95)	(0.8355, 19.8657)	(0.8439, 19.9326)	(0.8331, 19.8752)
(0.95, 0.9)	(0.8261, 19.7043)	(0.8502, 19.7877)	(0.8266, 19.6705)
(0.95, 0.85)	(0.8077, 19.3609)	(0.8493, 19.4905)	(0.8162, 19.1852)
(0.95, 0.8)	(0.7729, 18.6494)	(0.8346, 18.8933)	(0.7656, 18.3543)
(0.9, 1)	(0.8678, 19.8946)	(0.8531, 20.1115)	(0.9947, 19.8894)
(0.9, 0.95)	(0.8609, 19.7831)	(0.8662, 20.0015)	(0.8693, 19.7838)
(0.9, 0.9)	(0.8483, 19.5561)	(0.8743, 19.7852)	(0.8553, 19.4890)
(0.9, 0.85)	(0.8252, 19.0958)	(0.8729, 19.3608)	(0.8063, 19.0011)
(0.9, 0.8)	(0.7861, 18.1884)	(0.8559, 18.5507)	(0.7901, 17.8460)

harmonics with i_L and v_C increase with the decrease of α and β, respectively. Waveforms coming from these three methods show good agreements with each other, and the variation trend of these results coincides with the theoretical analysis in Sect. 10.2.3.

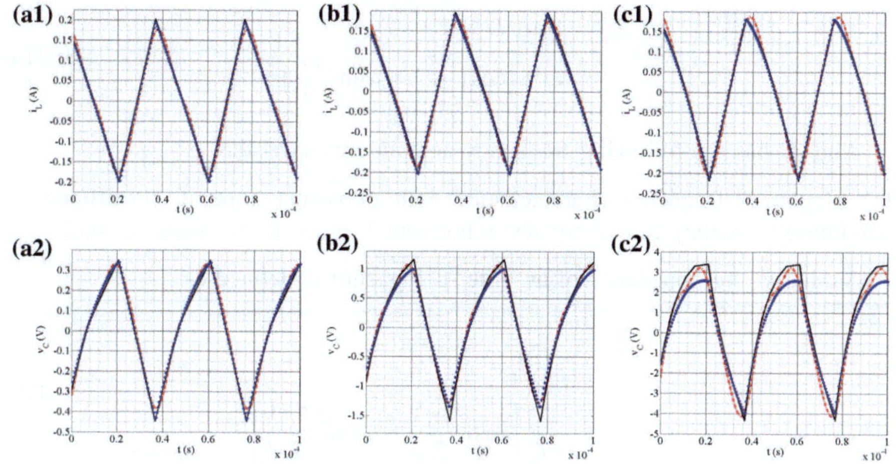

Fig. 10.5 Comparisons of steady-state ripples of i_L (upper row) and v_C (lower row) with the same inductor order $\alpha = 1$ and different capacitor order β, where (a1) (a2) for $\beta = 1$, (b1) (b2) for $\beta = 0.9$, (c1) (c2) for $\beta = 0.8$

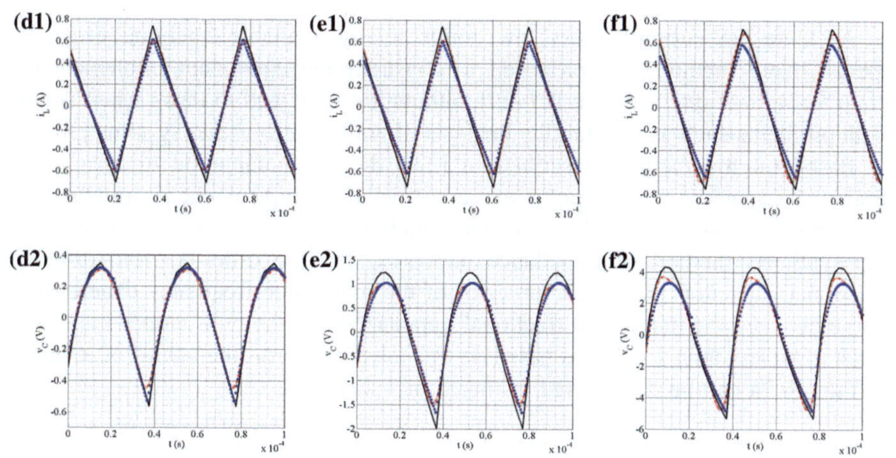

Fig. 10.6 Comparisons of steady-state ripples of i_L (upper row) and v_C (lower row) with the same inductor order $\alpha = 0.9$ and different capacitor order β, where (a1) (a2) for $\beta = 1$, (b1) (b2) for $\beta = 0.9$, (c1) (c2) for $\beta = 0.8$

10.2.5 Equivalent Circuit Implementation of Fractional-Order Devices

Theoretically, the definition of the impedance with the fractional-order element can be found in [26], where the impedance of an order-α inductor is defined as:

$$Z_L^{\alpha} = (j\omega)^{\alpha} L = \omega^{\alpha} L \left[\cos\left(\frac{\alpha\pi}{2}\right) + j\sin\left(\frac{\alpha\pi}{2}\right) \right] \quad (10.41a)$$

And Z_L^{α} has the following frequency-domain characteristics as:

(1) Magnitude-frequency characteristics: with a constant slope of 20α dB/dec.
(2) Phase-frequency characteristics: a horizontal line with the value of απ/2.

Similarly, the impedance of an order-β capacitor is defined as:

$$Z_c^{\beta} = \frac{1}{(j\omega)^{\beta} C} = \frac{1}{\omega^{\beta} C} \left[\cos\left(\frac{\beta\pi}{2}\right) - j\sin\left(\frac{\beta\pi}{2}\right) \right] \quad (10.41b)$$

And Z_C^{β} has the following frequency-domain characteristics as:

(1) Magnitude-frequency characteristics: with a constant slope of −20β dB/dec.
(2) Phase-frequency characteristics: a horizontal line with the value of −βπ/2.

As there are currently very few commercially available fractional-order elements, many methods are proposed to construct the equivalent implementation circuit of a fractional element [27–29], although it may lead to flaws when dealing with problems such as the stability analysis of nonlinear dynamic systems [30, 31].

Based on the method described in [32], the implementation circuits of the fractional-capacitor and inductor are constructed in typical chain structures, as shown in Fig. 10.7, where the parameters such as R_i, C_i and L_i can be obtained by the synthesis method of impedance networks. For the order-0.8 and order-0.9 fractional capacitor, the maximum error of the magnitude-frequency characteristic curve between the theoretical fractional impedance and the built circuit impedance is set to be less than 2 dB, and based on this tolerance error, the values of R_i and C_i in Fig. 10.7a can be determined, as listed in Table 10.4.

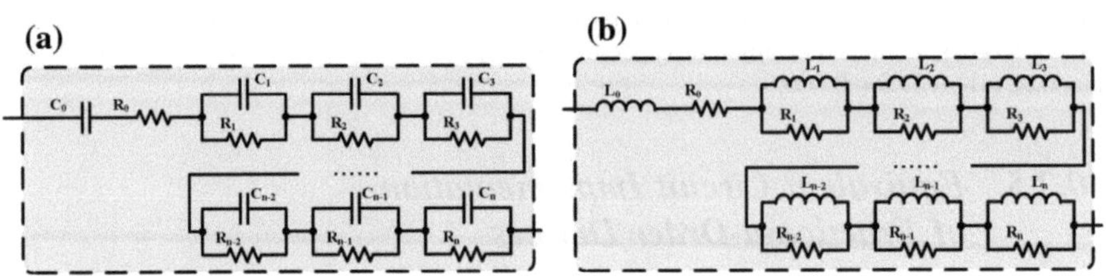

Fig. 10.7 The chain structure of fractional-order components: **a** the fractional-order capacitor, **b** the fractional-order inductor

Table 10.4 R_i and C_i values of the order-0.8 and order-0.9 chain fractance for fractional capacitor

β	0.8		0.9	
i	R_i (Ω)	C_i (μF)	R_i (Ω)	C_i (μF)
Type	Synthesis/actual	Synthesis/actual	Synthesis/actual	Synthesis/actual
1	18.6 m/20 m	0.40389/0.4	0.23487 m/0.25 m	2.5524/2.5
2	177.1 m/200 m	0.75315/0.78	23.4 m/20 m	4.2749/4.27
3	1.7662/1.8	1.3426/1.3	2.3392/2.55	7.1311/7.12
4	17.6598/20	2.3879/2.34	233.9241/225	11.895/11.89
5	176.5976/170	4.2463/4.2	23.476 k/23.47 k	19.772/19.7
6	1.7661 k/1.74 k	7.5505/7.5	5.8263 M/5.75 M	13.289/13.26
7	17.69 k/17.7 k	13.405/13.38	\	\
8	182.10 k/182 k	23.158/23	\	\
9	3.9267 M/3.91 M	19.097/19.09	\	\

As there are some commonly used values of resistances and capacitances, for example, capacitors of 33, 47, 68 μF and carbon film power resistors of 1.8, 7.5, 180 kΩ, and so on. To accomplish components with values listed in Table 10.4, resistances and capacitances are combined in a series-parallel combination structure.

Note that some resistors have milli-values, so we will choose constantan resistors or cement resistors in experiments with considering the power stress. The impedance of the 10 μF/(second)$^{1-\beta}$ capacitor under different order-β are listed in Table 10.5.

Additionally, the Bode diagrams of the fractances are depicted in Fig. 10.8, where the black solid lines represent the theoretical value calculated according to (10.41), and the red dash-dot and blue dash curves belong to the fractances obtained by synthesis and approximated by actual devices respectively. It could be seen, the magnitude-frequency characteristic curves meet well with each other in a certain frequency band, and the phase plot fluctuates in a certain range around the theoretical ones. Nonetheless, more precise results could be got by reducing the maximum error, but it would lead to an increase in the numbers of R_iC_i block in the chain structure.

According to the duality principle, the fractional-order inductors can be constructed. More specifically, as the impedance of the fractional-order inductor has a similar style with the admittance of a fractional-order capacitor, the reciprocal of a fractances transfer function could be regarded as the transfer function of a

Table 10.5 Comparison of impedances of the 10 μF/(second)$^{1-\beta}$ capacitor

Z_β	β	
	0.8	0.9
Ideal Impedance	2.1532 − 6.6269j	0.3295 − 2.0802j
Synthesis results	2.1566 − 6.6240j	0.3195 − 2.0976j
Actual selections	2.1829 − 6.7095j	0.3007 − 2.1352j

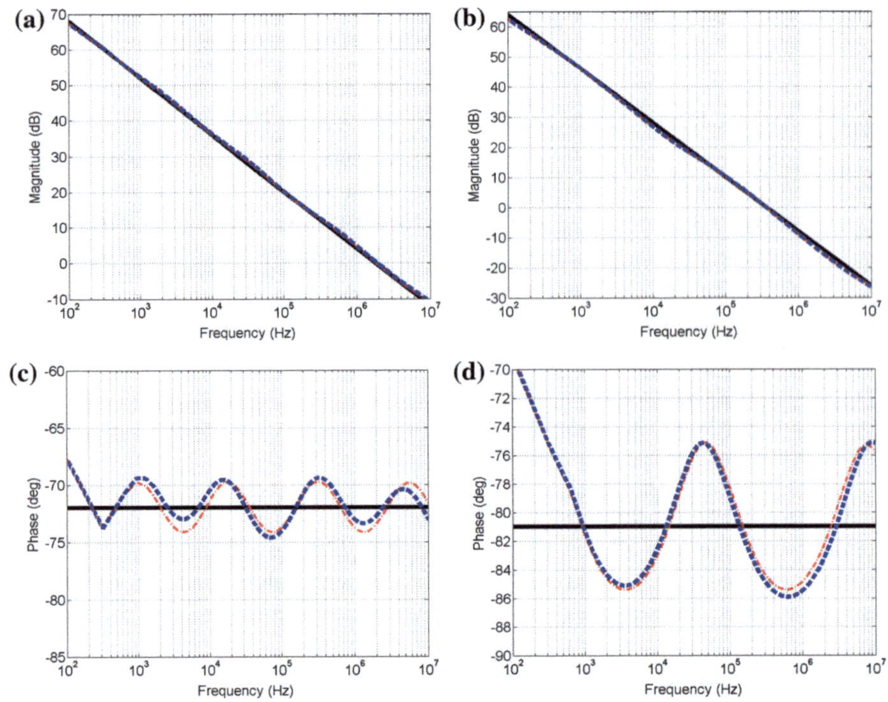

Fig. 10.8 Bode diagrams of the β-order capacitor: **a** and **c** are the magnitude and phase plots with $\beta = 0.8$ respectively, **b** and **d** are the magnitude and phase plots with $\beta = 0.9$ respectively

fractional-order inductor. So we can follow the same way described in [32] to construct the transfer functions of fractional-order inductors. When the maximum error between the approximate curves and the theoretical lines of frequency-domain characteristics is set to be 1 dB, the values of R_i and L_i for the inductors with the order $\alpha = 0.9$ and $\alpha = 0.8$ are listed in Table 10.6.

Note that some resistors' values are small, especially when the order $\alpha = 0.8$. It brings difficulties in implementations and it means that new ways of building fractional-order inductors should be found in future works. However, according to the prior discussions on the CCM-operating criterion, when the order with inductor is selected to be $\alpha = 0.8$, the converter would operate in DCM, so only the modeling simulation for this situation is provided, and the result of which is used as proof materials of the CCM-operating criterion.

Impedances of 477 μH inductor under different order-α are listed in Table 10.7. In addition, using values of R_i and L_i obtained from synthesis procedures and real selections, Bode diagrams of the generated fractional-order inductor are plotted in Fig. 10.9, where black solid lines represent the ideal, red dash-dot and blue dash lines belong to chain fractances by using parameters from synthesis procedures and real circuits, respectively.

Table 10.6 R_i and L_i values of the order-0.8 and order-0.9 fractional inductors

α	0.9		0.8	
i	R_i (Ω)	L_i (μH)	R_i (Ω)	L_i (μH)
Type	Synthesis/actual	Synthesis/actual	Synthesis/actual	Synethesis/actual
0	0	65.642/65	0	13.857
1	1.6397 k/1.635 k	18.634/18	82.8873	4.0363
2	167.0723/165	24.523/24	28.2491	5.8010
3	16.73/16.5	31.716/32	9.0663	7.8511
4	1.6732/1.5	40.967/40	2.8768	10.505
5	0.1673/0.165	52.911/51	0.9104	14.020
6	16.7 m/20 m	68.335/68	0.2880	18.699
7	1.7 m/2 m	88.221/90	0.0911	24.937
8	0.1663 m/1 m	113.33/113.4	0.0288	33.254
9	0.0155 m/1 m	136.09/136	0.0091	44.343
10	\	\	0.0029	59.126
11	\	\	0.9100 m	78.805
12	\	\	0.2872 m	104.86
13	\	\	0.0900 m	138.55
14	\	\	0.0273 m	177.53
15	\	\	0.0071 m	195.24

Table 10.7 Comparisons of the impedance of the 477 μH fractional inductor with different order-α

Z_α	α	
	0.8	0.9
Ideal impedance	3.5429 + 22.3690j	2.1154 + 6.5106j
Synthesis results	3.2088 + 22.0600j	2.1068 + 6.5222j
Real selections	3.1247 + 21.7475j	\

10.2.6 PSIM Simulations Based Equivalent Realization Circuits of Fractional Elements

Replacing the fractional devices in the Boost converter with there corresponding realization circuits, simulations are done in PSIM. Data obtained during simulations are reloaded to the workspace of MATLAB, and waveforms of state variables are plotted, as showed in Figs. 10.10 and 10.11.

From the results of modeling simulations, it could also be seen from Fig. 10.10 that the ripple amplitudes of i_L and v_C increase significantly with the decrease of order-α and order-β, respectively. These results meet well with both theoretical analysis and numerical simulations. Furthermore, to confirm the analysis of the

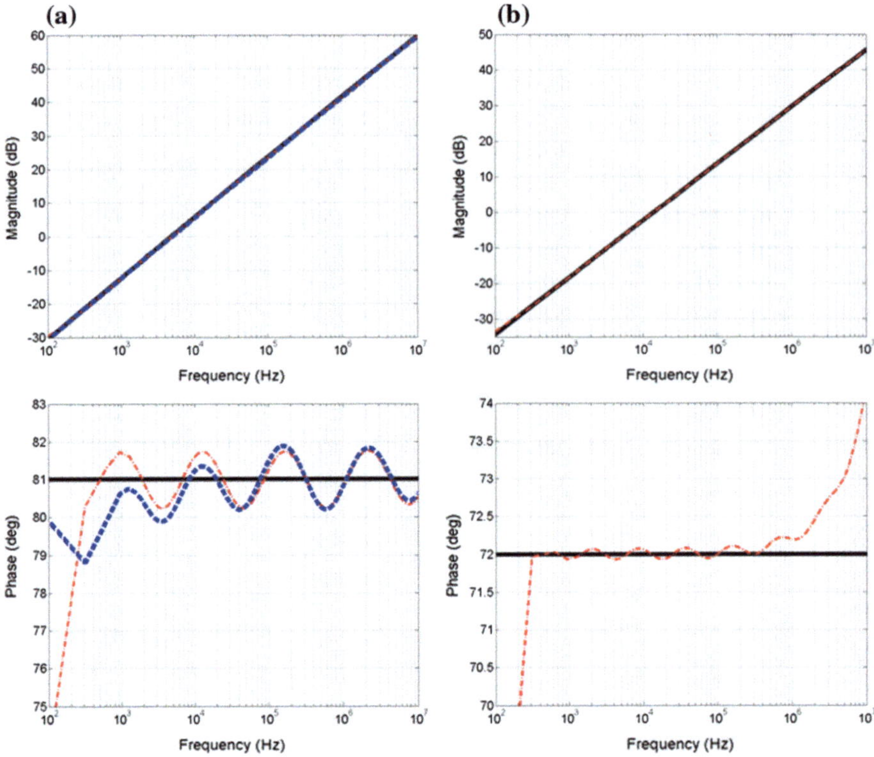

Fig. 10.9 Bode diagrams of the order-α inductor: **a** $\alpha = 0.9$, **b** $\alpha = 0.8$

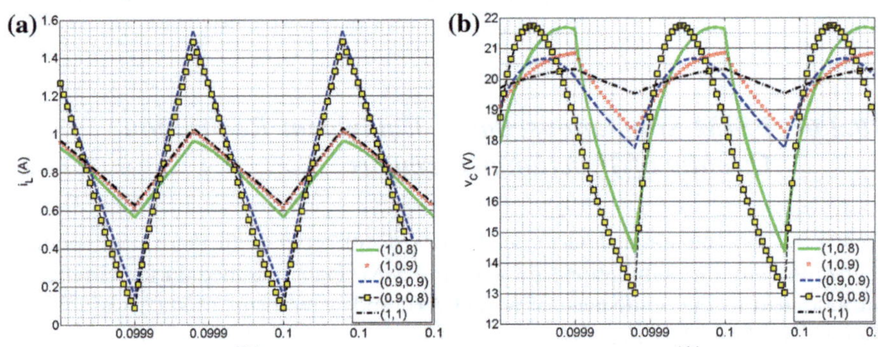

Fig. 10.10 Steady-state waveforms of state variables with different values of (α,β): **a** the inductor current, **b** the capacitor voltage

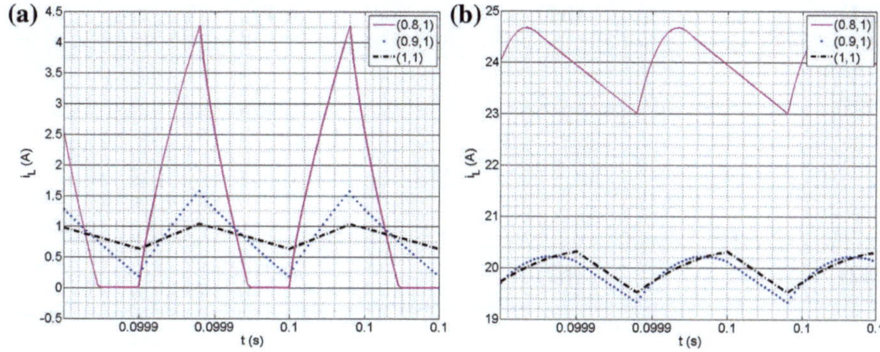

Fig. 10.11 Steady-state waveforms of state variables with different values of order-α: **a** the inductor current, **b** the capacitor voltage

CCM-operating boundary, the situation with (α, β) = (0.8, 1) is considered. The waveforms of state variables under three different values of (α, β) are put together, as depicted in Fig. 10.11. It can be seen that when α = 0.8, the converter would operate in DCM, which is in accord with the CCM-operating boundary in Fig. 10.2.

Moreover, these results indicate that, the circuits generated by the rational approximation method could play an alternative role of the ideal fractional-order elements in the steady state.

10.2.7 Experimental Results

To evaluate the performance of the proposed method, the fractional-order DC/DC Boost converter is built, the type of the MOSFET is IRF640B and the type of diode is MUR1620CT. To reduce the influences from parasitic parameters of the devices, the metalized polypropylene film capacitors are used to build the fractance. And, to reduce interference between power inductors in the chain structure, the Coilcraft MSS1278 and MSS1583 series shielded surface mount power inductors are used to build the fractional-order inductor. By using the actual parameters listed in Tables 10.4 and 10.6, experimental results are obtained, as shown in Figs. 10.3, 10.12 and 10.13.

Furthermore, the RMS and peak-to-peak values of state variables are listed in Table 10.8. Considering parasitic parameters, such as the maximum instantaneous forward voltage of the MOSTFET and the diode, the errors between experimental results and PSIM simulations are in an acceptable range.

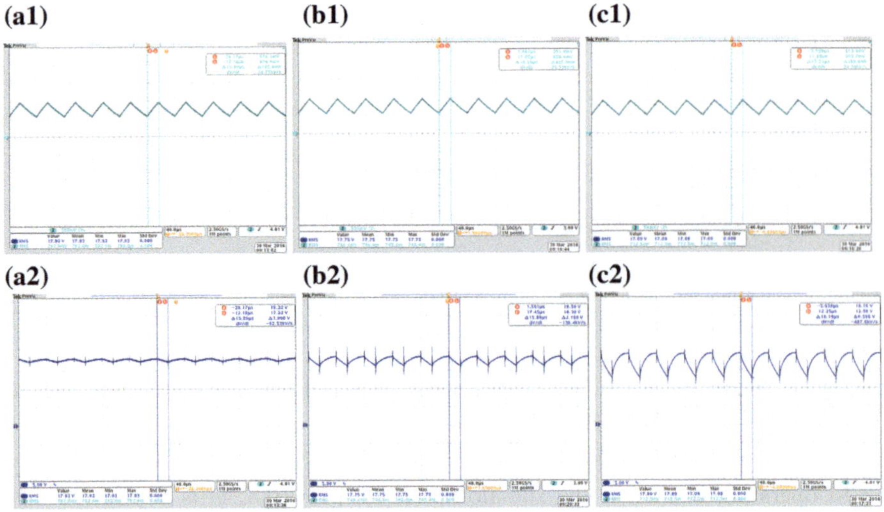

Fig. 10.12 Waveforms of steady-state ripples of i_L (upper row) and v_C (lower row) with the same inductor order $\alpha = 1$ and different capacitor order β, where (a1) (a2) for $\beta = 1$, (b1) (b2) for $\beta = 0.9$, (c1) (c2) for $\beta = 0.8$

Table 10.8 Comparisons between PSIM simulations and experiments

(α, β)	x_{RMS}		Δx_{pp}	
	v_{RMS} (V)	i_{RMS} (A)	Δv_{pp}	Δi_{pp}
(1, 1)	19.99/ 17.93	0.83/ 0.78	0.80/ 1.00	0.4025/ 0.3959
(1, 0.9)	19.89/ 17.75	0.82/ 0.75	2.55/ 2.20	0.4025/ 0.4079
(1, 0.8)	19.39/ 17.09	0.78/ 0.71	7.35/ 6.60	0.4025/ 0.3899
(0.9, 1)	19.99/ 17.57	0.96/ 0.84	0.83/ 0.84	1.3591/ 1.297
(0.9, 0.9)	19.54/ 17.43	0.94/ 0.83	2.91/ 2.80	1.3627/ 1.287
(0.9, 0.8)	18.38/ 16.40	0.89/ 0.81	7.65/ 7.87	1.3645/ 1.258

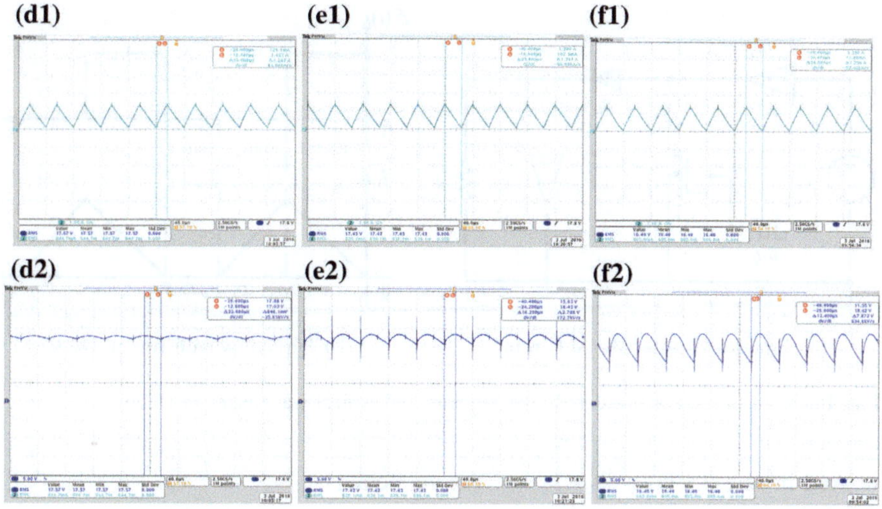

Fig. 10.13 Waveforms of steady-state ripples of i_L (upper row) and v_C (lower row) with the same inductor order $\alpha = 0.9$ and different capacitor order β, where (a1) (a2) for $\beta = 1$, (b1) (b2) for $\beta = 0.9$, (c1) (c2) for $\beta = 0.8$

10.3 Mathematical Model of Fractional-Order Boost Converter Operating in DCM

10.3.1 Equivalent System of DCM Non-integer Order Differential Equations

The fractional-order Boost DC/DC converter and its typical waveforms in DCM operation are shown in Fig. 10.14, where the switch S_T and diode S_D are ideal. The inductor L and capacitor C are both ideal fractional-order components. And their orders are α and β respectively. According to the states of S_T and S_D, the converter in DCM can be divided into three switching states in one steady-state period.

One chooses i_L and v_C as state variables, then the converter in state 1 ($0 \sim D_1T_s$) can be described by

$$\begin{cases} \frac{d^\alpha i_L}{dt^\alpha} = \frac{1}{L}V_S \\ \frac{d^\beta v_C}{dt^\beta} + \frac{v_C}{RC} = 0 \end{cases} \tag{10.42}$$

In state 2 ($D_1T_s \sim (D_1 + D_2)\,T_s$), the state variables are

$$\begin{cases} \frac{d^\alpha i_L}{dt^\alpha} + \frac{1}{L}v_C = \frac{1}{L}V_S \\ \frac{d^\beta v_C}{dt^\beta} + \frac{v_C}{RC} - \frac{1}{C}i_L = 0 \end{cases} \tag{10.43}$$

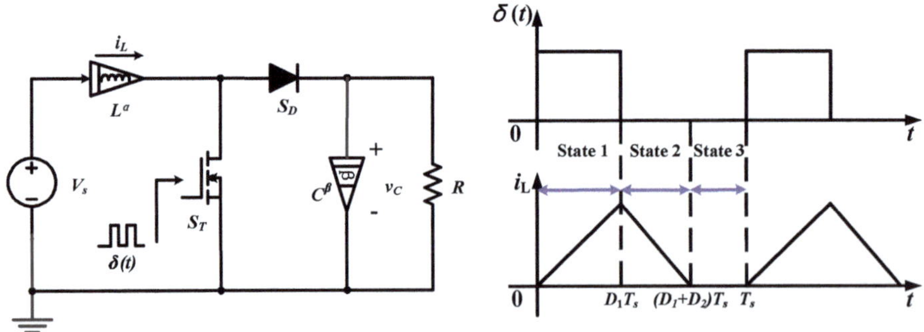

(a) Fractional-order Boost Circuit (b) Typical waveforms in DCM

Fig. 10.14 Ideal fractional-order Boost converter and typical waveforms in DCM

And in sate 3 ($(D_1 + D_2)\,T_s \sim T_s$), one obtains

$$\begin{cases} \frac{d^\alpha i_L}{dt^\alpha} = 0 \\ \frac{d^\beta v_C}{dt^\beta} + \frac{v_C}{RC} = 0 \end{cases} \tag{10.44}$$

Here we introduce the switching functions $\delta^{(q)}(t)$ (q = 1, 2, 3) to represent the three states with the definitions as follows

$$\delta^{(1)} = \begin{cases} 1, & 0 < t \le D_1 T \\ 0, & D_1 T < t \le T \end{cases} \tag{10.45}$$

$$\delta^{(3)} = \begin{cases} 1, & D_1 T < t \le (D_1 + D_2)T \\ 0, & 0 < t \le D_1 T, (D_1 + D_2)T < t \le T \end{cases} \tag{10.46}$$

$$\delta^{(2)} = \begin{cases} 0, & 0 < t \le (1 - D_3)T \\ 1, & (1 - D_3)T < t \le T \end{cases} \tag{10.47}$$

Obviously during one switching period, $\delta^{(1)} + \delta^{(2)} + \delta^{(3)} = 1$ and $D_1 + D_2 + D_3 = 1$. Let (10.43)*$\delta^{(1)}$ + (10.44)*$\delta^{(2)}$ + (10.45)*$\delta^{(3)}$, one can get the following vector state equation to describe the converter with DCM operation.

$$\boldsymbol{G}_0(p^\alpha, p^\beta)\boldsymbol{x} + \boldsymbol{G}_1\boldsymbol{f}^{(1)}(\boldsymbol{x}) + \boldsymbol{G}_2\boldsymbol{f}^{(2)}(\boldsymbol{x}) = \boldsymbol{U} \tag{10.48}$$

where the state vector $\boldsymbol{x} = [i_L\ v_C]^{Tr}$, the constant vector $\boldsymbol{U} = [V_S/L,\ 0]^{\mathrm{Tr}}$, and symbolic p^α and p^β represent the differential operator d^α/dt and d^β/dt respecitvely. The square matrices $\boldsymbol{G}_0(p^\alpha,p^\beta)$ and $\boldsymbol{G}_1$ and $\boldsymbol{G}_2$, as well as the nonlinear function $\boldsymbol{f}^{(1)}(\boldsymbol{x})$ and $\boldsymbol{f}^{(2)}(\boldsymbol{x})$ are given as followed

$$G_0(p^\alpha, p^\beta) = \begin{bmatrix} p^\alpha & \frac{1}{L} \\ \frac{-1}{C} & p^\beta + \frac{1}{RC} \end{bmatrix} G_1 = G_2 = \begin{bmatrix} 0 & \frac{-1}{L} \\ \frac{1}{C} & 0 \end{bmatrix} \quad (10.49)$$

$$\begin{cases} f^{(1)}(x) = \delta^{(1)}(x + E_1) \\ f^{(2)}(x) = \delta^{(2)}(x + E_2) \end{cases} \quad (10.50)$$

Here the constant vector $\mathbf{E}_1 = [0, 0]^{\mathrm{Tr}}$ and $\mathbf{E}_2 = [0, V_S]^{\mathrm{Tr}}$.

On the basis of [33, 34], this work further expands our modeling method for fractional-order DC/DC converters in DCM [35]. According to the principle of perturbation method, one assumes that the switching function $\delta^{(q)}(t)$ $(q = 1,2)$ in (10.45) and (10.47) can be expanded into the following series

$$\begin{cases} \delta^{(1)}(\mathrm{t}) = b_0 + \sum\limits_{k=1}^{\infty} (b_k e^{jk\tau} + \bar{b}_k e^{-jk\tau}) \\ \delta^{(2)}(\mathrm{t}) = c_0 + \sum\limits_{k=1}^{\infty} (c_k e^{jk\tau} + \bar{c}_k e^{-jk\tau}) \end{cases} \quad (10.51)$$

where $\tau = 2\pi f_s t$, $b_0 = D_2$, $c_0 = D_3$, and $\bar{b}_k$ and $\bar{c}_k$ are

$$\begin{cases} b_k = \frac{\sin(2k\pi D_1)}{2k\pi} - j\frac{1-\cos(2k\pi D_1)}{2k\pi} \\ c_k = \frac{\sin(2k\pi D_3)}{2k\pi} - j\frac{\cos(2k\pi D_3)-1}{2k\pi} \end{cases} \quad (10.52)$$

$\bar{b}_k$ and $\bar{c}_k$ is the conjugate complex of b_k and c_k. D_1 is the duty cycle, which is defined as the ratio of the turn-on time of S_T to the switching period T_s. D_2 is the duty cycle of state 2 and D_3 is the duty cycle of state 3. They are defined as

$$\begin{cases} D_2 = \frac{L\alpha\Gamma(\alpha)}{T^\alpha R D_1^\alpha}\left(1 + \sqrt{1 + \frac{4D_1^{\alpha+1}T^\alpha R}{2L\alpha\Gamma(\alpha)}}\right) \\ D_3 = 1 - D_1 - D_2 \end{cases} \quad (10.53)$$

in which Γ(.) is the gamma function [1]. Then one can expand $\delta^{(q)}(\mathrm{t})$ into following series

$$\begin{cases} \delta^{(1)}(\mathrm{t}) = \delta_0^{(1)} + \sum\limits_{i=1}^{n} \varepsilon^i \delta_i^{(1)} \\ \delta^{(2)}(\mathrm{t}) = \delta_0^{(2)} + \sum\limits_{i=1}^{n} \varepsilon^i \delta_i^{(2)} \end{cases} \quad (10.54)$$

in which $\delta_{\mathrm{i}}^{(1)}$, $\delta_{\mathrm{i}}^{(3)}$ are defined as

$$\begin{cases} \delta_0^{(1)} = b_0 + b_1 e^{j\tau} + c.c \\ \delta_i^{(1)} = b_{2i} e^{j2i\tau} + b_{2i+1} e^{j(2i+1)\tau} + c.c \end{cases} \quad \begin{cases} \delta_0^{(2)} = c_0 + c_1 e^{j\tau} + c.c \\ \delta_i^{(2)} = c_{2i} e^{j2i\tau} + c_{2i+1} e^{j(2i+1)\tau} + c.c \end{cases} \tag{10.55}$$

And one can expend the state vector $\boldsymbol{x}$ as

$$\boldsymbol{x} = \boldsymbol{x}_0 + \sum_{i=1}^{n} \varepsilon^i \boldsymbol{x}_i \tag{10.56}$$

where $\boldsymbol{x}_0$ and $\boldsymbol{x}_{\mathrm{i}}$ are called as the main wave and the ith order correction, and they consist of a set of numbers representing relative frequencies of harmonics. Here the indicator parameter ε^i is still introduced temporarily to indicate that the term $\boldsymbol{x}_{\mathrm{i}}$ is much smaller than the main wave $\boldsymbol{x}_0$, i.e., $\varepsilon^{i+1}\boldsymbol{x}_{i+1} < \varepsilon^i \boldsymbol{x}_i \ll \boldsymbol{x}_0$.

Substituting (10.54) and (10.56 into (10.50), and combining terms which have the same order of indicator parameter ε^i, one obtains

$$\boldsymbol{f}^{(q)}(\boldsymbol{x}) = \boldsymbol{f}_0^{(q)} + \sum_{i=1}^{\infty} \varepsilon^i \boldsymbol{f}_i^{(q)} = \boldsymbol{f}_{0m}^{(q)} + \varepsilon \boldsymbol{R}_1^{(q)} + \sum_{i=1}^{\infty} \left(\varepsilon^i \boldsymbol{f}_{im}^{(q)} + \varepsilon^{i+1} \boldsymbol{R}_{i+1}^{(q)} \right) \tag{10.57}$$

where $\boldsymbol{R}_{\mathrm{i+1}}^{(q)}$ includes the rest terms whose harmonic contents are different from $\boldsymbol{f}_i^{(q)}$, and

$$\begin{cases} \boldsymbol{f}_0^{(q)} = \delta_0^{(q)} (\boldsymbol{x}_0 + \boldsymbol{E}_q) \\ \boldsymbol{f}_1^{(q)} = \delta_0^{(q)} \boldsymbol{x}_1 + \delta_1^{(q)} (\boldsymbol{x}_0 + \boldsymbol{E}_q) \\ \boldsymbol{f}_2^{(q)} = \delta_0^{(q)} \boldsymbol{x}_2 + \delta_1^{(q)} \boldsymbol{x}_1 + \delta_2^{(q)} (\boldsymbol{x}_0 + \boldsymbol{E}_q) \\ \cdots \end{cases} \tag{10.58}$$

Then substituting (10.54), (10.56) and (10.57) into (10.48), and equating the terms with the same ε^i on the right-and left-hand sides, one can get the following equivalent linear fractional-order differential equations

$$\begin{cases} \boldsymbol{G}_0(p^\alpha, p^\beta)\boldsymbol{x}_0 + \boldsymbol{G}_1 \boldsymbol{f}_{0m}^{(1)} + \boldsymbol{G}_2 \boldsymbol{f}_{0m}^{(2)} = \boldsymbol{U} \\ \boldsymbol{G}_0(p^\alpha, p^\beta)\boldsymbol{x}_1 + \boldsymbol{G}_1(\boldsymbol{f}_{1m}^{(1)} + \boldsymbol{R}_1^{(1)}) + \boldsymbol{G}_2(\boldsymbol{f}_{1m}^{(2)} + \boldsymbol{R}_1^{(2)}) = 0 \\ \boldsymbol{G}_0(p^\alpha, p^\beta)\boldsymbol{x}_2 + \boldsymbol{G}_1(\boldsymbol{f}_{2m}^{(1)} + \boldsymbol{R}_2^{(1)}) + \boldsymbol{G}_2(\boldsymbol{f}_{2m}^{(2)} + \boldsymbol{R}_2^{(2)}) = 0 \\ \cdots \end{cases} \tag{10.59}$$

Each equation in (10.59) is linear and can be solved step by step using the method of harmonic balance, where the first equation is used to obtain the main wave $\boldsymbol{x}_0$, and following equations are used to get the corrections $\boldsymbol{x}_1$, $\boldsymbol{x}_2$, …, etc.

When the terms $\boldsymbol{x}_0$, $\boldsymbol{x}_1$, $\boldsymbol{x}_2$ and so on are found, the approximate steady-state solution of the state vector $\boldsymbol{x}$ can be acquired

$$\boldsymbol{x} = \boldsymbol{x}_0 + \boldsymbol{x}_1 + \boldsymbol{x}_2 + \ldots \tag{10.60}$$

10.3.2 Periodic Steady-State Solutions of Fractional-Order Boost Converter in DCM

A. *Solving the main wave* $\boldsymbol{x}_0$

Since DC/DC converters have low-pass properties, the main wave $\boldsymbol{x}_0$ can be chosen as dc value

$$\boldsymbol{x}_0 = \boldsymbol{a}_{00} = [\, I_{00} \quad V_{00} \,]^{Tr} \tag{10.61}$$

Then one gets $\boldsymbol{f}_{0\mathrm{m}}^{(\mathrm{q})}$ and $\boldsymbol{R}_1^{(\mathrm{q})}$

$$\begin{cases} \boldsymbol{f}_{0m}^{(1)} = \boldsymbol{a}_{00} b_0 = \boldsymbol{a}_{00} D_1 \\ \boldsymbol{f}_{0m}^{(2)} = (\boldsymbol{a}_{00} + \boldsymbol{E}_2) c_0 = (\boldsymbol{a}_{00} + \boldsymbol{E}_2) D_3 \end{cases} \tag{10.62a}$$

$$\begin{cases} \boldsymbol{R}_1^{(1)} = b_1 \boldsymbol{a}_{00} e^{j\tau} + c.c \\ \boldsymbol{R}_1^{(2)} = c_1 (\boldsymbol{a}_{00} + \boldsymbol{E}_2) e^{j\tau} + c.c \end{cases} \tag{10.62b}$$

The substitution of (10.61) and (10.62a) into the first equation in (10.59) leads to

$$\boldsymbol{G}_{00} \boldsymbol{a}_{00} + \boldsymbol{G}_1 \boldsymbol{a}_{00} D_1 + \boldsymbol{G}_2 (\boldsymbol{a}_{00} + \boldsymbol{E}_2) D_3 = \boldsymbol{U} \tag{10.63}$$

By solving this equation, one obtains

$$\boldsymbol{a}_{00} = (\boldsymbol{G}_{00} + \boldsymbol{G}_1 D_1 + \boldsymbol{G}_2 D_3)^{-1} \cdot (\boldsymbol{U} - \boldsymbol{G}_2 \boldsymbol{E}_2 D_3) \tag{10.64}$$

Here $\boldsymbol{G}_{00} = \boldsymbol{G}_0(0^{\alpha}, 0^{\beta})$, it can be obtained by setting $p = 0$ in (10.49). Then one can get:

$$\begin{cases} V_{00} = \frac{(1 - D_3) V_S}{D_2} \\ I_{00} = \frac{(1 - D_3) V_S}{R D_2^2} \end{cases} \tag{10.65}$$

B. *Solving the first order correction* $\boldsymbol{x}_1$

According to the harmonic content of $\boldsymbol{R}_1^{(\mathrm{q})}$, one can deduce that there is first-harmonic in $\boldsymbol{x}_1$. So $\boldsymbol{x}_1$ can be assumed as

$$\boldsymbol{x}_1 = \boldsymbol{a}_{11}e^{j\tau} + \bar{\boldsymbol{a}}_{11}e^{-j\tau} \tag{10.66}$$

in which $\boldsymbol{a}_{11} = [I_{11} \;\; V_{11}]^{\mathrm{Tr}}$, and $\bar{\boldsymbol{a}}_{11}$ means the complex-conjugate of $\boldsymbol{a}_{11}$. Substituting $\boldsymbol{x}_0$, $\boldsymbol{x}_1$, $\delta_0^{(1)}$, $\delta_1^{(1)}$ and $\delta_0^{(2)}$, $\delta_1^{(2)}$ into the second equation of (10.58), one can get

$$\begin{cases} \boldsymbol{f}_{1m}^{(1)} = b_0\boldsymbol{a}_{11}e^{j\tau} + c.c \\ \boldsymbol{f}_{1m}^{(2)} = c_0\boldsymbol{a}_{11}e^{j\tau} + c.c \end{cases} \tag{10.67}$$

and

$$\begin{cases} \boldsymbol{R}_2^{(1)} = (b_1\bar{\boldsymbol{a}}_{11} + \bar{b}_1\boldsymbol{a}_{11}) + (b_1\boldsymbol{a}_{11} + b_2\boldsymbol{a}_{00})e^{j2\tau} + b_3\boldsymbol{a}_{00}e^{j3\tau} + c.c \\ \boldsymbol{R}_2^{(2)} = (c_1\bar{\boldsymbol{a}}_{11} + \bar{c}_1\boldsymbol{a}_{11}) + (c_1\boldsymbol{a}_{11} + c_2(\boldsymbol{a}_{00} + \boldsymbol{E}_2))e^{j2\tau} + c_3(\boldsymbol{a}_{00} + \boldsymbol{E}_2)e^{j3\tau} + c.c \end{cases} \tag{10.68}$$

Then the substitution of (10.66), (10.67) and (10.62b) into the second equation in (10.59) brings

$$(\boldsymbol{G}_{01} + \boldsymbol{G}_1 b_0 + \boldsymbol{G}_2 c_0)\boldsymbol{a}_{11} + (\boldsymbol{G}_1 b_1 \boldsymbol{a}_{00} + \boldsymbol{G}_2(\boldsymbol{a}_{00} + \boldsymbol{E}_2)c_1) = 0 \tag{10.69}$$

By solving this equation, one obtains

$$\boldsymbol{a}_{11} = (\boldsymbol{G}_{01} + \boldsymbol{G}_1 b_0 + \boldsymbol{G}_2 c_0)^{-1} \cdot (-\boldsymbol{G}_1 b_1 \boldsymbol{a}_{00} - \boldsymbol{G}_2(\boldsymbol{a}_{00} + \boldsymbol{E}_2)c_1) \tag{10.70}$$

Here $\boldsymbol{G}_{01} = \boldsymbol{G}_0((j\omega)^{\alpha}, (j\omega)^{\beta})$, it can be obtained by setting $p = j\omega$ in (10.49). Thus, one can get:

$$\begin{cases} V_{11} = \dfrac{-D_2R[V_Sc_1 - V_{00}(b_1 + c_1)] - (jw)^{\alpha}RLI_{00}(b_1 + c_1)}{\Delta(j\omega)} \\ I_{11} = \dfrac{-[(jw)^{\beta}RC + 1][V_Sc_1 - V_{00}(b_1 + c_1)] + D_2RI_{00}(b_1 + c_1)}{\Delta(j\omega)} \end{cases} \tag{10.71}$$

where

$$\Delta(j\omega) = CLR(j\omega)^{\alpha+\beta} + L(j\omega)^{\alpha} + D_2^2R \tag{10.72}$$

C. *Solving the second order correction* x_3

According to the harmonic content of $\boldsymbol{R}_2$, one can deduce that there are DC components, second- and third-harmonics in $\boldsymbol{x}_2$. Then, the second order correction $\boldsymbol{x}_2$ can be assumed as

$$\boldsymbol{x}_2 = \boldsymbol{a}_{02} + \boldsymbol{a}_{22} e^{j2\tau} + \boldsymbol{a}_{32} e^{j3\tau} + c.c \tag{10.73}$$

in which $\boldsymbol{c.c}$ means the complex-conjugate of $\boldsymbol{a}_{22}$ and $\boldsymbol{a}_{32}$. Here $\boldsymbol{a}_{02} = [I_{02}\ V_{02}]^{\mathrm{Tr}}$ is the vector of dc values, it will give the corrections of $\boldsymbol{a}_{00}$. $\boldsymbol{a}_{22} = [I_{22}\ V_{22}]^{\mathrm{Tr}}$ and $\boldsymbol{a}_{32} = [I_{32}\ V_{32}]^{\mathrm{Tr}}$, they are the second- and third-harmonics of state variable $\boldsymbol{x}$ respectively.

Substituting $\boldsymbol{x}_0$, $\boldsymbol{x}_1$, $\boldsymbol{x}_2$ and $\delta_0^{(q)}$, $\delta_1^{(q)}$, $\delta_2^{(q)}$ into the third equation of (10.58), one acquires

$$\begin{cases} \boldsymbol{f}_{2m}^{(1)} = b_0\boldsymbol{a}_{02} + (b_0\boldsymbol{a}_{22} + \bar{b}_1\boldsymbol{a}_{32} + b_3\bar{\boldsymbol{a}}_{11})e^{j2\tau} + (b_0\boldsymbol{a}_{32} + b_1\boldsymbol{a}_{22} + b_2\boldsymbol{a}_{11})e^{j3\tau} + c.c \\ \boldsymbol{f}_{2m}^{(2)} = c_0\boldsymbol{a}_{02} + (c_0\boldsymbol{a}_{22} + \bar{c}_1\boldsymbol{a}_{32} + c_3\bar{\boldsymbol{a}}_{11})e^{j2\tau} + (c_0\boldsymbol{a}_{32} + c_1\boldsymbol{a}_{22} + c_2\boldsymbol{a}_{11})e^{j3\tau} + c.c \end{cases} \tag{10.74}$$

Substitution of (10.68), (10.73) and (10.74) into the third equation of (10.59) leads to

$$\boldsymbol{a}_{02} = (\boldsymbol{G}_{00} + \boldsymbol{G}_1 b_0 + \boldsymbol{G}_3 c_0)^{-1} \cdot (-\boldsymbol{G}_1(b_1\bar{\boldsymbol{a}}_{11} + \bar{b}_1\boldsymbol{a}_{11}) - \boldsymbol{G}_2(c_1\bar{\boldsymbol{a}}_{11} + \bar{c}_1\boldsymbol{a}_{11})) \tag{10.75a}$$

$$\begin{aligned} \boldsymbol{a}_{22} = {} & (\boldsymbol{G}_{02} + \boldsymbol{G}_1 b_0 + \boldsymbol{G}_2 c_0)^{-1} \cdot (-\boldsymbol{G}_1(b_1\boldsymbol{a}_{11} + b_2\boldsymbol{a}_{00} + b_3\bar{\boldsymbol{a}}_{11}) \\ & - \boldsymbol{G}_2(c_1\boldsymbol{a}_{11} + c_2(\boldsymbol{a}_{00} + \boldsymbol{E}_2) + c_3\bar{\boldsymbol{a}}_{11})) \end{aligned} \tag{10.75b}$$

$$\begin{aligned} \boldsymbol{a}_{32} = {} & (\boldsymbol{G}_{03} + \boldsymbol{G}_1 b_0 + \boldsymbol{G}_2 c_0)^{-1} \cdot (-\boldsymbol{G}_1(b_1\boldsymbol{a}_{22} + b_2\boldsymbol{a}_{11} + b_3(\boldsymbol{a}_{00} + \boldsymbol{E}_2)) \\ & - \boldsymbol{G}_2(c_1\boldsymbol{a}_{22} + c_2\boldsymbol{a}_{11} + c_3(\boldsymbol{a}_{00} + \boldsymbol{E}_2))) \end{aligned} \tag{10.75c}$$

The coefficient matrix $\boldsymbol{G}_{0\mathrm{k}}$ can be obtained by substituting the differential operator $p = jk\omega$ into (10.49), that is, $G_{0\mathrm{k}} = G_0((jk\omega)^{\alpha}, (jk\omega)^{\beta})$ $(k = 0, 2, 3)$. Thus, one obtains

$$\begin{cases} V_{02} = \dfrac{V_{11}(\bar{b}_1 + \bar{c}_1) + \overline{V}_{11}(b_1 + c_1)}{D_2} \\ I_{02} = \dfrac{V_{11}(\bar{b}_1 + \bar{c}_1) + \overline{V}_{11}(b_1 + c_1)}{RD_2^2} + \dfrac{I_{11}(\bar{b}_1 + \bar{c}_1) + \bar{I}_{11}(b_1 + c_1)}{D_2} \end{cases} \tag{10.76a}$$

$$\begin{cases} V_{22} = \dfrac{D_2R[V_{00}(b_2 + c_2) - c_2V_S + V_{11}(b_1 + c_1) + \overline{V}_{11}(b_3 + c_3)]}{\Delta(j2\omega)} \\ \qquad - \dfrac{(j2\omega)^{\alpha}RL[I_{00}(b_2 + c_2) + I_{11}(b_1 + c_1) + \bar{I}_{11}(b_3 + c_3)]}{\Delta(j2\omega)} \\ I_{22} = \dfrac{[(j2\omega)^{\beta}RC + 1][(b_2 + c_2)V_{00} - c_2V_S + V_{11}(b_1 + c_1) + \overline{V}_{11}(b_3 + c_3)]}{\Delta(j2\omega)} \\ \qquad + \dfrac{D_2R[I_{00}(b_2 + c_2) + I_{11}(b_1 + c_1) + \bar{I}_{11}(b_3 + c_3)]}{\Delta(j2\omega)} \end{cases} \tag{10.76b}$$

$$\begin{cases} V_{32} = \dfrac{D_2R[V_{22}(b_1+c_1)+V_{11}(b_2+c_2)+V_{00}(b_3+c_3)-c_3V_S]}{\Delta(j3\omega)} \\ \qquad -\dfrac{(j3\omega)^{\alpha}RL[I_{11}(b_2+c_2)+I_{22}(b_1+c_1)+I_{00}(b_3+c_3)]}{\Delta(j3\omega)} \\ I_{32} = \dfrac{[(j3\omega)^{\beta}RC+1][V_{00}(b_3+c_3)+V_{22}(b_1+c_1)+V_{11}(b_2+c_2)-c_3V_S]}{\Delta(j3\omega)} \\ \qquad +\dfrac{D_2R[I_{22}(b_1+c_1)+I_{11}(b_2+c_2)+I_{00}(b_3+c_3)]}{\Delta(j3\omega)} \end{cases} \tag{10.76c}$$

where $\Delta(jk\omega)$ has the similar form of (10.72).

Hence, the approximate periodic solutions of fractional-order Boost converter in DCM can be expressed as

$$x = [\, i_L \quad v_C \,]^{Tr} \approx (a_{00}+a_{02}) + (a_{11}e^{j\tau} + a_{22}e^{j2\tau} + a_{32}e^{j3\tau} + c.c) \tag{10.77}$$

10.3.3 Comparisons and Verification

The regulator in this work has a line voltage $V_s = 12$ V, and a load resistance $R = 100\ \Omega$. The inductance is $L = 477\ \mu$H, and the capacitance is $C = 10\ \mu\text{F}/(\text{s})^{1-\beta}$. The switching frequency is $f_s = 25$ kHz for a switching period $T_s = 1/f_s = 40\ \mu$s. The steady-state duty ratio is $D = 0.4$. This set of parameters ensures the regulator operates in DCM.

Introducing the above parameters into Eqs. (10.65), (10.71) and (10.76), one obtains $\boldsymbol{a}_{00}$, $\boldsymbol{a}_{02}$ and so on. According to the termination conditions of correcting procedure in Sect. 10.2, two times of corrections is enough for $\alpha = 0.9$ and $\beta = 0.8$. So one obtains

$$\begin{aligned} i_L &= 0.4744 - 0.4732\cos\tau + 0.4698\sin\tau - 0.0466\cos 2\tau \\ &\quad - 0.0674\sin 2\tau - 0.0015\cos 3\tau + 0.0194\sin 3\tau \end{aligned} \tag{10.78}$$

$$\begin{aligned} v_C &= 22.3482 - 1.0812\cos\tau - 3.2090\sin\tau + 0.2600\cos 2\tau \\ &\quad + 1.1341\sin 2\tau - 0.2402\cos 3\tau - 0.2173\sin 3\tau \end{aligned} \tag{10.79}$$

By adopting the frequency-domain approximation method in [29], the fractional-order capacitor and inductor are constructed, the structures of which are in typical chain topologies, as shown below (Fig. 10.15).

Following the same way in Sect. 10.2, by using synthesis method of impedance networks, when $L = 477\ \mu$H, $\alpha = 0.9$ and the maximum error between the approximate curve and the ideal one is less than 2 decibel, one can get the parameters as follows: $R_{11} = 1.635$ kΩ, $R_{12} = 165\ \Omega$, $R_{13} = 16.5\ \Omega$, $R_{14} = 1.65\ \Omega$, $R_{15} = 0.165\ \Omega$, $R_{16} = 20$ mΩ, $R_{17} = 2$ mΩ, $R_{18} = 1.6$ mΩ, $R_{19} = 1.5$ mΩ,

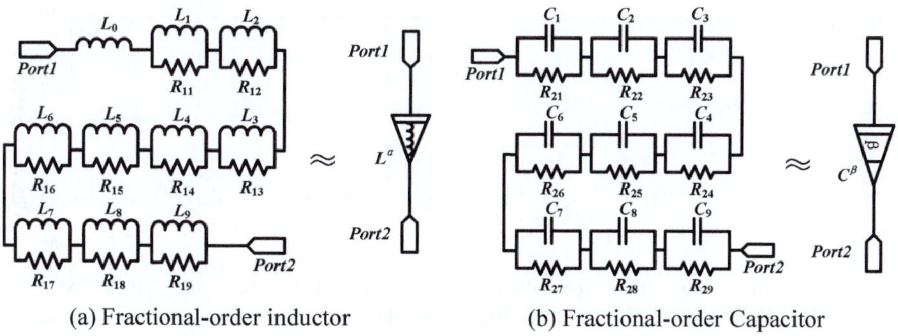

Fig. 10.15 Equivalent circuit of fractional-order elements

$L_0 = 65.6$ μH, $L_1 = 18.6$ μH, $L_2 = 24.5$ μH, $L_3 = 31.7$ μH, $L_4 = 40.9$ μH, $L_5 = 52.9$ μH, $L_6 = 68.3$ μH, $L_7 = 88.2$ μH, $L_8 = 113.4$ μH, $L_9 = 136$ μH.

According to the duality principle, parameters of the fractional order capacitor $C = 10\ \mu\text{F}/(\text{s})^{1-\beta}$ and $\beta = 0.8$ are as follows: $R_{21} = 20$ mΩ, $R_{22} = 200$ mΩ, $R_{23} = 1.8$ Ω, $R_{24} = 20$ Ω, $R_{25} = 176.6$ Ω, $R_{26} = 1.766$ kΩ, $R_{27} = 17.7$ Ω, $R_{28} = 182.1$ kΩ, $R_{29} = 3.927$ MΩ, $C_1 = 0.4$ μF, $C_2 = 0.78$ μF, $C_3 = 1.34$ μF, $C_4 = 2.34$ μF, $C_5 = 4.24$ μF, $C_6 = 7.5$ μF, $C_7 = 13.38$ μF, $C_8 = 23.1$ μF, $C_9 = 19.09$ μF.

The Bode diagrams of the constructed fractional-order components are depicted in Fig. 10.16, where the black solid lines represent the theoretical values, and the red dash–dotted and blue dashed curves belong to the devices obtained by synthesis and approximated by actual devices, respectively (Fig. 10.16).

Moreover, we use a frequency response analyzer (Venable Model 6305) to test the frequency-domain characteristics of the circuits. From Fig. 10.17, one finds that, by adopting structures and parameters above, the frequency-domain characteristics of fractional-order elements constructed meet well with theoretical design.

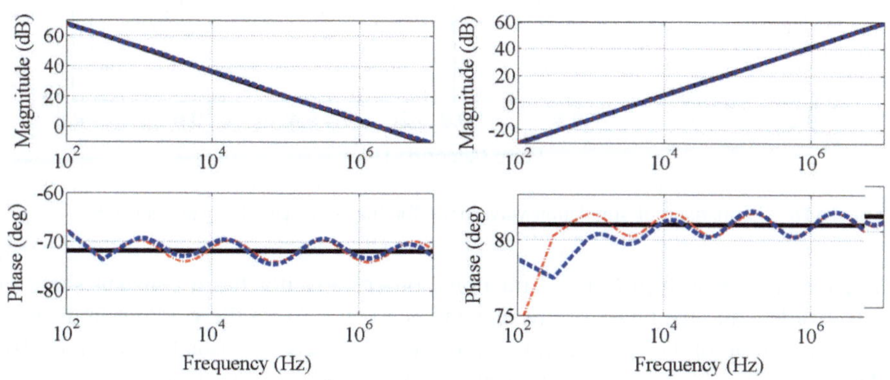

Fig. 10.16 Bode diagrams of fractional-order components

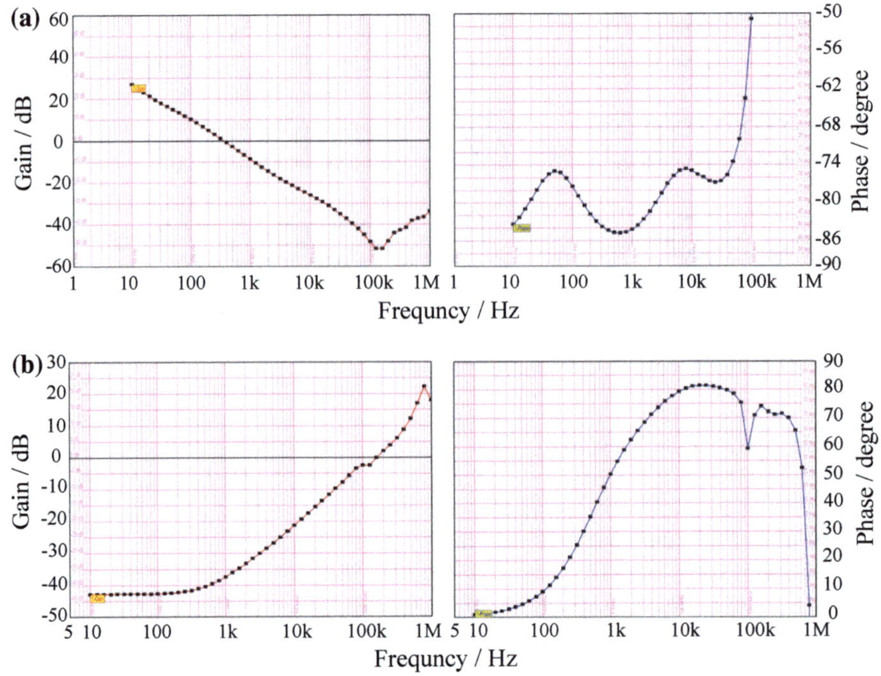

Fig. 10.17 Bode diagrams of approximate circuits, **a** for fractional-order capacitor, **b** for fractional-order inductor

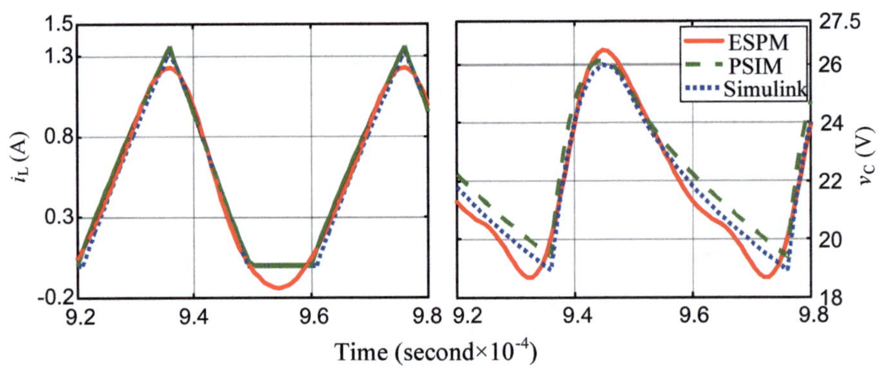

Fig. 10.18 Comparison of simulation waveforms for fractional DCM-Boost converter

The magnitude of $10\mu F/(s)^{1-0.8}$ fractional-order capacitor has a constant slope of −16 dB/dec, and the phase of this element is about −*76*° at the working frequency. The magnitude of 477 μH fractional-order inductor has a constant slope of 18 dB/dec, and the phase of this element is around *81*° at the working frequency (Fig. 10.17).

Table 10.9 Main features of state variables of the fractional DCM-Boost converter

		ESPM	Simulink	PSIM	Experiments
RMS	$i_L(A)$	0.672	0.647	0.687	0.614
	$v_C(V)$	22.492	22.877	22.939	20.5
i_L(A)	*Max*	1.230	1.332	1.363	1.30
	Min	−0.138	0	0	0
v_C(V)	*Max*	26.499	25.996	26.152	26.4
	Min	18.687	18.926	19.389	18.8

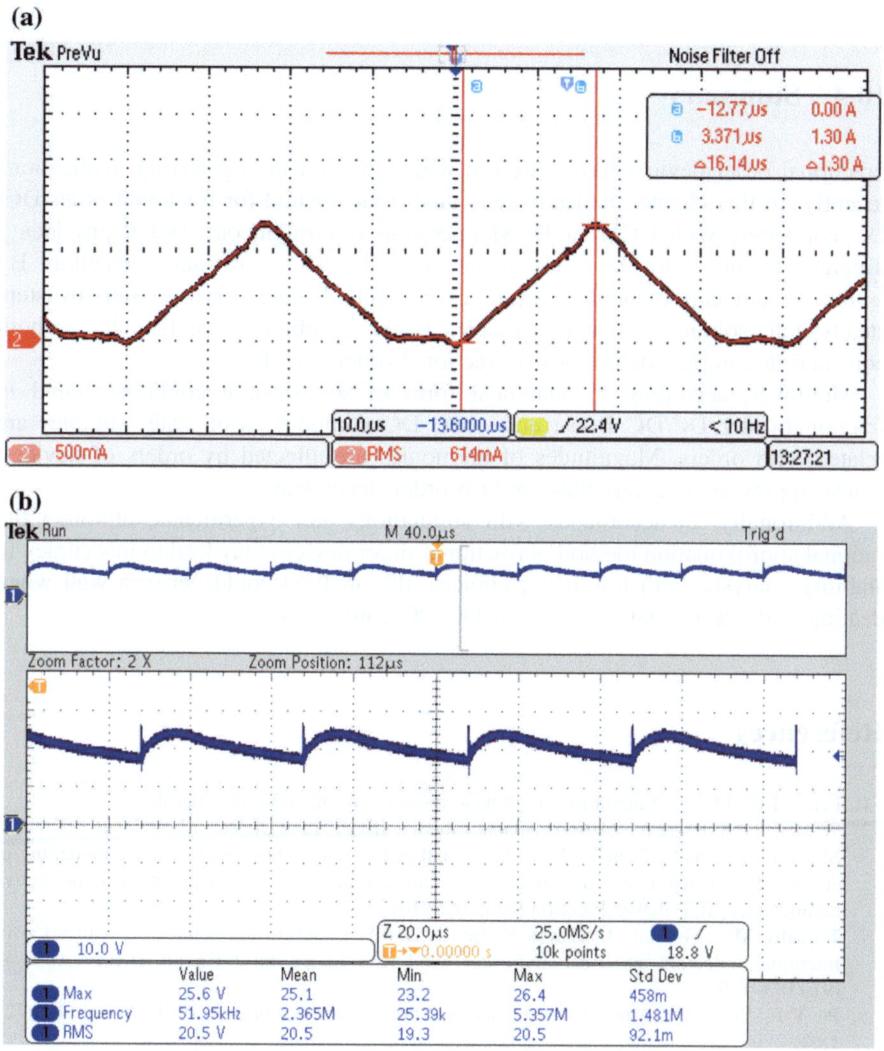

Fig. 10.19 Experimental waveforms of state variables of the fractional DCM-Boost converter: **a** inductor current, **b** capacitor voltage

Meanwhile, we implement both circuit based simulations in PSIM and numerical simulations based on Oustaloup's method in MATLAB/Simulink [25], and the results are put together with the approximate periodic solutions obtained by our method (Fig. 10.18).

It can be seen, the results from different ways are in a good agreement. Then we collect real-time experimental waveforms of state variables, as in Fig. 10.19. For comparison purposes, we list some main features of the waveforms in Table 10.9.

As can be seen from Fig. 10.19 and Table 10.9, the results from the ESPM show good agreement with those from simulations and experiments.

10.4 Summary

Fractional order devices have received wide attention in engineering applications recently. In this chapter, a steady-state analyzing method for fractional order DC/DC converters with CCM or DCM operation is brought out, and it provides a different insight into the non-integer order DC/DC converters behavior. By assuming and correcting the harmonics content of state variables step by step, steady-state solutions of state variables could be obtained analytically without considering complex definitions of fractional-order calculus.

Moreover, according to analytical form of solutions, it could be found in fractional-order DC/DC Boost converter, DC components of state variables are related with orders. Magnitudes of harmonics are affected by orders of devices, hence ripples of state variables are also order dependent.

Additionally, in accordance with simulations and experiments, although the rational approximation method of fractional-order devices may lead to inaccuracy in stability analysis, with few better choices, this method could perform well when dealing with steady-state analysis of DC/DC converters.

References

1. Petras I (2011) Fractional-order nonlinear systems. Springer-Verlag, Berlin
2. Westerlund S, Ekstam L (1994) Capacitor theory. IEEE Trans Dielectr Electr Insul 1:826–839
3. Malek H, Dadras S, Chen Y (2016) Fractional order equivalent series resistance modelling of electrolytic capacitor and fractional order failure prediction with application to predictive maintenance. IET Power Electron 9(8):1608–1613
4. Tripathy MC, Mondal D, Biswas K, Sen S (2015) Experimental studies on realization of fractional inductors and fractional-order band pass filters. Int J Circuit Theory Appl 43 (9):1183–1196
5. Pu Y-F, Yuan X (2016) Fracmemristor: fractional-order memristor. IEEE Access 4:1872–1888

6. Yang C, Yu H, Shang Y, Fei W (2015) Characterization of CMOS metamaterial transmission line by compact fractional-order equivalent circuit model. IEEE Trans Electron Devices 62 (9):3012–3018
7. Sabatier J, Cugnet M, Laruelle S, Grugeon S, Sahut B, Oustaloup A, Tarascon JM (2010) A fractional order model for lead-acid battery crankability estimation. Commun Nonlinear Sci Numer Simul 15(5):1308–1317
8. Freeborn TJ, Maundy B, Elwakil AS (2013) Measurement of super capacitor fractional-order model parameters from voltage-excited step response. IEEE J Emerg Sel Top Circuits Syst 3 (3):367–376
9. Tsirimokou G, Psychalinos C, Allagui A, Elwakil AS (2015) Simple non-impedance-based measuring technique for supercapacitors. Electron Lett 51(21):1699–1700
10. Biswas K, Sen S, Dutta PK (2006) Realization of a constant phase element and its performance study in a differentiator circuit. IEEE Trans Circuits Syst II-Express Briefs 53 (9):802–806
11. Mondal D, Biswas K (2013) Packaging of single-component fractional order element. IEEE Trans Device Mater Reliab 13(1):73–80
12. Elwakil AS (2010) Fractional-order circuits and systems: an emerging interdisciplinary research area. IEEE Circuits Syst Mag 10(4):40–50
13. Elwakil AS, Allagui A, Maundy BJ, Psychalinos C (2016) A low frequency oscillator using a super-capacitor. AEU-Int J Electron Commun 70(7):970–973
14. Hartley TT, Veillette RJ, Adams JL, Lorenzo CF (2015) Energy storage and loss in fractional-order circuit elements. IET Circuits Devices Syst 9(3):227–235
15. Diethelm K, Ford N, Freed A (2002) A predictor-corrector approach for the numerical solution of fractional differential equations. Nonlinear Dyn 29(1–4):3–22
16. Diethelm K (2004) The analysis of fractional differential equations. Springer-Verlag, Berlin
17. Podlubny I (1998) Fractional differential equations. Academic Press
18. Fa-Qiang W, Xi-Kui M (2013) Transfer function modeling and analysis of the open-loop Buck converter using the fractional calculus. Chin Phys B 22(3) (Mar 2013)
19. Wu C, Si G, Zhang Y, Yang N (2015) The fractional-order state space averaging modeling of the Buck Boost DC/DC converter in discontinuous conduction mode and the performance analysis. Nonlinear Dyn 79(1):689–703
20. Tseng C, Pei S, Hsia S (2000) Computation of fractional derivatives using Fourier transform and digital FIR differentiator. Sig Process 80(1):151–159
21. Qiu S, Filanovsky I (2000) Harmonic analysis of PWM converters. IEEE Trans Circuits Syst I-Regul Papers 47(9):1340–1349
22. Zhu H, Ding M, Li Y (2011) Gibbs phenomenon for fractional Fourier series. IET Signal Proc 5(8):728–738
23. Erickson RW, Maksimovic D (2001) Fundamentals of power electronics. Springer, US
24. Oustaloup A, Levron F, Mathieu B, Nanot F (2000) Frequency band complex noninteger differentiator: characterization and synthesis. IEEE Trans Circuits Syst I-Fundam Theory Appl 47(1):25–39
25. Sierociuk D, Podlubny I, Petras I (2013) Experimental evidence of variable-order behavior of ladders and nested ladders. IEEE Trans Control Syst Technol 21(2):459–466
26. Valsa J, Vlach J (2013) RC models of a constant phase element. Int J Circuit Theory Appl 41 (1):59–67
27. Tripathy MC, Mondal D, Biswas K, Sen S (2015) Experimental studies on realization of fractional inductors and fractional-order band pass filters. Int J Circuit Theory Appl 43 (9):1183–1196
28. Sarafraz MS, Tavazoei MS (2015) Realizability of fractional-order impedances by passive electrical networks composed of a fractional capacitor and RLC components. IEEE Trans Circuits Syst I-Regul Papers 62(12):2829–2835
29. Tavakoli-Kakhki M, Haeri M, Tavazoei MS (2011) Notes on the state space realizations of rational order transfer functions. IEEE Trans Circuits Syst I-Regul Papers 58(5):1099–1108

30. Tavazoei MS (2015) Comments on "Chaotic characteristics analysis and circuit implementation for a fractional-order system". IEEE Trans Circuits Syst I-Regul Papers 62(1):329–332
31. Charef A, Sun H, Tsao Y, Onaral B (1992) Fractal system as represented by singularity function. IEEE Trans Autom Control 37:1465–1470
32. Chen YF, Chen X, Zhang B, Qiu DY (2015) A new analyzing scheme for non-integer order DC-DC converters. In: 41st annual conference of the IEEE industrial electronics society, pp 1633–1638
33. Wu Y, Hu AP, Budgett D, Malpas SC, Dissanayake T (2011) Efficient power-transfer capability analysis of the TET system using the equivalent small parameter method. IEEE Trans Biomed Circuits Syst 5(3):272–282
34. Chen Xi, Chen Yanfeng, Zhang Bo, Qiu Dongyuan (2017) A modeling and analysis method for fractional-order DC–DC converters. IEEE Trans Power Electron 32(9):7034–7044
35. Chen YF, Chen X, Zhang B, Qiu DY (2017) A symbolic analysis method for fractional-order boost converter in discontinuous conduction mode. In: 43rd annual conference of the IEEE industrial electronics society, pp 1403–1408

MIX
Papier aus verantwortungsvollen Quellen
Paper from responsible sources
FSC® C105338

If you have any concerns about our products,
you can contact us on
ProductSafety@springernature.com

In case Publisher is established outside the EU,
the EU authorized representative is:
Springer Nature Customer Service Center GmbH
Europaplatz 3, 69115 Heidelberg, Germany

Printed by Libri Plureos GmbH
in Hamburg, Germany